Advanced Electrical Installation Work

FIFTH EDITION

TREVOR LINSLEY
Senior Lecturer
Blackpool and The Fylde College

AMSTERDAM • BOSTON • HEIDELBERG • LONDON • NEW YORK • OXFORD
PARIS • SAN DIEGO • SAN FRANCISCO • SINGAPORE • SYDNEY • TOKYO
Newnes is an imprint of Elsevier

Newnes is an imprint of Elsevier
The Boulevard, Langford Lane, Kidlington, Oxford OX5 1GB, UK
30 Corporate Drive, Suite 400, Burlington, MA 01803, USA

First published by Arnold 1998
Reprinted by Butterworth-Heinemann 2001, 2003 (twice), 2004 (twice)
Fourth edition 2005
Fifth edition 2008
Reprinted 2009

British Library Cataloguing in Publication Data
A catalogue record for this book is available from the British Library

Library of Congress Cataloging-in-Publication Data
A catalog record for this book is available from the Library of Congress

ISBN–13: 978-0-7506-8752-2

For information on all Newnes publications
visit our website at www.elsevierdirect.com

Printed and bound in *China*

09 10 11 12 10 9 8 7 6 5 4 3 2

Advanced Electrical Installation Work

To Joyce, Samantha and Victoria

Contents

Preface

The 5th Edition of *Advanced Electrical Installation Work* has been completely rewritten in 10 Chapters to closely match the 10 Outcomes of the City and Guilds qualification. The technical content has been revised and updated to the requirements of the new 17th Edition of the IEE Regulations BS 7671: 2008. Improved page design with new illustrations gives greater clarity to each topic.

This book of electrical installation theory and practice will be of value to the electrical trainee working towards:

- The City and Guilds 2330 Level 3 Certificate in Electrotechnical Technology, Installation Route.
- The City and Guilds 2356 Level 2 NVQ in Installing Electrotechnical Systems.
- The SCOTVEC and BTEC Electrical Utilisation Units at Levels II and III.
- Those taking Engineering NVQ and modern Apprenticeship Courses.

Advanced Electrical Installation Work provides a sound basic knowledge of electrical practice which other trades in the construction industry will find of value, particularly those involved in multi-skilling activities.

The book incorporates the requirements of the latest Regulations, particularly:

- 17th Edition IEE Wiring Regulations.
- British Standards BS 7671: 2008.
- Part P of the Building Regulations, Electrical Safety in Dwellings: 2006.
- Hazardous Waste Regulations: 2005.
- Work at Height Regulations: 2005.

Trevor Linsley
2008

Acknowledgements

I would like to acknowledge the assistance given by the following manufacturers and professional organizations in the preparation of this book:

- The Institution of Engineering and Technology for permission to reproduce Regulations and Tables from the 17th Edition IEE Regulations.
- The British Standards Institution for permission to reproduce material from BS 7671: 2008.
- Crabtree Electrical Industries for technical information and data.
- RS Components Limited for technical information and photographs.
- Stocksigns Limited for technical information and photographs.
- Wylex Electrical Components for technical information and photographs.
- Jason Vann Smith MIET MIEEE MBCS MACM for the photographs used in the page design.

I would like to thank the many College Lecturers who responded to the questionnaire from Elsevier the publishers, regarding the proposed new edition of this book. Their recommendations have been taken into account in producing this improved 5th Edition.

I would also like to thank the editorial and production staff at Elsevier the publishers for their enthusiasm and support. They were able to publish this 5th Edition within the very short timescale created by the publication of the 17th Edition of the IEE Regulations.

Finally, I would like to thank Joyce, Samantha and Victoria for their support and encouragement.

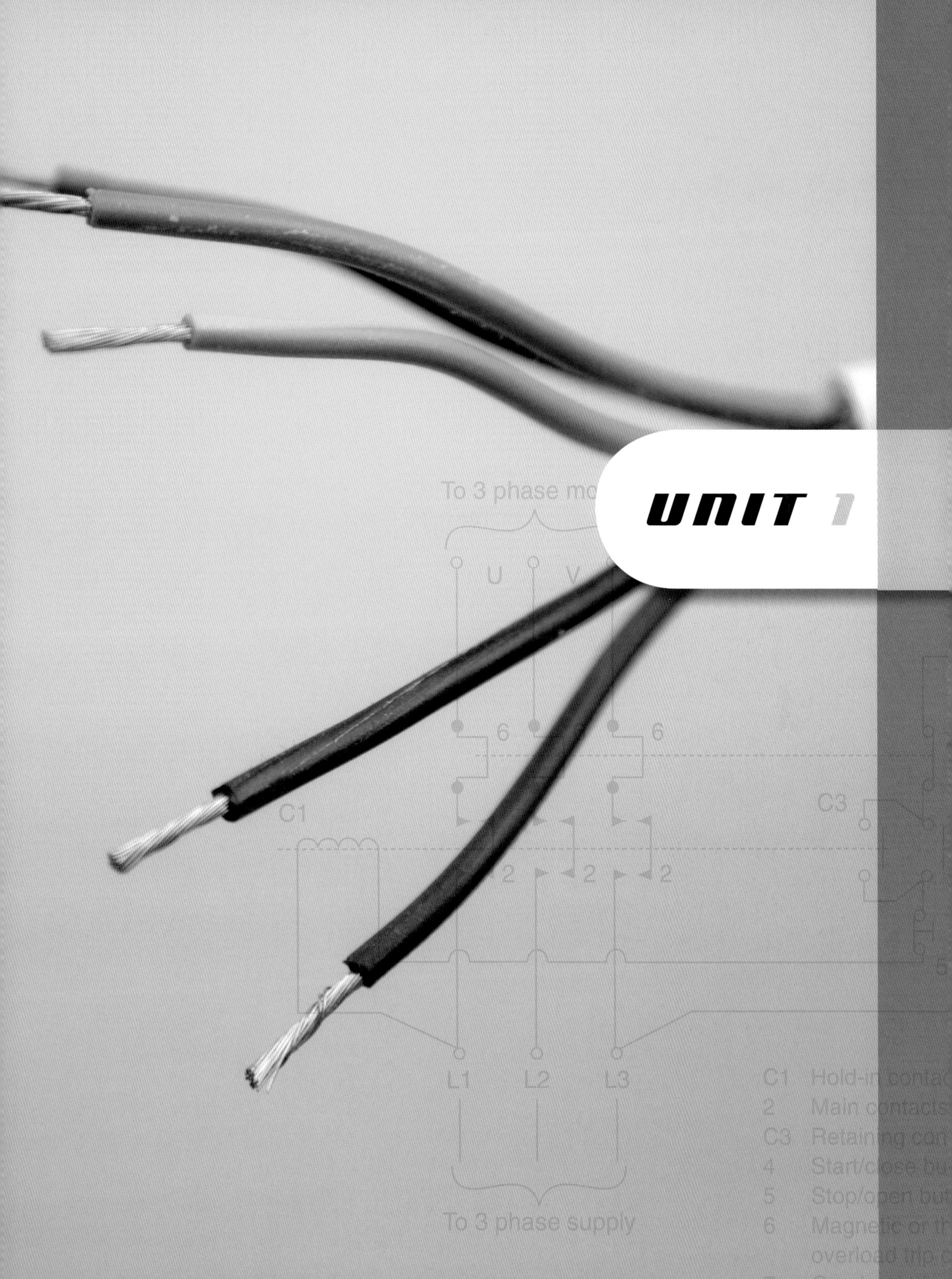
UNIT 1
U
V
6
6
C1
C3
2
2
2
L1
L2
L3
To 3 phase supply

CH 1

Statutory regulations and safe working procedures

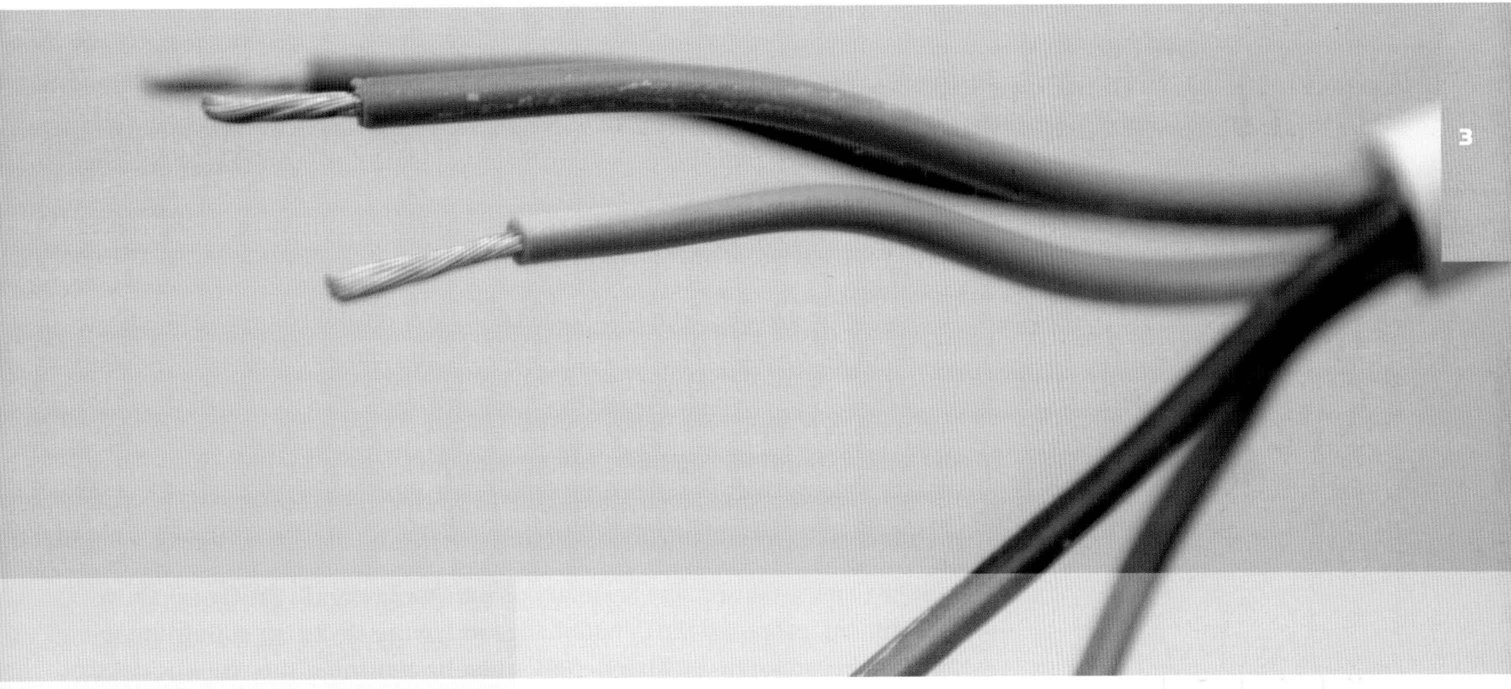

Unit 1 - Application of health and safety and electrical principles – Outcome 1

Underpinning knowledge: when you have completed this chapter you should be able to:

- identify the Safety Regulations relevant to the electrotechnical industry
- identify environmental legislation relevant to the electrotechnical industry
- state employer and employee responsibilities
- state the human and environmental conditions leading to workplace accidents
- describe a procedure for reporting accidents
- recognize workplace safety signs
- carry out a risk assessment
- list the changing work patterns within the industry

Introduction

This first chapter of Advanced Electrical Installation work covers the health and safety core skills required by the City and Guilds Level 3 Certificate in Electrotechnical Technology. That is the Health and Safety Laws and Regulations that underpin the electrotechnical industry.

Let me begin by looking at the background to the modern Health and Safety Regulations and the electricity supply and wiring regulations.

Electricity generation as we know it today began when Michael Faraday conducted the famous ring experiment in 1831. This experiment, together with many other experiments of the time, made it possible for Lord Kelvin and Sebastian de Ferranti to patent in 1882 the designs for an electrical machine called the Ferranti–Thompson dynamo, which enabled the generation of electricity on a commercial scale.

In 1887 the London electric supply corporation was formed with Ferranti as chief engineer. This was one of the many privately owned electricity generating stations supplying the electrical needs of the United Kingdom. As the demand for electricity grew, more privately owned generating stations were built until eventually the government realized that electricity was a national asset which would benefit from nationalization.

In 1926 the Electricity Supply Act placed the responsibility for generation in the hands of the Central Electricity Board. In England and Wales the Central Electricity Generating Board (CEGB) had the responsibility for the generation and transmission of electricity on the supergrid. In Scotland, generation was the joint responsibility of the North of Scotland Hydro-Electricity Board and the South of Scotland Electricity Board. In Northern Ireland electricity generation was the responsibility of the Northern Ireland Electricity Service.

In 1988 Cecil Parkinson, the Secretary of State for Energy in the Conservative government, proposed the denationalization of the electricity supply industry; this became law in March 1991, thereby returning the responsibility for generation, transmission and distribution to the private sector. It was anticipated that this action, together with new legislation over the security of supplies, would lead to a guaranteed quality of provision, with increased competition leading eventually to cheaper electricity.

During the period of development of the electricity services, particularly in the early days, poor design and installation led to many buildings being damaged by fire and the electrocution of human beings and livestock. It was the insurance companies which originally drew up a set of rules and guidelines of good practice in the interest of reducing the number of claims made upon them. The first rules were made by the American Board of Fire Underwriters and were quickly followed by the Phoenix Rules of 1882. In the same year the first edition of the Rules and Regulations for the Prevention of Fire Risk arising from Electrical Lighting was issued by the Institute of Electrical Engineers.

The current edition of these regulations is called the Requirements for Electrical Installations, IEE Wiring Regulations (BS 7671: 2008), and since

July 2008 we have been using the 17th edition. All the rules have been revised, updated and amended at regular intervals to take account of modern developments, and the 17th edition brought the UK Regulations into harmony with those of the rest of Europe.

The laws and regulations affecting the electrotechnical industry have steadily increased over the years. There is a huge amount of legislation from the European law-makers in Brussels. These laws and regulations will permeate each and every sector of the electrotechnical industry and reform and modify our future work patterns and behaviour.

In this section I want to deal with the laws and regulations that affect our industry under three general headings because there are a large number of them, and it may help us to appreciate the reasons for them.

(i) First of all I want to look at the laws concerned with health and safety at work, making the working environment safe.

(ii) Then I want to go on to the laws that protect our environment from, for example, industrial waste and pollution.

(iii) Finally, I will look at employment legislation and the laws which protect us as individual workers, people and citizens in Chapter 3 of this book.

The Health and Safety at Work Act 1974

Many governments have passed laws aimed at improving safety at work but the most important recent legislation has been the Health and Safety at Work Act 1974. The purpose of the act is to provide the legal framework for stimulating and encouraging high standards of health and safety at work; the act puts the responsibility for safety at work on both workers and managers.

The Health and Safety at Work Act is an 'Enabling Act' that allows the Secretary of State to make further laws, known as regulations, without the need to pass another Act of Parliament. Regulations are law, passed by Parliament and are usually made under the Health and Safety at Work Act 1974. This applies to regulations based on European directives as well as new UK Regulations. The way it works is that the Health and Safety at Work Act established the Health and Safety Commission (HSC) and gave it the responsibility of drafting new regulations and enforcing them through its executive arm known as the Health and Safety Executive (HSE) or through the local Environmental Health Officers (EHO). The HSC has equal representation from employers, trade unions and special interest groups. Their role is to set out the regulations as goals to be achieved. They describe what must be achieved in the interests of safety, but not how it must be done.

Definition

Under the Health and Safety at Work Act an *employer* has a duty to care for the health and safety of employees.

Under the Health and Safety at Work Act an **employer** has a duty to care for the health and safety of employees (Section 2 of the Act). To do this he has a *responsibility* to ensure that:

- the working conditions and standard of hygiene are appropriate;

Safety First

Laws

The Health and Safety at Work Act provides the legal framework for stimulating and encouraging health and safety at work. It is:

- the most important,
- the most far reaching,
- single piece of legislation.

Definition

Employees have a duty to care for their own health and safety and that of others who may be affected by their actions.

- the plant, tools and equipment are properly maintained;
- safe systems of work are in place;
- safe methods of handling, storing and transporting goods and materials are used;
- there is a system for reporting accidents in the workplace;
- the company has a written Health & Safety Policy statement;
- the necessary safety equipment – such as personal protective equipment (PPE), dust and fume extractors and machine guards – are available and properly used;
- the workers are trained to use equipment and plant safely.

Employees have a duty to care for their own health and safety and that of others who may be affected by their actions (Section 7 of the Act). To do this they must:

- take reasonable care to avoid injury to themselves or others as a result of their work activity;
- co-operate with their employer, helping him or her to comply with the requirements of the act;
- not interfere with or misuse anything provided to protect their health and safety.

Failure to comply with the Health and Safety at Work Act is a criminal offence and any infringement of the law can result in heavy fines, a prison sentence or both.

ENFORCEMENT

Laws and rules must be enforced if they are to be effective. The system of control under the Health and Safety at Work Act comes from the HSE which is charged with enforcing the law. The HSE is divided into a number of specialist inspectorates or sections which operate from local offices throughout the United Kingdom. From the local offices the inspectors visit individual places of work.

The HSE inspectors have been given wide-ranging powers to assist them in the enforcement of the law. They can:

1. enter premises unannounced and carry out investigations, take measurements or photographs;
2. take statements from individuals;
3. check the records and documents required by legislation;
4. give information and advice to an employee or employer about safety in the workplace;
5. demand the dismantling or destruction of any equipment, material or substance likely to cause immediate serious injury;

6. issue an improvement notice which will require an employer to put right, within a specified period of time, a minor infringement of the legislation;
7. issue a prohibition notice which will require an employer to stop immediately any activity likely to result in serious injury, and which will be enforced until the situation is corrected;
8. prosecute all persons who fail to comply with their safety duties, including employers, employees, designers, manufacturers, suppliers and the self-employed.

SAFETY DOCUMENTATION

Under the Health and Safety at Work Act, the employer is responsible for ensuring that adequate instruction and information is given to employees to make them safety-conscious. Part 1, Section 3 of the Act instructs all employers to prepare a written health and safety policy statement and to bring this to the notice of all employees. Your employer must let you know who your safety representatives are and the new health and safety poster shown in Fig. 1.1 has a blank section into which the names and contact information of your specific representatives can be added. This is a large laminated poster, 595 × 415 mm suitable for wall or notice board display.

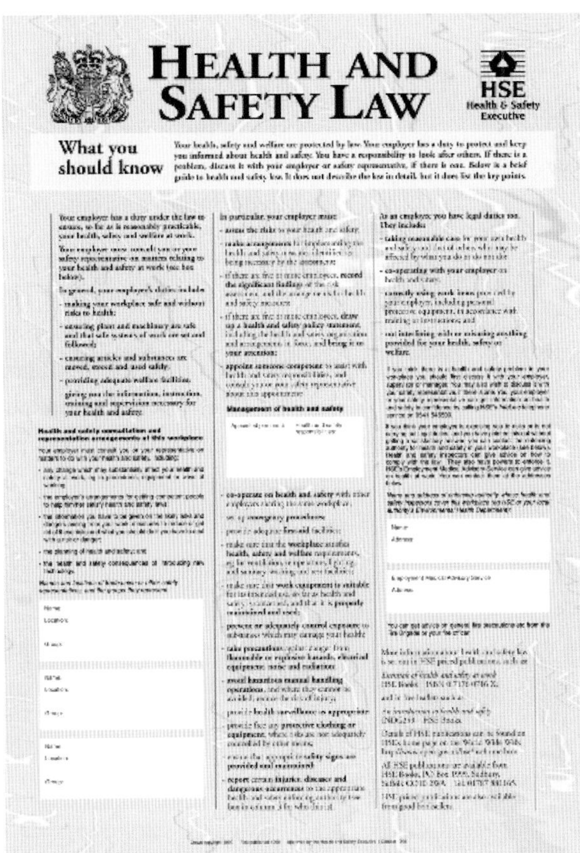

FIGURE 1.1
New Health and Safety Law poster. *Source*: HSE © Crown copyright material is reproduced with the permission of the Controller of HMSO and Her Majesty's Stationery Office, Norwich.

All workplaces employing five or more people must display the type of poster shown in Fig. 1.1 after 30 June 2000.

To promote adequate health and safety measures the employer must consult with the employees' safety representatives. In companies which employ more than 20 people this is normally undertaken by forming a safety committee which is made up of a safety officer and employee representatives, usually nominated by a trade union. The safety officer is usually employed full-time in that role. Small companies might employ a safety supervisor, who will have other duties within the company, or alternatively they could join a 'safety group'. The safety group then shares the cost of employing a safety adviser or safety officer, who visits each company in rotation. An employee who identifies a dangerous situation should initially report to his site safety representative. The safety representative should then bring the dangerous situation to the notice of the safety committee for action which will remove the danger. This may mean changing company policy or procedures or making modifications to equipment. All actions of the safety committee should be documented and recorded as evidence that the company takes seriously its health and safety policy.

Safety First

Information

- Have you seen the new Health and Safety Law poster like Fig 1.1?
  - in your place of work
  - at the college.
- Were the blank sections filled in?

The Management of Health and Safety at Work Regulations 1999

The Health and Safety at Work Act 1974 places responsibilities on employers to have robust Health and Safety systems and procedures in the workplace. Directors and managers of any company who employ more than five employees can be held personally responsible for failures to control health and safety.

The Management of Health and Safety at Work Regulations 1999 tell us that employers must systematically examine the workplace, the work activity and the management of safety in the establishment through a process of '*risk assessments*'. A record of all significant risk assessment findings must be kept in a safe place and be available to an HSE inspector if required. Information based on these findings must be communicated to relevant staff and if changes in work behaviour patterns are recommended in the interests of safety, then they must be put in place. The process of risk assessment is considered in detail later in this chapter.

Risks, which may require a formal assessment in the electrotechnical industry, might be:

- working at heights;
- using electrical power tools;
- falling objects;
- working in confined places;
- electrocution and personal injury;

- working with 'live' equipment;
- using hire equipment;
- *manual handling*: pushing, pulling, lifting;
- *site conditions*: falling objects, dust, weather, water, accidents and injuries.

And any other risks which are particular to a specific type of work place or work activity.

Provision and Use of Work Equipment Regulations 1998

These regulations tidy up a number of existing requirements already in place under other regulations such as the Health and Safety at Work Act 1974, the Factories Act 1961 and the Offices, Shops and Railway Premises Act 1963.

The Provision and Use of Work Equipment Regulations 1998 places a general duty on employers to ensure minimum requirements of plant and equipment. If an employer has purchased good quality plant and equipment, which is well maintained, there is little else to do. Some older equipment may require modifications to bring it in line with modern standards of dust extraction, fume extraction or noise, but no assessments are required by the regulations other than those generally required by the Management Regulations 1999 discussed previously.

The Control of Substances Hazardous to Health Regulations 2002

The original Control of Substances Hazardous to Health (COSHH) Regulations were published in 1988 and came into force in October 1989. They were re-enacted in 1994 with modifications and improvements, and the latest modifications and additions came into force in 2002.

The COSHH Regulations control people's exposure to hazardous substances in the workplace. Regulation 6 requires employers to assess the risks to health from working with hazardous substances, to train employees in techniques which will reduce the risk and provide PPE so that employees will not endanger themselves or others through exposure to hazardous substances. Employees should also know what cleaning, storage and disposal procedures are required and what emergency procedures to follow. The necessary information must be available to anyone using hazardous substances as well as to visiting HSE inspectors.

Hazardous substances include:

1. any substance which gives off fumes causing headaches or respiratory irritation;
2. man-made fibres which might cause skin or eye irritation (e.g. loft insulation);
3. acids causing skin burns and breathing irritation (e.g. car batteries, which contain dilute sulphuric acid);

4. solvents causing skin and respiratory irritation (strong solvents are used to cement together PVC conduit fittings and tube);
5. fumes and gases causing asphyxiation (burning PVC gives off toxic fumes);
6. cement and wood dust causing breathing problems and eye irritation;
7. exposure to asbestos – although the supply and use of the most hazardous asbestos material is now prohibited, huge amounts were installed between 1950 and 1980 in the construction industry and much of it is still in place today. In their latest amendments the COSHH Regulations focus on giving advice and guidance to builders and contractors on the safe use and control of asbestos products. These can be found in Guidance Notes EH 71.

Where PPE is provided by an employer, employees have a duty to use it to safeguard themselves.

PPE at Work Regulations 1992

PPE is defined as all equipment designed to be worn, or held, to protect against a risk to health and safety. This includes most types of protective clothing, and equipment such as eye, foot and head protection, safety harnesses, life jackets and high-visibility clothing.

Under the Health and Safety at Work Act, employers must provide free of charge any PPE and employees must make full and proper use of it. Safety signs such as those shown in Fig. 1.2 are useful reminders of the type of PPE to be used in a particular area. The vulnerable parts of the body which may need protection are the head, eyes, ears, lungs, torso, hands and feet and, additionally, protection from falls may need to be considered. Objects falling from a height present the major hazard against which head protection is provided. Other hazards include striking the head against projections and hair becoming entangled in machinery. Typical methods of protection include helmets, light duty scalp protectors called 'bump caps' and hairnets.

Safety First

PPE

- What type of PPE do you use at work?
- Make a list in the margin of the book.

The eyes are very vulnerable to liquid splashes, flying particles and light emissions such as ultraviolet light, electric arcs and lasers. Types of eye protectors include safety spectacles, safety goggles and face shields. Screen based workstations are being used increasingly in industrial and commercial locations by all types of personnel. Working with VDUs (visual display units) can cause eye strain and fatigue and, therefore, work patterns should be varied and operators are entitled to free eye tests.

Noise is accepted as a problem in most industries and surprisingly there has been very little control legislation. The HSE have published a 'Code of Practice' and 'Guidance Notes' HSG 56 for reducing the exposure of employed persons to noise. A continuous exposure limit of below 90 dB for an 8-hour working day is recommended by the code.

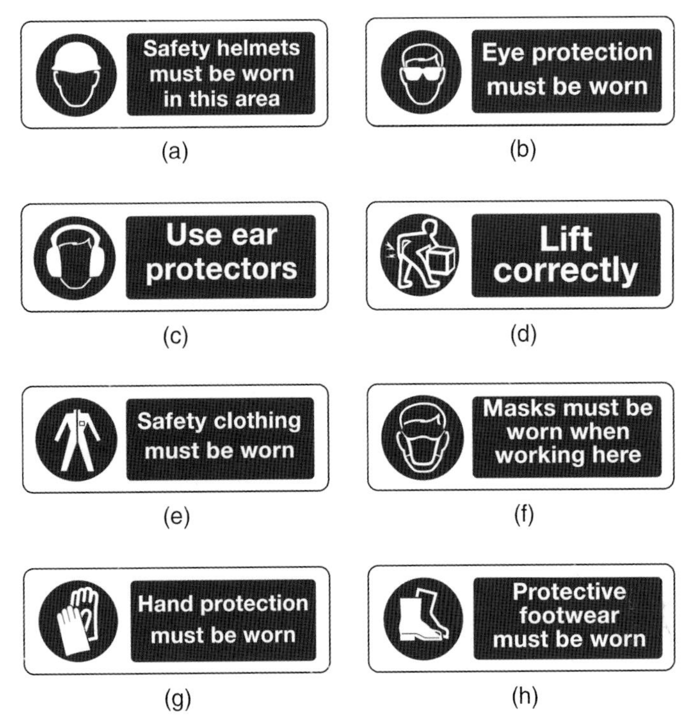

FIGURE 1.2
Safety signs showing type of PPE to be worn.

Noise may be defined as any disagreeable or undesirable sound or sounds, generally of a random nature, which do not have clearly defined frequencies. The usual basis for measuring noise or sound level is the decibel scale. Whether noise of a particular level is harmful or not also depends upon the length of exposure to it. This is the basis of the widely accepted limit of 90 dB of continuous exposure to noise for 8 hours per day.

A peak sound pressure of above 200 pascals or about 120 dB is considered unacceptable and 130 dB is the threshold of pain for humans. If a person has to shout to be understood at 2 m, the background noise is about 85 dB. If the distance is only 1 m, the noise level is about 90 dB. Continuous noise at work causes deafness, makes people irritable, affects concentration, causes fatigue and accident proneness and may mask sounds which need to be heard in order to work efficiently and safely.

It may be possible to engineer out some of the noise, for example, by placing a generator in a separate sound-proofed building. Alternatively, it may be possible to provide job rotation, to rearrange work locations or provide acoustic refuges.

Where individuals must be subjected to some noise at work it may be reduced by ear protectors. These may be disposable ear plugs, reusable ear plugs or ear muffs. The chosen ear protector must be suited to the user and suitable for the type of noise and individual personnel should be trained in its correct use.

Breathing reasonably clean air is the right of every individual, particularly at work. Some industrial processes produce dust which may present a potentially serious hazard. The lung disease asbestosis is caused by the inhalation of asbestos dust or particles and the coal dust disease pneumoconiosis, suffered by many coal miners, has made people aware of the dangers of breathing in contaminated air.

Some people may prove to be allergic to quite innocent products such as flour dust in the food industry or wood dust in the construction industry. The main effect of inhaling dust is a measurable impairment of lung function. This can be avoided by wearing an appropriate mask, respirator or breathing apparatus as recommended by the company's health and safety policy and indicated by local safety signs.

A worker's body may need protection against heat or cold, bad weather, chemical or metal splash, impact or penetration and contaminated dust. Alternatively, there may be a risk of the worker's own clothes causing contamination of the product, as in the food industry. Appropriate clothing will be recommended in the company's health and safety policy. Ordinary working clothes and clothing provided for food hygiene purposes are not included in the PPE at Work Regulations.

Hands and feet may need protection from abrasion, temperature extremes, cuts and punctures, impact or skin infection. Gloves or gauntlets provide protection from most industrial processes but should not be worn when operating machinery because they may become entangled in it. Care in selecting the appropriate protective device is required; for example, barrier creams provide only a limited protection against infection.

Boots or shoes with in-built toe caps can give protection against impact or falling objects and, when fitted with a mild steel sole plate, can also provide protection from sharp objects penetrating through the sole. Special slip resistant soles can also be provided for employees working in wet areas.

Whatever the hazard to health and safety at work, the employer must be able to demonstrate that he or she has carried out a risk analysis, made recommendations which will reduce that risk and communicated these recommendations to the workforce. Where there is a need for PPE to protect against personal injury and to create a safe working environment, the employer must provide that equipment and any necessary training which might be required and the employee must make full and proper use of such equipment and training.

RIDDOR

RIDDOR stands for Reporting of Injuries, Diseases and Dangerous Occurrences Regulation 1995, which is sometimes referred to as RIDDOR 95, or just RIDDOR for short. The HSE requires employers to report some work related accidents or diseases so that they can identify where and how risks arise, investigate serious accidents and publish statistics and data to help reduce accidents at work.

What needs reporting? Every work related death, major injury, dangerous occurrence, disease or any injury which results in an absence from work of over 3 days.

Where an employee or member of the public is killed as a result of an accident at work the employer or his representative must report the accident to the Environmental Health Department of the Local Authority by telephone that day and give brief details. Within 10 days this must be followed up by a completed accident report form (Form No. F2508). Major injuries sustained as a result of an accident at work include amputations, loss of sight (temporary or permanent), fractures to the body other than to fingers, thumbs or toes and any other serious injury. Once again, the Environmental Health Department of the Local Authority must be notified by telephone on the day that the serious injury occurs and the telephone call followed up by a completed Form F2508 within 10 days. Dangerous occurrences are listed in the regulations and include the collapse of a lift, an explosion or injury caused by an explosion, the collapse of a scaffold over 5 m high, the collision of a train with any vehicle, the unintended collapse of a building and the failure of fairground equipment.

Depending upon the seriousness of the event, it may be necessary to immediately report the incident to the Local Authority. However, the incident must be reported within 10 days by completing Form F2508. If a doctor notifies an employer that an employee is suffering from a work related disease then form F2508A must be completed and sent to the Local Authority. Reportable diseases include certain poisonings, skin diseases, lung disease, infections and occupational cancer. The full list is given within the pad of report forms.

An accident at work resulting in an over 3 day injury, that is, an employee being absent from work for over 3 days as a result of an accident at work, requires that accident report form F2508 be sent to the Local Authority within 10 days.

An over 3 day injury is one which is not major but results in the injured person being away from work for more than 3 days not including the day the injury occurred.

Who are the reports sent to? They are sent to the Environmental Health Department of the Local Authority or the area HSE offices (see the Appendix L of this book for area office addresses). Accident report forms F2508 can also be obtained from them or by ringing the HSE Infoline or by ringing the incident contact centre on telephone number 0845 300 9923.

For most businesses, a reportable accident, dangerous occurrence or disease is a very rare event. However, if a report is made, the company must keep a record of the occurrence for 3 years after the date on which the incident happened. The easiest way to do this would probably be to file a photo copy of the completed accident report form F2508, but a record may be kept in any form which is convenient.

The Control of Major Accidents and Hazards Regulations 1999

The Control of Major Accidents and Hazards (COMAH) Regulations came into force on 1 April 1999. Their main aim is to prevent any major accidents involving dangerous substances such as chlorine, liquefied petroleum gas (LPG), explosives and arsenic pentoxide that would cause serious harm to people or damage the environment. The COMAH Regulations regard risks to the environment just as seriously as harm to people.

These regulations apply mainly to the chemical industry but also apply to some storage facilities and nuclear sites.

Operators who fall within the scope of these regulations must 'take all measures necessary to prevent major accidents and limit their consequences to people and the environment'. This sets high standards of control but by requiring operators to put in place measures for both prevention and mitigation, which means to make less serious, there is the recognition that all risks cannot be completely eliminated. Operators must, therefore, be able to show that they have taken 'all measures necessary' to prevent an accident occurring.

The COMAH Regulations are enforced by the HSE and the Environment Agency.

Dangerous Substances and Explosive Atmospheres Regulations 2002

The Dangerous Substances and Explosive Atmospheres Regulations (DSEAR) came into force on 9 December 2002 and complement the Management of Health and Safety at Work Regulations 1999. They are designed to implement the safety requirements of the Chemical Agents and Explosive Atmospheres Directive.

DSEAR deals with any dangerous substance that has the potential to create a risk to persons from energetic or energy releasing events such as fires or explosions. Dangerous substances include petrol, LPG, paint, solvents and combustible or explosive dust produced in machining and sanding operations, flour mills and distilleries.

Many of these substances will also create a health risk, for example, solvents are toxic as well as being flammable. However, DSEAR does not address the health risk, only the fire and explosion risk. The potential health risk is dealt with under the COSHH Regulations discussed earlier in this chapter.

The DSEAR Regulations follow the modern risk assessment-based approach. Technical and organizational measures are required to eliminate or reduce risks as far as is reasonably practicable. There is a requirement to provide equipment and procedures to deal with accidents and emergencies and also to provide information and training for employees.

So what sort of industries does DSEAR apply to? DSEAR is concerned with the harmful effects from burns, pressure injuries from explosions and asphyxiation arising from fires and explosions. Typical industries might be those concerned with the storage of petrol as a fuel for vehicles, agricultural and horticultural storage and the movement of bulk powders for the food industry, the storing of waste dust in a range of manufacturing industries, dust produced in the mining of coal, storage and transportation of paint and LPG.

The Construction (Design and Management) Regulations 1994

The Construction (Design and Management) Regulations (CDM) are aimed at improving the overall management of health, safety and welfare throughout all stages of the construction project.

Definition

'Duty holder', someone who has a duty of care for health, safety and welfare matters on site. This phrase recognizes the level of responsibility which electricians are expected to take on a part of their job in order to control electrical safety in the work environment.

The person requesting that construction work commence, the client, must first of all appoint a **'duty holder'**, someone who has a duty of care for health, safety and welfare matters on site. This person will be called a 'planning supervisor'. The planning supervisor must produce a 'pre-tender' health and safety plan and co-ordinate and manage this plan during the early stages of construction.

The client must also appoint a principal contractor who is then required to develop the health and safety plan made by the planning supervisor, and keep it up to date during the construction process to completion.

The degree of detail in the health and safety plan should be in proportion to the size of the construction project and recognize the health and safety risks involved on that particular project. Small projects will require simple straightforward plans, large projects, or those involving significant risk, will require more detail. The CDM Regulations will apply to most large construction projects but they do not apply to the following:

- Construction work, other than demolition work, that does not last longer than 30 days and does not involve more than four people.
- Construction work carried out inside commercial buildings such as shops and offices, which does not interrupt the normal activities carried out on those premises.
- Construction work carried out for a domestic client.
- The maintenance and removal of pipes or lagging which forms a part of a heating or water system within the building.

The Construction (Health, Safety and Welfare) Regulations 1996

An electrical contractor is a part of the construction team, usually as a sub-contractor, and therefore the regulations particularly aimed at the construction industry also influence the daily work procedures and environment of an electrician. The most important recent piece of legislation are the Construction Regulations.

The temporary nature of construction sites makes them one of the most dangerous places to work. These regulations are made under the Health and Safety at Work Act 1974 and are designed specifically to promote safety at work in the construction industry. Construction work is defined as any building or civil engineering work, including construction, assembly, alterations, conversions, repairs, upkeep, maintenance or dismantling of a structure.

The general provision sets out minimum standards to promote a good level of safety on site. Schedules specify the requirements for guardrails, working platforms, ladders, emergency procedures, lighting and welfare facilities. Welfare facilities set out minimum provisions for site accommodation: washing facilities, sanitary conveniences and protective clothing. There is now a duty for all those working on construction sites to wear head protection, and this includes electricians working on site as sub-contractors.

Safety First

Head protection

- The Construction Regulations require everyone working on a construction site to wear head protection
- This includes electricians.

Building Regulations – Part P 2006

The Building Regulations lay down the design and build standards for construction work in buildings in a series of Approved Documents. The scope of each Approved Document is given below:

Part A structure
Part B fire safety
Part C site preparation and resistance to moisture
Part D toxic substances
Part E resistance to the passage of sound
Part F ventilation
Part G hygiene
Part H drainage and waste disposal
Part J combustion appliances and fuel storage systems
Part K protection from falling, collision and impact
Part L conservation of fuel and power
Part M access and facilities for disabled people
Part N glazing – safety in relation to impact, opening and cleaning
Part P electrical safety.

Part P of the Building Regulations was published on 22 July 2004, bringing domestic electrical installations in England and Wales under building regulations control. This means that anyone carrying out domestic electrical installation work from 1 January 2005 must comply with Part P of the Building Regulations. An ammended document was published in an attempt at greater clarity and this came into effect on 6 April 2006.

If the electrical installation meets the requirements of the IEE Regulations BS 7671, then it will also meet the requirements of Part P of the Building Regulations, so no change there. What is going to change under Part P is this new concept of 'notification' to carry out electrical work.

NOTIFIABLE ELECTRICAL WORK

Any work to be undertaken by a firm or individual who is *not* registered under an 'approved competent person scheme' must be notified to the Local Authority Building Control Body before work commences. That is, work that involves:

- the provision of at least one new circuit,
- work carried out in kitchens,

- work carried out in bathrooms,
- work carried out in special locations such as swimming pools and hot air saunas.

Upon completion of the work, the Local Authority Building Control Body will test and inspect the electrical work for compliance with Part P of the Building Regulations.

NON-NOTIFIABLE ELECTRICAL WORK

Work carried out by a person or firm registered under an authorized Competent Persons Self-Certification Scheme or electrical installation work that does not include the provision of a new circuit. This includes work such as:

- replacing accessories such as socket outlets, control switches and ceiling roses;
- replacing a like for like cable for a single circuit which has become damaged by, for example, impact, fire or rodent;
- re-fixing or replacing the enclosure of an existing installation component provided the circuits protective measures are unaffected;
- providing mechanical protection to existing fixed installations;
- adding lighting points (light fittings and switches) to an existing circuit, provided that the work is not in a kitchen, bathroom or special location;
- installing or upgrading the main or supplementary equipotential bonding provided that the work is not in a kitchen, bathroom or special location.

All replacement work is non-notifiable even when carried out in kitchens, bathrooms and special locations, but certain work carried out in kitchens, bathrooms and special locations may be notifiable, even when carried out by an authorized competent person. The IEE have published a guide called the *Electricians' Guide* to the Building Regulations which brings clarity to this subject. In specific cases the Local Authority Building Control Officer or an approved Inspector will be able to confirm whether Building Regulations apply.

Failure to comply with the Building Regulations is a criminal offence and Local Authorities have the power to require the removal or alteration of work that does not comply with these requirements.

Electrical work carried out by DIY home-owners will still be permitted after the introduction of Part P. Those carrying out notifiable DIY work must first submit a building notice to the Local Authority before the work begins. The work must then be carried out to the standards set by the IEE Wiring Regulations BS 7671 and a building control fee paid for such work to be inspected and tested by the Local Authority.

COMPETENT PERSONS SCHEME

The Competent Persons Self-Certification Scheme is aimed at those who carry out electrical installation work as the primary activity of their business. The government has approved schemes to be operated by BRE Certification Ltd., British Standards Institution, ELECSA Ltd., NICEIC Certification Services Ltd., and Napit Certification Services Ltd. All the different bodies will operate the scheme to the same criteria and will be monitored by the Department for Communities and Local Government, formally called the Office of the Deputy Prime Minister.

Those individuals or firms wishing to join the Competent Persons Scheme will need to demonstrate their competence, if necessary, by first undergoing training. The work of members will then be inspected at least once each year. There will be an initial registration and assessment fee and then an annual membership and inspection fee.

The Electricity Safety, Quality and Continuity Regulations 2002

The Electricity Safety, Quality and Continuity Regulations replaces the Electricity Supply Regulations 1988. They are statutory regulations which are enforceable by the laws of the land. They are designed to ensure a proper and safe supply of electrical energy up to the consumer's terminals.

These regulations impose requirements upon the regional electricity companies regarding the installation and use of electric lines and equipment. The regulations are administered by the Engineering Inspectorate of the Electricity Division of the Department of Energy and will not normally concern the electrical contractor except that it is these regulations which lay down the earthing requirement of the electrical supply at the meter position.

The regional electricity companies must declare the supply voltage and maintain its value between prescribed limits or tolerances.

The government agreed on 1 January 1995 that the electricity supplies in the United Kingdom would be harmonized with those of the rest of Europe. Thus the voltages used previously in low voltage supply systems of 415 and 240V have become 400V for three-phase supplies and 230V for single-phase supplies. The permitted tolerances to the nominal voltage have also been changed from ±6% to +10% and −6%.

The next proposed change is for the tolerance levels to be adjusted to ±10% of the declared nominal voltage. (IEE Regulations Appendix 2:14).

The frequency is maintained at an average value of 50 Hz over 24 hours so that electric clocks remain accurate.

Regulation 29 gives the area boards the power to refuse to connect a supply to an installation which in their opinion is not constructed, installed and protected to an appropriately high standard. This regulation would only be enforced if the installation did not meet the requirements of the IEE Regulations for Electrical Installations.

The Electricity at Work Regulations 1989

This legislation came into force in 1990 and replaced earlier regulations such as the Electricity (Factories Act) Special Regulations 1944. The regulations are made under the Health and Safety at Work Act 1974, and enforced by the HSE. The purpose of the regulations is to 'require precautions to be taken against the risk of death or personal injury from electricity in work activities'.

Section 4 of the Electricity at Work Regulations (EWR) tells us that 'all systems must be constructed so as to prevent danger ..., and be properly maintained. ... Every work activity shall be carried out in a manner which does not give rise to danger. ... In the case of work of an electrical nature, it is preferable that the conductors be made dead before work commences'.

The EWR do not tell us specifically how to carry out our work activities and ensure compliance, but if proceedings were brought against an individual for breaking the EWR, the only acceptable defence would be 'to prove that all reasonable steps were taken and all diligence exercised to avoid the offence' (Regulation 29).

An electrical contractor could reasonably be expected to have 'exercised all diligence' if the installation was wired according to the IEE Wiring Regulations and this is confirmed in the 17th Edition at Regulation 114 (see below). However, electrical contractors must become more 'legally aware' following the conviction of an electrician for manslaughter at Maidstone Crown Court in 1989. The Court accepted that an electrician had caused the death of another man as a result of his shoddy work in wiring up a central heating system. He received a 9 month suspended prison sentence. This case has set an important legal precedent, and in future any tradesman or professional who causes death through negligence or poor workmanship risks prosecution and possible imprisonment.

DUTY OF CARE

The Health and Safety at Work Act and the Electricity at Work Regulations (EWR) make numerous references to employer and employees having a 'duty of care' for the health and safety of others in the work environment. In this context the EWR refer to a person as a 'duty holder'. This phrase recognizes the level of responsibility which electricians are expected to take on a part of their job in order to control electrical safety in the work environment.

Definition

Everyone has a *duty of care* but not everyone is a duty holder. The person who exercises 'control over the whole systems, equipment and conductors' and is the Electrical Company's representative on site, is the duty holder.

Everyone has a **duty of care** but not everyone is a duty holder. The regulations recognize the amount of control that an individual might exercise over the whole electrical installation. The person who exercises 'control over the whole systems, equipment and conductors' and is the Electrical Company's representative on site, is *the duty holder*. He might be a supervisor or

Definition

'Reasonably practicable' or 'absolute' If the requirement of the regulation is absolute, then that regulation must be met regardless of cost or any other consideration. If the regulation is to be met 'so far as is reasonably practicable' then risks, cost, time trouble and difficulty can be considered.

manager, but he will have a duty of care on behalf of his employer for the electrical, health, safety and environmental issues on that site.

Duties referred to in the regulations may have the qualifying terms **'reasonably practicable'** or *'absolute'*. If the requirement of the regulation is absolute, then that regulation must be met regardless of cost or any other consideration. If the regulation is to be met 'so far as is reasonably practicable' then risks, cost, time trouble and difficulty can be considered.

Often there is a cost effective way to reduce a particular risk and prevent an accident occurring. For example, placing a fire-guard in front of the fire at home when there are young children in the family is a reasonably practicable way of reducing the risk of a child being burned.

If a regulation is not qualified with 'so far as is reasonably practicable' then it must be assumed that the regulation is absolute. In the context of the EWR, where the risk is very often death by electrocution, the level of duty to prevent danger more often approaches that of an absolute duty of care.

The IEE Wiring Regulations 17th Edition to BS 7671: 2008

REQUIREMENTS FOR ELECTRICAL INSTALLATIONS

The Institution of Electrical Engineers Requirements for Electrical Installations (the IEE Regulations) are non-statutory regulations. They relate principally to the design, selection, erection, inspection and testing of electrical installations, whether permanent or temporary, in and about buildings generally and to agricultural and horticultural premises, construction sites and caravans and their sites. Paragraph 7 of the introduction to the EWR says: 'the IEE Wiring Regulations is a code of practice which is widely recognized and accepted in the United Kingdom and compliance with them is likely to achieve compliance with all relevant aspects of the EWR'. The IEE Regulations confirm this relationship at Regulation 114 which states that compliance with the IEE Regulations may be used in a Court of Law to claim compliance with a statutory requirement such as the EWR. The IEE Wiring Regulations only apply to installations operating at a voltage up to 1000V a.c. They do not apply to electrical installations in mines and quarries, where special regulations apply because of the adverse conditions experienced there.

The current edition of the IEE Wiring Regulations, is the 17th edition. The main reason for incorporating the IEE Wiring Regulations into British Standard BS 7671: 2008 was to create harmonization with European standards.

To assist electricians in their understanding of the regulations a number of guidance notes have been published. The guidance notes which I will frequently make reference to in this book are those contained in the *On Site Guide*. Eight other guidance notes booklets are also currently available. These are:

1. *Selection and Erection;*
2. *Isolation and Switching;*
3. *Inspection and Testing;*

4. *Protection against Fire;*
5. *Protection against Electric Shock;*
6. *Protection against Overcurrent;*
7. *Special Locations;*
8. *Earthing and Bonding.*

These guidance notes are intended to be read in conjunction with the regulations.

The IEE Wiring Regulations are the electricians' bible and provide the authoritative framework of information for anyone working in the electrotechnical industry.

ENVIRONMENTAL LAWS AND REGULATIONS

Environmental laws protect the environment in which we live by setting standards for the control of pollution to land, air and water.

If a wrong is identified in the area in which we now think of as 'environmental' it can be of two kinds.

1. An offence in common law which means damage to property, nuisance or negligence leading to a claim for damages.
2. A statutory offence against one of the laws dealing with the protection of the environment. These offences are nearly always 'crimes' and punished by fines or imprisonment rather than by compensating any individual.

The legislation dealing with the environment has evolved for each part – air, water, land noise, radioactive substances where an organization's activities impact upon the environmental laws they are increasingly adopting environmental management systems which comply with ISO 14001. Let us now look at some of the regulations and try to see the present picture at the beginning of the new millennium.

Environmental Protection Act 1990

In the context of environmental law, the Environmental Protection Act 1990 was a major piece of legislation. The main sections of the act are:

Part 1	Integrated pollution control by HM Inspectorate of Pollution, and air pollution control by Local Authorities
Part 2	Wastes on land
Part 3	Statutory nuisances and clean air
Part 4	Litter
Part 5	Radioactive Substances Act 1960
Part 6	Genetically modified organisms
Part 7	Nature conservation
Part 8	Miscellaneous, including contaminated land.

The Royal Commission of 1976 identified that a reduction of pollutant to one medium, air, water or land, then led to an increase of pollutant to another. It, therefore, stressed the need to take an integrated approach to pollution control. The processes subject to an integrated pollution control are:

- Air emissions.
- Processes which give rise to significant quantities of special waste, that is, waste defined in law in terms of its toxicity or flammability.
- Processes giving rise to emissions to sewers or 'Red List' substances. These are 23 substances including mercury, cadmium and many pesticides, which are subject to discharge consent to the satisfaction of the Environment Agency.

Where a process is under integrated control the Inspectorate is empowered to set conditions to ensure that the best practicable environmental option (BPEO) is employed to control pollution. This is the cornerstone of the Environmental Protection Act.

Pollution Prevention and Control Regulations 2000

The system of Pollution Prevention and Control is replacing that of Integrated Pollution Control established by the Environmental Protection Act 1990, thus bringing environmental law into the new millennium and implementing the European Directive (EC/96/61) on integrated pollution prevention and control. The new system will be fully implemented by 2007.

Pollution Prevention and Control is a regime for controlling pollution from certain industrial activities. This regime introduces the concept of Best Available Technique (BAT) for reducing and preventing pollution to an acceptable level.

Industrial activities are graded according to their potential to pollute the environment:

- A(1) installations are regulated by the Environment Agency.
- A(2) installations are regulated by the Local Authorities.
- Part B installations are also regulated by the Local Authority.

All three systems require the operators of certain industrial installations to obtain a permit to operate. Once an operator has submitted a permit application, the regulator then decides whether to issue a permit. If one is issued it will include conditions aimed at reducing and preventing pollution to acceptable levels. A(1) installations are generally perceived as having the greatest potential to pollute the environment. A(2) installations and Part B installations would have the least potential to pollute.

The industries affected by these regulations are those dealing with petrol vapour recovery, incineration of waste, mercury emissions from crematoria, animal rendering, non-ferrous foundry processes, surface treating of metals and plastic materials by powder coating, galvanizing of metals and the manufacture of certain specified composite wood-based boards.

Clean Air Act 1993

We are all entitled to breathe clean air but until quite recently the only method of heating houses and workshops was by burning coal, wood or peat in open fires. The smoke from these fires created air pollution and the atmosphere in large towns and cities was of poor quality. On many occasions in the 1950s the burning of coal in London was banned because the city was grinding to a halt because of the combined effect of smoke and fog, called smog. Smog was a very dense fog in which you could barely see more than a metre in front of you and which created serious breathing difficulties. In the new millennium we are no longer dependent upon coal and wood to heat our buildings, smokeless coal has been created and the gaseous products of combustion are now diluted and dispersed by new chimney design regulations. Using well engineered combustion equipment together with the efficient arrestment of small particles in commercial chimneys of sufficient height, air pollution has been much reduced. This is what the Clean Air Act set out to achieve and it has been largely successful.

Definition

The *Clean Air Act* applies to all small and medium sized companies operating furnaces, boilers, or incinerators.

The **Clean Air Act** applies to all small and medium sized companies operating furnaces, boilers, or incinerators. Compliance with the Act does not require an application for authorization and so companies must make sure that they do not commit an offence. In general the emission of dark smoke from any chimney is unacceptable. The emission of dark smoke from any industrial premises is also unacceptable. This might be caused by, for example, the burning of old tyres or old cable.

In England, Scotland and Wales it is not necessary for the Local Authority to have witnessed the emission of dark smoke before taking legal action. Simply the evidence of burned materials, which potentially give rise to dark smoke when burned, is sufficient. In this way the law aims to stop people creating dark smoke under the cover of darkness.

Definition

A *public nuisance* is 'an act unwarranted by law or an omission to discharge a legal duty which materially affects the life, health, property, morals or reasonable comfort or convenience of Her Majesty's subjects'.

A **public nuisance** is 'an act unwarranted by law or an omission to discharge a legal duty which materially affects the life, health, property, morals or reasonable comfort or convenience of Her Majesty's subjects'. This is a criminal offence and Local Authorities can prosecute, defend or appear in proceedings that affect the inhabitants of their area.

Controlled Waste Regulations 1998

Under these Regulations we have a 'Duty of Care to handle, recover and dispose of all waste responsibly'. This means that all waste must be handled, recovered and disposed of by individuals or businesses that are authorized to do so under a system of signed Waste Transfer Notes.

The Environmental Protection (Duty of Care) Regulations 1991 state that as a business you have a duty to ensure that any waste you produce is handled safely and in accordance with the law. This is the 'Duty of Care' and applies to anyone who produces, keeps, carries, treats or disposes of waste from business or industry.

You are responsible for the waste that you produce, even after you have passed it on to another party such as a Skip Hire company, a Scrap Metal merchant, recycling company or local council. The Duty of Care has no time limit and extends until the waste has either been finally and properly disposed of or fully recovered.

So what does this mean for your company?

- Make sure that waste is only transferred to an authorized company.
- Make sure that waste being transferred is accompanied by the appropriate paperwork showing what was taken, where it was to be taken and by whom.
- Segregate the different types of waste that your work creates.
- Label waste skips and waste containers so that it is clear to everyone what type of waste goes into that skip.
- Minimize the waste that you produce and do not leave waste behind for someone else to clear away. Remember there is no time limit on your Duty of Care for waste.

Occupiers of domestic properties are exempt from the Duty of Care for the household waste that they produce. However, they do have a Duty of Care for the waste produced by, for example, a tradesperson working at a domestic property.

Definition

Special waste is covered by the Special Waste Regulations 1996 and is waste that is potentially hazardous or dangerous and which may, therefore, require special precautions during handling, storage, treatment or disposal. Examples of special waste are asbestos, lead-acid batteries, used engine oil, solvent-based paint, solvents, chemical waste and pesticides.

Special waste is covered by the Special Waste Regulations 1996 and is waste that is potentially hazardous or dangerous and which may, therefore, require special precautions during handling, storage, treatment or disposal. Examples of special waste are asbestos, lead-acid batteries, used engine oil, solvent-based paint, solvents, chemical waste and pesticides. The disposal of special waste must be carried out by a competent person, with special equipment and a licence.

New Hazardous Waste Regulations were introduced in July 2005 and under these Regulations electric discharge lamps and tubes such as fluorescent, sodium, metal halide and mercury vapour are classified as hazardous waste. While each lamp only contains a very small amount of mercury, vast numbers are used and disposed of each year, resulting in a significant environmental threat. The environmentally responsible way to dispose of lamps and tubes is to recycle them and this process is now available through the electrical wholesalers.

Electrotechnical companies produce relatively small amounts of waste and even smaller amounts of special waste. Most companies buy in the expertise of specialist waste companies these days and build these costs into the contract.

Waste Electrical and Electronic Equipment EU Directive 2007

The Waste Electrical and Electronic Equipment (WEEE) Regulations will ensure that Britain complies with its EU obligation to recycle waste from electrical products. The Regulation came into effect in July 2007 and from that date any company which makes, distributes or trades in electrical or electronic goods such as household appliances, sports equipment and even torches and toothbrushes will have to make arrangements for recycling these goods at the end of their useful life. Batteries will be covered separately by yet another forthcoming EU directive.

Some sectors are better prepared for the new regulations than others. Mobile phone operators, O2, Orange, Virgin and Vodafone, along with retailers such as Currys and Dixons, have already joined together to recycle their mobile phones collectively. In Holland the price of a new car now includes a charge for the recycling costs.

Further Information is available on the DTI and DEFRA website under WEEE.

Radioactive Substances Act 1993

These regulations apply to the very low ionizing radiation sources used by specialized industrial contractors. The radioactive source may be sealed or unsealed. Unsealed sources are added to a liquid in order to trace the direction or rate of flow of that liquid. Sealed radioactive sources are used in radiography for the non-destructive testing of materials or in liquid level and density gauges.

This type of work is subject to the Ionising Radiations Regulations 1999 (IRR), which impose comprehensive duties on employers to protect people at work against exposure to ionizing radiation. These regulations are enforced by the HSE, while the Radioactive Substances Act (RSA) is enforced by the Environmental Agency.

The RSA 1993 regulates the keeping, use, accumulation and disposal of radioactive waste, while the IRR 1999 regulates the working and storage conditions when using radioactive sources. The requirements of RSA 1993 are in addition to and separate from IRR 1999 for any industry using radioactive sources. These regulations also apply to offshore installations and to work in connection with pipelines.

Dangerous Substances and Preparations and Chemicals Regulations 2000

Chemical substances that are classified as carcinogenic, mutagenic or toxic, or preparations which contain those substances, constitute a risk to the general public because they may cause cancer, genetic disorders and birth defects, respectively.

These Regulations were introduced to prohibit the supply of these dangerous drugs to the general public, to protect consumers from contracting fatal diseases through their use.

The Regulations require that new labels be attached to the containers of these drugs which identify the potential dangers and indicate that they are restricted to professional users only.

The Regulations implement Commission Directive 99/43/EC, known as the 17th Amendment, which brings the whole of Europe to an agreement that these drugs must not be sold to the general public, this being the only way of offering the highest level of protection for consumers.

The Regulations will be enforced by the Local Authority Trading Standards Department.

Noise Regulations

Before 1960 noise nuisance could only be dealt with by common law as a breach of the peace under various Acts or local by-laws. In contrast, today there are many statutes, Government circulars, British Standards and EU Directives dealing with noise matters. Environmental noise problems have been around for many years. During the eighteenth century, in the vicinity of some London hospitals, straw was put on the roads to deaden the sound of horses' hooves and the wheels of carriages. Today we have come a long way from this self-regulatory situation.

Definition

'A *statutory nuisance* must materially interfere with the enjoyment of one's dwelling. It is more than just irritating or annoying and does not take account of the undue sensitivity of the receiver'.

In the context of the *Environmental Protection Act 1990*, noise or vibration is a **statutory nuisance** if it is prejudicial to health or is a nuisance. However, nuisance is not defined and has exercised the minds of lawyers, magistrates and judges since the concept of nuisance was first introduced in the 1936 Public Health Act. There is a wealth of case law but a good working definition might be 'A statutory nuisance must materially interfere with the enjoyment of one's dwelling. It is more than just irritating or annoying and does not take account of the undue sensitivity of the receiver'.

The line that separates nuisance from no nuisance is very fine and non-specific. Next door's intruder alarm going off at 3 a.m. for an hour or more is clearly a statutory nuisance, whereas one going off a long way from your home would not be a nuisance. Similarly, an all night party with speakers in the garden would be a nuisance, whereas an occasional party finishing at say midnight would not be a statutory nuisance.

At Stafford Crown Court on 1 November 2004, Alton Towers, one of the country's most popular Theme Parks, was ordered by a judge to reduce noise levels from its 'white knuckle' rides. In the first judgment of its kind, the judge told the Park's owners that neighbouring residents must not be interrupted by noise from rides such as Nemesis, Air, Corkscrew, Oblivion or from loudspeakers or fireworks.

The owners of Alton Towers, Tussauds Theme Parks Ltd., were fined the maximum sum of £5000 and served with a Noise Abatement Order for being guilty of breaching the 1990 Environmental Protection Act. Mr Richard Buxton, for the prosecution, said that the £5000 fine reflected the judge's view that Alton Towers had made little or no effort to reduce the noise nuisance.

Many nuisance complaints under the Act are domestic and are difficult to assess and investigate. Barking dogs, stereos turned up too loud, washing machines running at night to use 'low cost' electricity, television, DIY activities are all difficult to assess precisely as statutory nuisance. Similarly, sources of commercial noise complaints are also varied and include deliveries of goods during the night, general factory noises, refrigeration units, noise from public houses and clubs are all common complaints.

Industrial noise can be complex and complaints difficult to resolve both legally and technically. Industrial noise assessment is aided by BS 4142 but no guidance exists for other noise nuisance. The Local Authority has a duty to take reasonable steps to investigate all complaints and to take appropriate action.

The Noise and Statutory Nuisance Act 1993

This Act extended the statutory nuisance provision of the Environmental Protection Act 1990 to cover noise from vehicles, machinery or equipment in the streets. The definition of equipment includes musical instruments but the most common use of this power is to deal with car alarms and house intruder alarms being activated for no apparent reason and which then continue to cause a nuisance for more than 1 hour.

In the case of a car alarm a notice is fixed to the vehicle and an officer from the Local Authority spends 1 hour trying to trace the owner with help from the police and their National Computer system. If the alarm is still sounding at the end of this period, then the Local Authority Officer can break into the vehicle and silence the alarm. The vehicle must be left as secure as possible but if this cannot be done then it can be removed to a safe compound after the police have been notified. Costs can be recovered from the registered keeper.

Home intruder alarms that have been sounding for 1 hour can result in a 'Notice' being served on the occupier of the property, even if he or she is absent from the property at the time of the offence. The Notice can be served by putting it through a letterbox. A Local Authority Officer can then immediately silence the alarm without going into the property. However, these powers are *adoptive* and some Local Authorities have indicated that they will not adopt them because Sections 7–9 of the Act makes provision for incorporating the 'Code of Practice relating to Audible Intruder Alarms' into the statute. The two key points of the Code are the installation of a 20 minute cut-off of the external sounder and the notification to the police and Local Authority of two key holders who can silence the alarm.

Noise Act 1996

This Act clarifies the powers which may be taken against work which is in default under the nuisance provision of the Environmental Protection Act 1990. It provides a mechanism for permanent deprivation, return of seized equipment and charges for storage.

The Act also includes an *adoptive* provision making night time noise between 23:00 and 07:00 hours a criminal offence if the noise exceeds a certain level to be prescribed by the Secretary of State. If a notice is not complied with, a fixed penalty may be paid instead of going to court.

Noise at Work Regulations 1989

The Noise at Work Regulations, unlike the previous vague or limited provisions, apply to all work places and require employers to carry out assessments of the noise levels within their premises and to take appropriate action where necessary. The 1989 Regulations came into force on 1 January 1990 implementing in the United Kingdom the EC Directive 86/188/EEC 'The Protection of Workers from Noise'.

Three action levels are defined by the Regulations:

1. The first action level is a daily personal noise exposure of 85 dB, expressed as 85 dB(A).
2. The second action level is a daily personal noise exposure of 90 dB(A).
3. The third defined level is a peak action level of 140 dB(A) or 200 Pa of pressure which is likely to be linked to the use of cartridge operated tools, shooting guns or similar loud explosive noises. This action level is likely to be most important where workers are subjected to a small number of loud impulses during an otherwise quiet day.

The Noise at Work Regulations are intended to reduce hearing damage caused by loud noise. So, what is a loud noise? If you cannot hear what someone is saying when they are 2 m away from you or if they have to shout to make themselves heard, then the noise level is probably above 85 dB and should be measured by a competent person.

At the first action level an employee must be provided with ear protection (ear muffs or ear plugs) on request. At the second action level the employer must reduce, so far as is reasonably practicable, other than by providing ear protection, the exposure to noise of that employee.

Hearing damage is cumulative, it builds up, leading eventually to a loss of hearing ability. Young people, in particular, should get into the routine of avoiding noise exposure before their hearing is permanently damaged. The damage can also take the form of permanent tinnitus (ringing noise in the ears) and an inability to distinguish words of similar sound such as bit and tip.

Vibration is also associated with noise. Direct vibration through vibrating floors or from vibrating tools, can lead to damage to the bones of the feet or hands. A condition known as 'vibration white finger' is caused by an impaired blood supply to the fingers, associated with vibrating hand tools.

Employers and employees should not rely too heavily on ear protectors. In practice, they reduce noise exposure far less than is often claimed, because they may be uncomfortable or inconvenient to wear. To be effective, ear protectors need to be worn all the time when in noisy places. If left off for even a short time, the best protectors cannot reduce noise exposure effectively.

Protection against noise is best achieved by controlling it at source. Wearing ear protection must be a last resort. Employers should:

- Design machinery and processes to reduce noise and vibration (mounting machines on shock absorbing materials can dampen out vibration).
- When buying new equipment, where possible, choose quiet machines. Ask the supplier to specify noise levels at the operator's working position.
- Enclose noisy machines in sound absorbing panels.
- Fit silencers on exhaust systems.
- Install motor drives in a separate room away from the operator.
- Inform workers of the noise hazard and get them to wear ear protection.
- Reduce a worker's exposure to noise by job rotation or provide a noise refuge.

New regulations introduced in 2006 reduce the first action level to 80 dB(A) and the second level to 85 dB(A) with a peak action level of 98 dB(A) or 140 Pa of pressure. Every employer must make a 'noise' assessment and provide workers with information about the risks to hearing if the noise level approaches the first action level. He must do all that is reasonably practicable to control the noise exposure of his employees and clearly mark ear protection zones. Employees must wear personal ear protection whilst in such a zone.

The EHO (Environmental Health Officer)

The responsibilities of the EHO are concerned with reducing risks and eliminating the dangers to human health associated with the living and working environment. They are responsible for monitoring and ensuring the maintenance of standards of environmental and public health, including food safety, workplace health and safety, housing, noise, odour, industrial waste, pollution control and communicable diseases in accordance with the law. Although they have statutory powers with which to enforce the relevant regulations, the majority of their work involves advising and educating in order to implement public health policies.

The majority of EHO are employed by Local Authorities, who are the agencies concerned with the protection of public health. Increasingly, however, officers are being employed by the private sector, particularly those concerned with food, such as large hotel chains, airlines and shipping companies.

Your Local Authority EHO would typically have the responsibility of enforcing the environmental laws discussed above. Their typical work activities are to:

- ensure compliance with the Health and Safety at Work Act 1974, the Food Safety Act 1990 and the Environmental Protection Act 1990;

- carry out Health and Safety investigations, food hygiene inspections and food standards inspections;
- investigate public health complaints such as illegal dumping of rubbish, noise complaints and inspect contaminated land;
- investigate complaints from employees about their workplace and carry out accident investigations;
- investigate food poisoning outbreaks;
- obtain food samples for analysis where food is manufactured, processed or sold;
- visit housing and factory accommodation to deal with specific incidents such as vermin infestation and blocked drains;
- test recreational water, such as swimming pool water and private water supplies in rural areas;
- inspect and licence pet shops, animal boarding kennels, riding stables and zoos;
- monitor air pollution in heavy traffic areas and remove abandoned vehicles;
- work in both an advisory capacity and as enforcers of the law, educating managers of premises on issues which affect the safety of staff and members of the public.

In carrying out these duties, officers have the right to enter any workplace without giving notice, although notice may be given if they think it appropriate. They may also talk to employees, take photographs and samples and serve an Improvement Notice, detailing the work which must be carried out if they feel that there is a risk to health and safety that needs to be dealt with.

Enforcement Law Inspectors

If the laws relating to work, the environment and people are to be effective, they must be able to be enforced. The system of control under the Health and Safety at Work Act comes from the HSE or the Local Authority. Local Authorities are responsible for retail and service outlets such as shops, garages, offices, hotels, public houses and clubs. The HSE are responsible for all other work premises including the Local Authorities themselves. Both groups of inspectors have the same powers. They are allowed to:

- enter premises, accompanied by a police officer if necessary;
- examine, investigate and require the premises to be left undisturbed;
- take samples and photographs as necessary, dismantle and remove equipment;
- require the production of books or documents and information;

- seize, destroy or render harmless any substance or article;
- issue enforcement notices and initiate prosecutions.

Definition

An *improvement notice* identifies a contravention of the law and specifies a date by which the situation is to be put right.

Definition

A *prohibition notice* is used to stop an activity which the inspector feels may lead to serious injury.

There are two types of enforcement notices, an **'improvement notice'** and a **'prohibition notice'**.

An improvement notice identifies a contravention of the law and specifies a date by which the situation is to be put right. An appeal may be made to an Employment Tribunal within 21 days.

A prohibition notice is used to stop an activity which the inspector feels may lead to serious injury. The notice will identify which legal requirement is being contravened and the notice takes effect as soon as it is issued. An appeal may be made to the Employment Tribunal but the notice remains in place and work is stopped during the appeal process.

Cases may be heard in the Magistrates' or Crown Courts.

Magistrates' Court (Summary Offences) for health and safety offences, employers may be fined up to £20,000 and employees or individuals up to £5000. For failure to comply with an enforcement notice or a court order, anyone may be imprisoned for up to 6 months.

Crown Court (Indictable Offences) for failure to comply with an enforcement notice or a court order, fines are unlimited in the Crown Court and may result in imprisonment for up to 2 years.

Actions available to an inspector upon inspection of premises:

- Take no action – the law is being upheld.
- Give verbal advice – minor contraventions of the law identified.
- Give written advice – omissions have been identified and a follow up visit will be required to ensure that they have been corrected.
- Serve an improvement notice – a contravention of the law has, or is taking place and the situation must be remedied by a given date. A follow up visit will be required to ensure that the matter has been corrected.
- Serve a prohibition notice – an activity has been identified which may lead to serious injury. The law has been broken and the activity must stop immediately;
- Prosecute – the law has been broken and the employer prosecuted.

On any visit one or more of the above actions may be taken by the inspector.

In-house safety representatives

The HSE and the EHO are the health and safety professionals. The day that one of these inspectors arrives to look at the health and safety systems and procedures that your company has in place is a scary day! Most companies are very conscientious about their health and safety responsibilities and want to comply with the law. Many of the regulations demand that the Health and Safety systems and procedures are regularly reviewed and monitored and that employees are informed and appropriately trained. To meet the requirements there is a need for 'competent persons' to be appointed to the various roles within the company structure to support the company directors in their management of the Health and Safety Policy. The number of people involved, and whether health and safety is their only company role, will depend upon the size of the company and the type of work being carried out. To say that 'everyone is responsible for health and safety' is very misleading and would definitely not impress a visiting HSE inspector. There is no equality of responsibility under the law between those who provide direction and create policy and those who are employed to carry out instructions. Company directors and employers have substantially more responsibilities than employees as far as the Health and Safety at Work Act is concerned. There therefore needs to be an appropriate structure and nominated 'competent persons' within the company to manage health and safety at work.

At the top of the health and safety structure there will need to be a senior manager. Like all management functions, establishing control and maintaining it day in day out is crucial to effective health and safety management. Senior managers must take proactive responsibility for controlling issues that could lead to ill health or injury. A nominated senior manager at the top of the organization must oversee policy implementation and monitoring.

Health and safety responsibilities must then be assigned to line managers and health and safety expertise must be available to them to help them achieve the requirements of the Health and Safety at Work Act and the Regulations made under the Act. The purpose of a health and safety organization within a company is to harness the collective enthusiasm, skill and effort of the whole workforce, with managers taking key responsibility and providing clear direction. The prevention of accidents and ill health through management systems of control then becomes the focus rather than looking for individuals to blame after an accident has happened. Two key personnel in this type of system might hold the job title '*Safety Officer*' and '*Safety Representative*'.

Definition

The *Safety Officer* will be the specialist member of staff, having responsibility for health and safety within the company. He or she will report to the senior manager responsible for health and safety.

The Safety Officer will probably hold a Health and Safety qualification such as NEBOSH (National Examination Board in Occupational Safety and Health) and will:

- monitor the internal Health and Safety systems,
- carry out risk assessments,
- maintain accident reports and records,
- arrange or carry out in-house training,
- update systems as Regulations change.

The **Safety Officer** will be the specialist member of staff, having responsibility for health and safety within the company. He or she will report to the senior manager responsible for health and safety and together they will develop strategies for implementing and maintaining the company's health and safety policies.

The Safety Officer will probably hold a health and safety qualification such as NEBOSH (National Examination Board in Occupational Safety and Health) and will:

- monitor the internal health and safety systems,
- carry out risk assessments,

- maintain accident reports and records,
- arrange or carry out in-house training,
- update systems as Regulations change.

If an accident occurs, the Safety Officer would lead the investigation, identify the cause and advise the senior manager responsible for health and safety on possible improvements to the system.

Definition

The *Safety Representative* will be the person who represents a small section of the workforce on the Safety Committee. The role of the Safety Representative will be to bring to the Safety Committee the health and safety concerns of colleagues and to take back to colleagues, information from the Committee.

The **Safety Representative** will be the person who represents a small section of the workforce on the Safety Committee. The role of the Safety Representative will be to bring to the Safety Committee the health and safety concerns of colleagues and to take back to colleagues, information from the Committee. The office of Safety Representative is often held by the Trade Union representative, since it is a similar role, representing colleagues on management committees. If the company does not have a Safety Committee then the Safety Representative will liaise with the Safety Officer, informing him of the training and other health and safety requirements of colleagues.

The Safety Officer and Safety Representative hold important positions within a company, informing both employers and employees on health and safety matters and helping the company meet its obligation to 'consult with employees' under the Health and Safety Regulations.

Regular monitoring and reviewing of systems and procedures is an essential part of any Health and Safety system. Similarly, monitoring and evaluating systems systematically is an essential part of many quality management systems. In Chapter 3 we will look at quality systems.

Try This

Safety Officer

- Is there someone in your company responsible for safety?
- What is his name?
- What does he do?

Safe working procedures

The principles which were laid down in the many Acts of Parliament and the Regulations that we have already looked at in this chapter, control our working environment. They make our workplace safer, but despite all this legislation, workers continue to be injured and killed at work or die as a result of a work-related injury. The number of deaths has consistently averaged about 200 each year for the past 8 years. These figures only relate to employees. If you include the self-employed and members of the public killed in work-related accidents, the numbers almost double.

In addition to the deaths, about 28,000 people have major accidents at work and about 130,000 people each year, receive minor work-related injuries which keep them off work for more than 3 days.

It is a mistake to believe that these things only happen in dangerous occupations such as deep sea diving, mining and quarrying, fishing industry, tunnelling and fire-fighting or that it only happens in exceptional circumstances such as would never happen in your workplace. This is not the case. Some basic thinking and acting beforehand, could have prevented most of these accident statistics, from happening.

Definition

Human errors include behaving badly or foolishly, being careless and not paying attention to what you should be doing at work.

Definition

Environmental conditions include unguarded or faulty machinery.

Safety First

Safety Procedures

- Hazard Risk Assessment *is an essential part* of any Health and Safety management system.
- The aim of the planning process is to minimize risk.
- HSE publication HSG(65).

CAUSES OF ACCIDENTS

Most accidents are caused by either human error or environmental conditions. **Human errors** include behaving badly or foolishly, being careless and not paying attention to what you should be doing at work, doing things that you are not competent to do or have not been trained to do. You should not work when tired or fatigued and should never work when you have been drinking alcohol or taking drugs.

Environmental conditions include unguarded or faulty machinery, damaged or faulty tools and equipment, poorly illuminated or ventilated workplaces and untidy, dirty or overcrowded workplaces.

The most common causes of accidents

These are:

- slips, trips and falls;
- manual handling, that is moving objects by hand;
- using equipment, machinery or tools;
- storage of goods and materials which then become unstable;
- fire;
- electricity;
- mechanical handling.

Accident prevention measures

To control the risk of an accident we usually:

- eliminate the cause;
- substitute a procedure or product with less risk;
- enclose the dangerous situation;
- put guards around the hazard;
- use safe systems of work;
- supervise, train and give information to staff;
- if the hazard cannot be removed or minimized then provide PPE.

Let us now look at the application of one of the procedures that make the workplace a safer place to work but first of all I want to explain what I mean when I use the words hazard and risk.

HAZARD AND RISK

A **hazard** is something with the 'potential' to cause harm, for example, chemicals, electricity or working above ground.

A **risk** is the 'likelihood'of harm actually being done.

Competent persons are often referred to in the Health and Safety at Work Regulations, but who is 'competent'? For the purposes of the Act, a competent person is anyone who has the necessary technical skills, training and expertise to safely carry out the particular activity. Therefore, a **competent person** dealing with a hazardous situation reduces the risk.

Definitions

A *hazard* is something with the 'potential' to cause harm, for example, chemicals, electricity or working above ground.

A *risk* is the 'likelihood' of harm actually being done.

Definition

A *competent person* is anyone who has the necessary technical skills, training and expertise to safely carry out the particular activity.

Think about your workplace and at each stage of what you do, think about what might go wrong. Some simple activities may be hazardous. Here are some typical activities where accidents might happen.

Typical activity	Potential hazard
Receiving materials	Lifting and carrying
Stacking and storing	Falling materials
Movement of people	Slips, trips and falls
Building maintenance	Working at heights or in confined spaces
Movement of vehicles	Collisions

How high are the risks? Think about what might be the worst result, is it a broken finger or someone suffering permanent lung damage or being killed? How likely is it to happen? How often is that type of work carried out and how close do people get to the hazard? How likely is it that something will go wrong?

How many people might be injured if things go wrong. Might this also include people who do not work for your company?

Employers of more than five people must document the risks at work and the process is known as Hazard Risk Assessment.

HAZARD RISK ASSESSMENT – THE PROCESS

The Management of Health and Safety at Work Regulations 1999 tells us that employers must systematically examine the workplace, the work activity and the management of safety in the establishment through a process of risk assessments. A record of all significant risk assessment findings must be kept in a safe place and be made available to an HSE Inspector if required. Information based on the risk assessment findings must be communicated to relevant staff and if changes in work behaviour patterns are recommended in the interests of safety, then they must be put in place.

So risk assessment must form a part of any employer's robust policy of health and safety. However, an employer only needs to 'formally' assess the significant risks. He is not expected to assess the trivial and minor types of household risks. Staff are expected to read and to act upon these formal risk assessments and they are unlikely to do so enthusiastically if the file is full of trivia. An assessment of risk is nothing more than a careful examination of what, in your work, could cause harm to people. It is a record that shows whether sufficient precautions have been taken to prevent harm.

The HSE recommends five steps to any risk assessment.

Step 1

Look at what might reasonably be expected to cause harm. Ignore the trivial and concentrate only on significant hazards that could result in serious harm or injury. Manufacturers data sheets or instructions can also help you spot hazards and put risks in their true perspective.

Step 2

Decide who might be harmed and how. Think about people who might not be in the workplace all the time – cleaners, visitors, contractors or maintenance personnel. Include members of the public or people who share the workplace. Is there a chance that they could be injured by activities taking place in the workplace.

Step 3

Evaluate what is the risk arising from an identified hazard. Is it adequately controlled or should more be done? Even after precautions have been put in place, some risk may remain. What you have to decide, for each significant hazard, is whether this remaining risk is low, medium or high. First of all, ask yourself if you have done all the things that the law says you have got to do. For example, there are legal requirements on the prevention of access to dangerous machinery. Then ask yourself whether generally accepted industry standards are in place, but do not stop there – think for yourself, because the law also says that you must do what is reasonably practicable to keep the workplace safe. Your real aim is to make all risks small by adding precautions, if necessary.

If you find that something needs to be done, ask yourself:

(i) Can I get rid of this hazard altogether?

(ii) If not, how can I control the risk so that harm is unlikely?

Only use PPE when there is nothing else that you can reasonably do.

If the work that you do varies a lot, or if there is movement between one site and another, select those hazards which you can reasonably foresee, the ones that apply to most jobs and assess the risks for them. After that, if you spot any unusual hazards when you get on site, take what action seems necessary.

Step 4

Record your findings and say what you are going to do about risks that are not adequately controlled. If there are fewer than five employees you do not need to write anything down but if there are five or more employees, the significant findings of the risk assessment must be recorded. This means writing down the more significant hazards and assessing if they are adequately controlled and recording your most important conclusions. Most employers have a standard risk assessment form which they use such as that shown in Fig. 1.3 but any format is suitable. The important thing is to make a record.

HAZARD RISK ASSESSMENT	FLASH-BANG ELECTRICAL CO.
For Company name or site: ---------------------- Address: ---------------------- ----------------------	Assessment undertaken by: ---------------------- Signed: ---------------------- Date: ----------------------
STEP 5 Assessment review date: ----------------------	
STEP 1 List the hazards here ---------------------- ---------------------- ---------------------- ---------------------- ---------------------- ---------------------- ---------------------- ---------------------- ---------------------- ---------------------- ----------------------	STEP 2 Decide who might be harmed ---------------------- ---------------------- ---------------------- ---------------------- ---------------------- ---------------------- ---------------------- ---------------------- ---------------------- ---------------------- ----------------------
STEP 3 Evaluate (what is) the risk – is it adequately controlled? State risk level as low, medium or high ---------------------- ---------------------- ---------------------- ---------------------- ---------------------- ---------------------- ---------------------- ---------------------- ----------------------	STEP 4 Further action – what else is required to control any risk identified as medium or high? ---------------------- ---------------------- ---------------------- ---------------------- ---------------------- ---------------------- ---------------------- ---------------------- ----------------------

FIGURE 1.3
Hazard risk assessment standard form.

There is no need to show how the assessment was made, providing you can show that:

1. a proper check was made,
2. you asked those who might be affected,
3. you dealt with all obvious and significant hazards,
4. the precautions are reasonable and the remaining risk is low,
5. you informed your employees about your findings.

Risk assessments need to be *suitable* and *sufficient*, not perfect. The two main points are:

1. Are the precautions reasonable?
2. Is there a record to show that a proper check was made?

File away the written Assessment in a dedicated file for future reference or use. It can help if an HSE Inspector questions the company's precautions or if the company becomes involved in any legal action. It shows that the company has done what the law requires.

Step 5

Review the assessments from time to time and revise them if necessary.

COMPLETING A RISK ASSESSMENT

When completing a risk assessment such as that shown in Fig. 1.3, do not be over complicated. In most firms in the commercial, service and light industrial sector, the hazards are few and simple. Checking them is common sense but necessary.

Step 1

List only hazards which you could reasonably expect to result in significant harm under the conditions prevailing in your workplace. Use the following examples as a guide:

- Slipping or tripping hazards (e.g. from poorly maintained or partly installed floors and stairs).
- Fire (e.g. from flammable materials you might be using, such as solvents).
- Chemicals (e.g. from battery acid).
- Moving parts of machinery (e.g. blades).
- Rotating parts of handtools (e.g. drills).
- Accidental discharge of cartridge operated tools.
- High pressure air from airlines (e.g. air powered tools).
- Pressure systems (e.g. steam boilers).

- Vehicles (e.g. fork lift trucks).
- Electricity (e.g. faulty tools and equipment).
- Dust (e.g. from grinding operations or thermal insulation).
- Fumes (e.g. from welding).
- Manual handling (e.g. lifting, moving or supporting loads).
- Noise levels too high (e.g. machinery).
- Poor lighting levels (e.g. working in temporary or enclosed spaces).
- Low temperatures (e.g. working outdoors or in refrigeration plant).
- High temperatures (e.g. working in boiler rooms or furnaces).

Step 2

Decide who might be harmed, do not list individuals by name. Just think about groups of people doing similar work or who might be affected by your work:

- Office staff
- Electricians
- Maintenance personnel
- Other contractors on site
- Operators of equipment
- Cleaners
- Members of the public.

Pay particular attention to those who may be more vulnerable, such as:

- staff with disabilities,
- visitors,
- young or inexperienced staff,
- people working in isolation or enclosed spaces.

Step 3

Calculate what is the risk – is it adequately controlled? Have you already taken precautions to protect against the hazards which you have listed in Step 1. For example:

- Have you provided adequate information to staff.
- Have you provided training or instruction.

Do the precautions already taken

- meet the legal standards required,

- comply with recognized industrial practice,
- represent good practice,
- reduce the risk as far as is reasonably practicable.

If you can answer 'yes' to the above points then the risks are adequately controlled, but you need to state the precautions you have put in place. You can refer to company procedures, company rules, company practices, etc., in giving this information. For example, if we consider there might be a risk of electric shock from using electrical power tools, then the risk of a shock will be *less* if the company policy is to PAT test all power tools each year and to fit a label to the tool showing that it has been tested for electrical safety. If the stated company procedure is to use battery drills whenever possible, or 110V drills when this is not possible, and to *never* use 230V drills, then this again will reduce the risk. If a policy such as this is written down in the company Safety Policy Statement, then you can simply refer to the appropriate section of the Safety Policy Statement and the level of risk will be low.

Step 4

Further action – what more could be done to reduce those risks which were found to be inadequately controlled?

You will need to give priority to those risks that affect large numbers of people or which could result in serious harm. Senior managers should apply the principles below when taking action, if possible in the following order:

1. Remove the risk completely.
2. Try a less risky option.
3. Prevent access to the hazard (e.g. by guarding).
4. Organize work differently in order to reduce exposure to the hazard.
5. Issue PPE.
6. Provide welfare facilities (e.g. washing facilities for removal of contamination and first aid).

Any hazard identified by a risk assessment as *high risk* must be brought to the attention of the person responsible for health and safety within the company. Ideally, in Step 4 of the Risk Assessment you should be writing, 'No further action is required. The risks are under control and identified as low risk'.

The assessor may use as many standard Hazard Risk Assessment forms, such as that shown in Fig. 1.3, as the assessment requires. Upon completion they should be stapled together or placed in a plastic wallet and stored in the dedicated file.

You might like to carry out a risk assessment on a situation you are familiar with at work, or at college using the standard form of Fig. 1.3, or your employer's standard forms.

Accident reports

Every accident must be reported to an employer and minor accidents reported to a supervisor, safety officer or first aider and the details of the accident and treatment given suitably documented. A first aid logbook or accident book such as that shown in Fig. 1.4 containing first aid treatment record sheets could be used to effectively document accidents which occur in the workplace and the treatment given. Failure to do so may influence the payment of compensation at a later date if an injury leads to permanent disability. To comply with the Data Protection Regulations, from 31 December 2003 all First Aid Treatment Logbooks or Accident Report books must contain perforated sheets which can be removed after completion and filed away for personal security.

If the accident results in death, serious injury or an injury that leads to an absence from work of more than 3 days, then your employer must report the accident to the local office of the HSE. The quickest way to do this is to call the Incident Control Centre on 0845 300 9923. They will require the following information:

- The name of the person injured.
- A summary of what happened.
- A summary of events prior to the accident.
- Information about the injury or loss sustained.
- Details of witnesses.
- Date and time of accident.
- Name of the person reporting the incident.

The Incident Control Centre will forward a copy of every report they complete to the employer for them to check and hold on record. However, good practice would recommend an employer or his representative make an extensive report of any serious accident that occurs in the workplace. In addition to recording the above information, the employer or his representative should:

FIGURE 1.4
First aid logbook/Accident book with data protection compliant removable sheets.

- Sketch diagrams of how the accident occurred, where objects were before and after the accident, where the victim fell, etc.
- Take photographs or video that show how things were after the accident, for example, broken stepladders, damaged equipment, etc.
- Collect statements from witnesses. Ask them to write down what they saw.
- Record the circumstances surrounding the accident. Was the injured person working alone – in the dark – in some other adverse situation or condition – was PPE being worn – was PPE recommended in that area?

The above steps should be taken immediately after the accident has occurred and after the victim has been sent for medical attention. The area should be made safe and the senior management informed so that any actions to prevent a similar occurrence can be put in place. Taking photographs and obtaining witnesses' statements immediately after an accident happens, means that evidence may still be around and memories still sharp.

Safety signs

The rules and regulations of the working environment are communicated to employees by written instructions, signs and symbols. All signs in the working environment are intended to inform. They should give warning of possible dangers and must be obeyed. At first there were many different safety signs but British Standard BS 5499 Part 1 and the Health and Safety (Signs and Signals) Regulations 1996 have introduced a standard system which gives health and safety information with the minimum use of words. The purpose of the regulations is to establish an internationally understood system of safety signs and colours which draw attention to equipment and situations that do, or could, affect health and safety. Text-only safety signs became illegal from 24 December 1998. From that date, all safety signs have had to contain a pictogram or symbol such as those shown in Fig. 1.5. Signs fall into four categories: prohibited activities; warnings; mandatory instructions and safe conditions.

PROHIBITION SIGNS

These are must not do signs. These are circular white signs with a red border and red cross bar, and are given in Fig. 1.6. They indicate an activity which *must not* be done.

FIGURE 1.5
Text-only safety signs do not comply.

WARNING SIGNS

Warning signs give safety information. These are triangular yellow signs with a black border and symbol, and are given in Fig. 1.7. They *give warning* of a hazard or danger.

MANDATORY SIGNS

These are must do signs. These are circular blue signs with a white symbol, and are given in Fig. 1.8. They *give instructions* which must be obeyed.

ADVISORY OR SAFE CONDITION SIGNS

These are square or rectangular green signs with a white symbol, and are given in Fig. 1.9. They *give information* about safety provision.

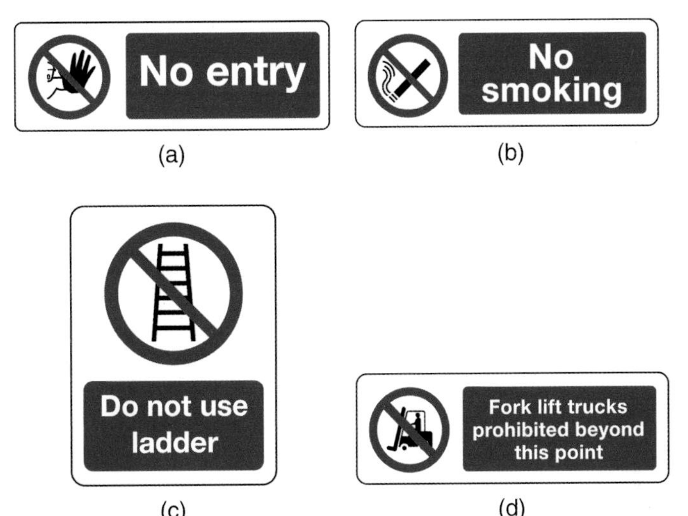

FIGURE 1.6
Prohibition signs. (These are MUST NOT DO signs.)

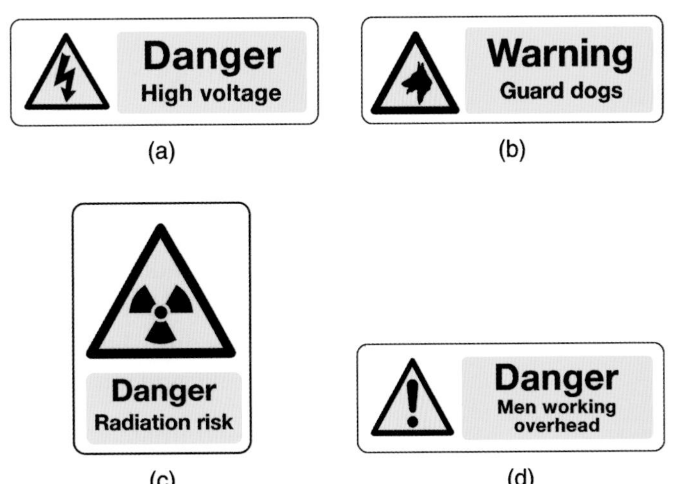

FIGURE 1.7
Warning signs. (These give safety information.)

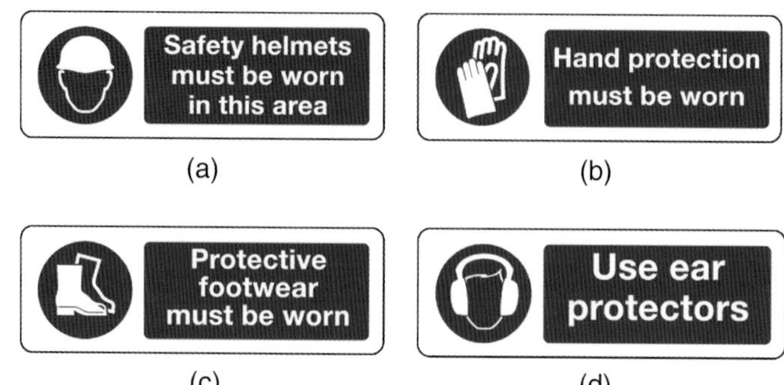

FIGURE 1.8
Mandatory signs. (These are MUST DO signs.)

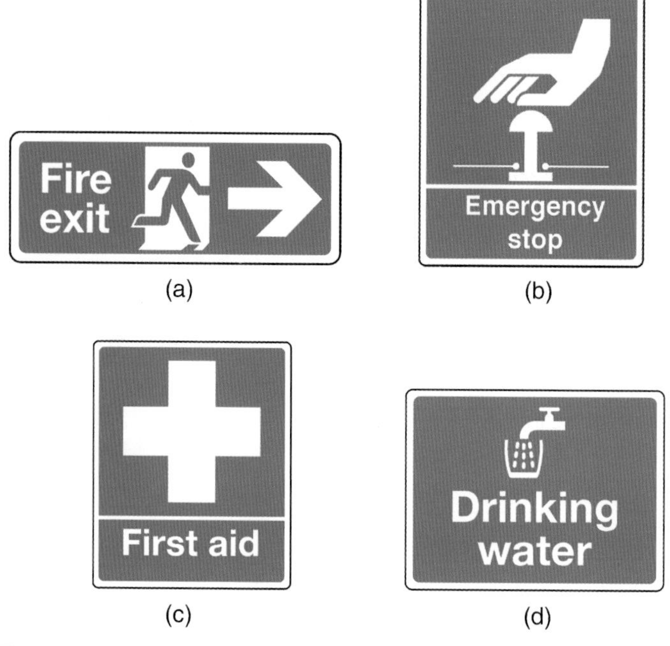

FIGURE 1.9
Advisory or safe condition signs. (These also give safety information.)

Good customer relationships

Remember that it is the customers who actually pay the wages of everyone employed in your company. You should always be polite and listen carefully to their wishes. They may be elderly or of a different religion or cultural background than you. In a domestic situation, the playing of loud music on a radio will not be approved of. Treat the property in which you are working with the utmost care. When working in houses, shops and offices use dust sheets to protect floor coverings and furnishings. Clean up periodically and make a special effort when the job is completed.

Dress appropriately: an unkempt or untidy appearance will encourage the customer to think that your work will be of poor quality.

The electrical installation in a building is often carried out alongside other trades. It makes good sense to help other trades where possible and to develop good working relationships with other employees. The customer will be most happy if the workers give an impression of working together as a team for the successful completion of the project. The customer will be most impressed by the workers punctuality, professional attitude, dedication to the job in hand and completion of the work in the agreed time.

Finally, remember that the customer will probably see more of the electrician and the electrical trainee than of the managing director of your firm and, therefore, the image presented by you will be assumed to reflect the policy of the company. You are, therefore, your company's most important representative. Always give the impression of being capable and in command of the situation, because this gives customers confidence in the company's ability to meet their needs. However, if a problem does occur which is outside your previous experience and you do not feel confident to solve it successfully, then contact your supervisor for professional help and guidance. It is not unreasonable for a young member of the company's team to seek help and guidance from those employees with more experience. This approach would be preferred by most companies rather than having to meet the cost of an expensive blunder.

Safety First

Safety signs

Which safety signs do you see around you?

- When you are at work?
- When you are at college?

Key Fact

Customer relationships

- Good customer relationships are important.
- Always be polite.
- Do not switch on a radio.
- Do not use a mobile phone during the 'customer's time'.
- Always be punctual.
- Always behave in a professional manner.

Legal contracts

Before work commences, some form of legal contract should be agreed between the two parties, that is, those providing the work (e.g. the subcontracting electrical company) and those asking for the work to be carried out (e.g. the main building company) or an individual customer or client.

A contract is a formal document which sets out the terms of agreement between the two parties. A standard form of building contract typically contains four sections:

1. The articles of agreement – this names the parties, the proposed building and the date of the contract period.
2. The contractual conditions – this states the rights and obligations of the parties concerned, for example, whether there will be interim payments for work or a penalty if work is not completed on time.
3. The appendix – this contains details of costings, for example, the rate to be paid for extras as daywork, who will be responsible for defects, how much of the contract tender will be retained upon completion and for how long.
4. The supplementary agreement – this allows the electrical contractor to recoup any value-added tax paid on materials at interim periods.

In signing the contract, the electrical contractor has agreed to carry out the work to the appropriate standards in the time stated and for the agreed cost. The other party, say the main building contractor, is agreeing to pay the price stated for that work upon completion of the installation.

If a dispute arises the contract provides written evidence of what was agreed and will form the basis for a solution.

For smaller electrical jobs, a verbal contract may be agreed, but if a dispute arises there is no written evidence of what was agreed and it then becomes a matter of one person's word against another's.

Changing work patterns

The electrotechnical industries cover a large range of activities and occupations from panel building, instrumentation, maintenance, cable jointing, highway electrical systems to motor re-winding, alarm and security systems, building management systems and computer installations. Electricians are often employed in the electrical contracting industry, installing wiring systems and equipment in houses, hospitals, schools, shops and offices. Electricians are also employed directly by factories, local councils, large commercial organizations, hospitals and the armed services where their skills are in demand. Employment opportunities for electrically trained people are enormous. There are about 21,000 electrical contracting companies registered in the United Kingdom. These companies employ from less than 10 people to the big multi-national companies, although the majority are small companies of less than 10 people. Then there are the small self-employed electrical businesses and those who work for the Local Authority, hospitals or armed forces who do not get counted as electrical personnel but as blue collar workers or soldiers.

The new technology of recent times has created many new opportunities for electrically competent personnel from installing satellite dishes, computer networks, extension telephone sockets for Internet connections to dichroic reflector miniature spotlight installations, intruder alarms and external illumination of garden areas.

New editions of the Regulations create work opportunities in domestic and public buildings bringing them up to the latest safety requirements.

A structured apprenticeship gives a broad range of experience opportunities and the achievement of the appropriate City and Guilds qualifications will lead to qualified electrician status with good electrical core skills.

New technologies present new opportunities to build on these core skills. New editions of the Regulations present new training opportunities. The acquisition of new skills gives the opportunity to transfer these new skills to new employers. Flexible workers with a range of skills can work in different disciplines in different parts of the electrotechnical industry in different parts of the country. Flexible workers are an attractive proposition to a prospective employer.

Electricians trained in the installation of conduit systems can easily transfer their skills to those we think of as belonging to a plumber or heating engineer in mechanical services. For those employed in the maintenance of fluid systems, instrumentation, monitoring and control will be required, and this may present opportunities of further responsibility or an increase in salary or status within a company.

Maintenance work demands that a craftsman has a range of skills and the flexibility to use them. If an electric motor was found to be faulty, then to replace it would require mechanical engineering and fitting skills as well as electrical skills and the one man who can do that job has multiple skills and can demand more pay.

Increased leisure opportunities have seen a huge increase in fitness centres containing lots of electrical equipment. The overuse and misuse of equipment means that it breaks down more frequently. When does it break down? At the most inconvenient time of course! The fitness centre manager wants the equipment fixed reasonably quickly, even if it is Sunday. They are at work so why isn't the electrician!

I live close to a seaside resort. All the people involved in the holiday seasonal work, work hard long hours, usually from Easter until the end of the summer. Everything then closes down and they then get on with their planned maintenance work. However, things also go wrong during the holiday season and the electrician is expected to support them when they need him.

This leads to a demand for flexible working hours or a flexible working week. Some of the small electrical companies have a rota system so that at least one member of staff is on cover for breakdowns and emergencies over a weekend. If the rota is shared out, then each individual only need cover, say one in four weekends, and as a result receives extra pay.

The foreseeable future for those employed in the electrotechnical industries is that they will require a firm practical and academic foundation. New technologies will require that we continue to learn new skills and new ideas will create new business opportunities for electrical companies. Regulations and laws will be updated to improve health and safety and to meet the demands of industry along with new training opportunities for employees to keep up to date with new requirements. Employees will also need to be flexible, not only in relation to what they can do but when they can do it. Very few people these days work regular fixed hours.

Check your Understanding

When you have completed these questions check out the answers at the back of the book.

Note: more than one multiple choice answer may be correct.

1. Under the Health and Safety at Work Act an employer has a duty of care to his employees. Identify an employer's duties from the list below.
   a. provide appropriate PPE
   b. wear appropriate PPE
   c. have plant and equipment properly maintained
   d. take reasonable care to avoid injury.

2. Under the Health and Safety at Work Act an employee has a duty of care to his employer. Identify an employee's responsibilities from the list below.
   a. provide appropriate PPE
   b. wear appropriate PPE
   c. have plant and equipment properly maintained
   d. take reasonable care to avoid injury.

3. The Health and Safety Laws are enforced by:
   a. The Local Trading Standards
   b. The IEE Regulations
   c. The ECA (Electrical Contractors Association)
   d. The HSE (Health and Safety Executive).

4. Every company that employs more than **five** people must have a:
   a. Health and Safety Policy statement
   b. pension plan for employees
   c. Health and Safety Law poster displayed
   d. means of ensuring Health and Safety awareness among its employees.

5. Safety signs showing the type of PPE are coloured:
   a. blue and white
   b. green and white
   c. red and white
   d. yellow and white.

6. Warning signs are coloured:
   a. blue and white
   b. green and white
   c. red and white
   d. yellow and white.

7. Prohibition signs are coloured:
   a. blue and white
   b. green and white
   c. red and white
   d. yellow and white.
8. Advisory or Safe Condition signs are coloured:
   a. blue and white
   b. green and white
   c. red and white
   d. yellow and white.
9. Someone who has the necessary technical skills training and experience to safely carry out a particular activity is said to be a:
   a. legal contract
   b. risk
   c. hazard
   d. competent person.
10. Something with the potential to cause harm is called a:
   a. legal contract
   b. risk
   c. hazard
   d. competent person.
11. The chance of harm actually being done as a result of a work activity we call a:
   a. legal contract
   b. risk
   c. hazard
   d. competent person.
12. Before any work begins, some form of agreement between the two parties should be made regarding cost, completion date and what work is to be done. This is called a:
   a. legal contract
   b. risk
   c. hazard
   d. competent person.

13. List **five** Safety Regulations which are very relevant to the electrotechnical industry.
14. List **three** pieces of environmental legislation which has an impact on electrotechnical activities.
15. List the responsibilities of an employer and employee under the Health and Safety at Work Act.
16. List the human and environmental conditions which lead to accidents in the workplace.
17. Produce a rough sketch and show the colour of the **four** types of Safety Sign, that is: Warning, Advisory, Mandatory and Prohibition signs.
18. Everyone has a duty of care but not everyone is a duty holder. Briefly describe the meaning of 'duty of care' and 'duty holder'.
19. Which **five** laws or regulations have the most impact upon your work in the electrotechnical industry?
20. The Electricity at Work Regulations describe things that must be done as 'reasonably practicable' or 'absolute'. Briefly describe the meaning of 'reasonably practicable' and 'absolute'.
21. State the advantages of having a written contract between the electrical company and their customers rather than a verbal contract.
22. Make a list of the things that could be done to ensure a good customer relationship.
23. The electrotechnical industry is expanding to meet the changing needs of customers who want to use the latest technology in their businesses. Make a list of the different types of electrical activity that your company provides to its customers.

Safe working practices and emergency procedures

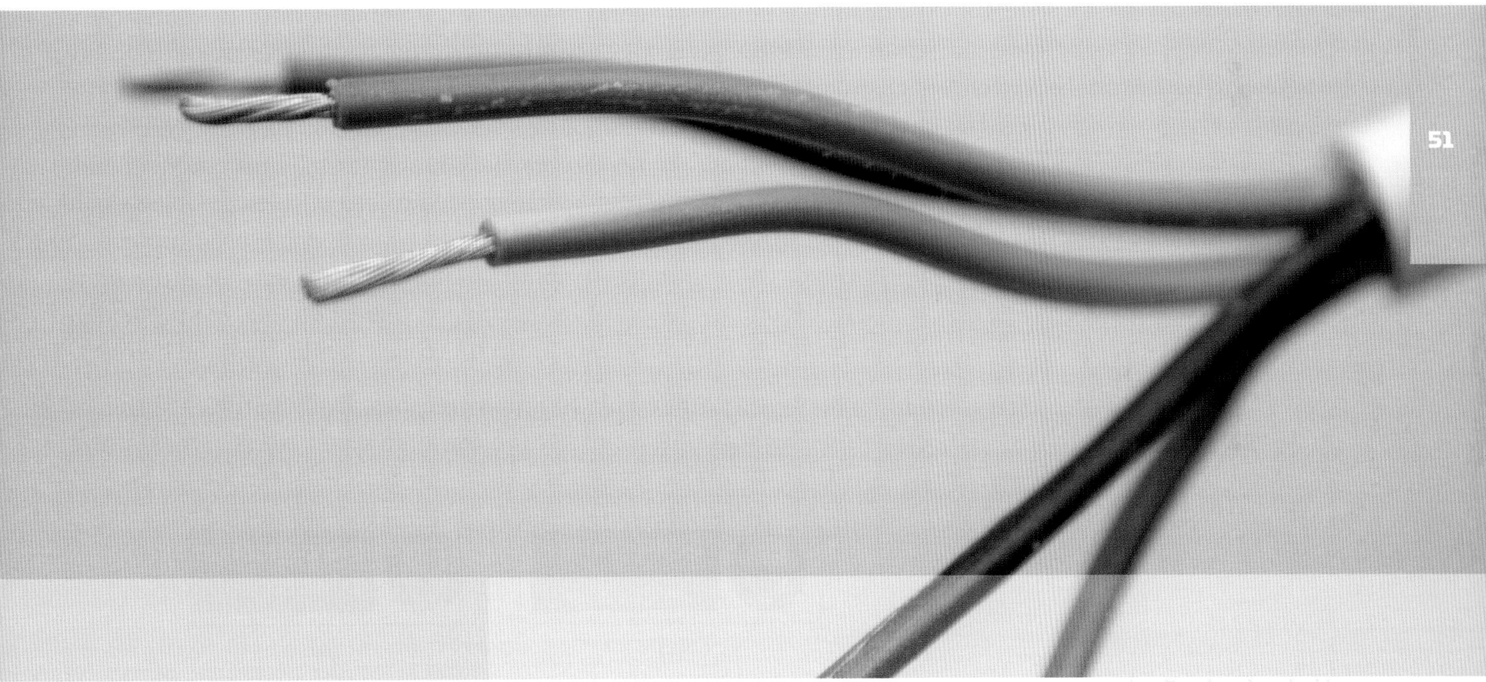

Unit 1 – Application of health and safety and electrical principles – Outcome 2

Underpinning knowledge: when you have completed this chapter you should be able to:

- outline 'permit-to-work' and 'secure electrical isolation procedures'
- state the requirements for the provision of first aid equipment
- state fire prevention methods and evacuation procedures
- describe work situations where you must not work alone
- state the emergency action following electric shock
- define asphyxiation, dangerous occurrence and hazardous malfunctions

Definition

A *hazard* is something with the potential to cause harm, for example, chemicals, electricity, working above ground.

Definition

A *risk* is the likelihood or chance of harm actually being done by the hazard.

Definition

PPE is defined as all equipment designed to be worn, or held, to protect against a risk to health and safety.

Safe working procedures to prevent injury

Where a particular hazard exists in the working environment, an employer must carry out a risk assessment and establish procedures which will reduce or eliminate the risk. When the risk cannot be completely removed, an employer must provide personal protective equipment (PPE) to protect his employees from a risk to health and safety.

Personal protective equipment

PPE is defined as all equipment designed to be worn, or held, to protect against a risk to health and safety. This includes most types of protective clothing, and equipment such as eye, foot and head protection, safety harnesses, life jackets and high-visibility clothing.

Under the Health and Safety at Work Act, employers must provide free of charge any **PPE** and employees must make full and proper use of it. Safety signs such as those shown in Fig. 2.1 are useful reminders of the type of PPE to be used in a particular area. The vulnerable parts of the body which may need protection are the head, eyes, ears, lungs, torso, hands and feet and, additionally, protection from falls may need to be considered. Objects falling from a height present the major hazard against which head protection is provided. Other hazards include striking the head against projections and hair becoming entangled in machinery. Typical methods of protection include helmets, light duty scalp protectors called 'bump caps' and hairnets.

The eyes are very vulnerable to liquid splashes, flying particles and light emissions such as ultraviolet light, electric arcs and lasers. Types of eye

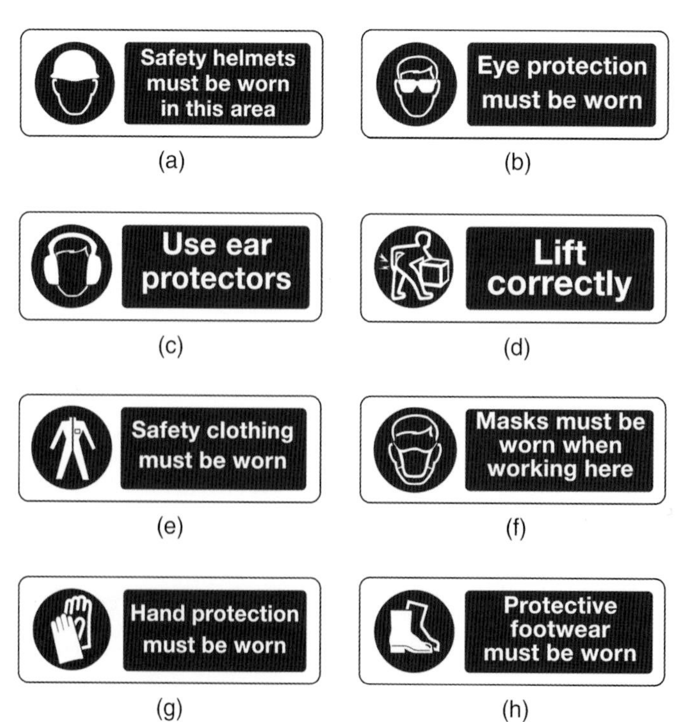

FIGURE 2.1
Safety signs showing type of PPE to be worn.

protectors include safety spectacles, safety goggles and face shields. Screen based workstations are being used increasingly in industrial and commercial locations by all types of personnel. Working with VDUs (visual display units) can cause eye strain and fatigue and therefore, every display screen operator is entitled to a free eye test.

Noise is accepted as a problem in most industries and we looked in some detail at the Noise Regulations a little earlier in this book under the Environmental Laws section.

Noise may be defined as any disagreeable or undesirable sound or sounds, generally of a random nature, which do not have clearly defined frequencies. The usual basis for measuring noise or sound level is the decibel scale. Whether noise of a particular level is harmful or not also depends upon the length of exposure to it. This is the basis of the widely accepted limit of 85 dB of continuous exposure to noise for 8 hours per day.

Where individuals must be subjected to some noise at work it may be reduced by ear protectors. These may be disposable ear plugs, reusable ear plugs or ear muffs. The chosen ear protector must be suited to the user and suitable for the type of noise and individual personnel should be trained in its correct use.

Breathing reasonably clean air is the right of every individual, particularly at work. Some industrial processes produce dust which may present a potentially serious hazard. The lung disease asbestosis is caused by the inhalation of asbestos dust or particles and the coal dust disease pneumoconiosis, suffered by many coal miners, has made people aware of the dangers of breathing in contaminated air.

Some people may prove to be allergic to quite innocent products such as flour dust in the food industry or wood dust in the construction industry. The main effect of inhaling dust is a measurable impairment of lung function. This can be avoided by wearing an appropriate mask, respirator or breathing apparatus as recommended by the company's health and safety policy and indicated by local safety signs such as those shown in Fig. 2.2.

A worker's body may need protection against heat or cold, bad weather, chemical or metal splash, impact or penetration and contaminated dust. Alternatively, there may be a risk of the worker's own clothes causing contamination of the product, as in the food industry. Appropriate clothing will be recommended in the company's health and safety policy. Ordinary

(a)

(b)

FIGURE 2.2
Breathing protection signs.

working clothes and clothing provided for food hygiene purposes are not included in the PPE at Work Regulations.

Hands and feet may need protection from abrasion, temperature extremes, cuts and punctures, impact or skin infection. Gloves or gauntlets provide protection from most industrial processes but should not be worn when operating machinery because they may become entangled in it. Care in selecting the appropriate protective device is required; for example, barrier creams provide only a limited protection against infection.

Boots or shoes with in-built toe caps can give protection against impact or falling objects and, when fitted with a mild steel sole plate, can also provide protection from sharp objects penetrating through the sole. Special slip resistant soles can also be provided for employees working in wet areas.

Whatever the hazard to health and safety at work, the employer must be able to demonstrate that he or she has carried out a risk assessment, made recommendations which will reduce that risk and communicated these recommendations to the workforce. Where there is a need for PPE to protect against personal injury and to create a safe working environment, the employer must provide that equipment and any necessary training which might be required and the employee must make full and proper use of such equipment and training.

WORKING ALONE

Some working situations are so potentially hazardous that not only must PPE be worn but you must also never work alone and safe working procedures must be in place before your work begins to reduce the risk.

It is unsafe to work in isolation in the following situations:

- when working above ground,
- when working below ground,
- when working in confined spaces,
- when working close to unguarded machinery,
- when a fire risk exists,
- when working close to toxic or corrosive substances.

WORKING ABOVE GROUND

We looked at this topic as it applies to electrotechnical personnel in Chapter 8 of *Basic Electrical Installation Work* 5th Edition under the sub-heading 'Safe Working above Ground'. The new Work at Height Regulations 2005 tells us that a person is at height if that person could be injured by falling from it. The Regulations require that:

- We should avoid working at height if at all possible.
- No work should be done at height which can be done on the ground. For example, equipment can be assembled on the ground then taken up to height, perhaps for fixing.

- Ensure the work at height is properly planned.
- Take account of any risk assessments carried out under Regulation 3 of the Management of Health and Safety at Work Regulations.

WORKING BELOW GROUND

Working below ground might be working in a cellar or an unventilated basement with only one entrance/exit. There is a risk that this entrance/exit might become blocked by materials, fumes or fire. When working in trenches there is always the risk of the sides collapsing if they are not adequately supported by temporary steel sheets. There is also the risk of falling objects so always:

- wear a hard hat,
- never go into an unsupported excavation,
- erect barriers around the excavation,
- provide good ladder access,
- ensure the work is properly planned,
- take account of the risk assessment before starting work.

WORKING IN CONFINED SPACES

When working in confined spaces there is always the risk that you may become trapped or overcome by a lack of oxygen or by gas, fumes, heat or an accumulation of dust. Examples of confined spaces are:

- storage tanks and silos on farms,
- enclosed sewer and pumping stations,
- furnaces,
- ductwork.

In my experience, electricians spend a lot of time on their knees in confined spaces because many electrical cable systems run out of sight away from public areas of a building.

The Confined Spaces Regulations 1997 require that:

- A risk assessment is carried out before work commences.
- If there is a serious risk of injury in entering the confined space then the work should be done on the outside of the vessel.
- Follow a safe working procedure such as a 'permit-to-work procedure' which is discussed later in this chapter, and put adequate emergency arrangements in place before work commences.

WORKING NEAR UNGUARDED MACHINERY

There is an obvious risk in working close to unguarded machinery and indeed, most machinery will be guarded but in some production processes and with overhead travelling cranes, this is not always possible. To reduce the risks associated with these hazards:

- have the machinery stopped during your work activity if possible,
- put temporary barriers in place,
- make sure that the machine operator knows that you are working on the equipment,
- identify the location of emergency stop buttons,
- take account of the risk assessment before work commences.

A RISK OF FIRE

When working in locations containing stored flammable materials such as petrol, paraffin, diesel or bottled gas, there is always the risk of fire. To minimize the risk:

- take account of the risk assessment before work commences,
- keep the area well ventilated,
- locate the fire extinguishers,
- secure your exit from the area,
- locate the nearest fire alarm point,
- follow a safe working procedure and put adequate emergency arrangements in place before work commences.

Safety First

Working alone

- Never work alone in:
  - confined spaces
  - storage tanks
  - enclosed ductwork.

Secure electrical isolation

Electric shock occurs when a person becomes part of the electrical circuit. The level or intensity of the shock will depend upon many factors, such as age, fitness and the circumstances in which the shock is received. The lethal level is approximately 50 mA, above which muscles contract, the heart flutters and breathing stops. A shock above the 50 mA level is therefore fatal unless the person is quickly separated from the supply. Below 50 mA only an unpleasant tingling sensation may be experienced or you may be thrown across a room or shocked enough to fall from a roof or ladder, but the resulting fall may lead to serious injury.

To prevent people receiving an electric shock accidentally, all circuits contain protective devices. All exposed metal is earthed, fuses and miniature circuit breakers (MCBs) are designed to trip under fault conditions and residual current devices (RCDs) are designed to trip below the fatal level.

Construction workers and particularly electricians do receive electric shocks, usually as a result of carelessness or unforeseen circumstances. As an electrician working on electrical equipment you must always make sure that the equipment is switched off or electrically isolated before commencing work. Every circuit must be provided with a means of isolation (IEE Regulation 132.15). When working on portable equipment or desktop units it is often simply a matter of unplugging the equipment from the adjacent supply. Larger pieces of equipment, and electrical machines may require isolating at the local isolator switch before work commences. To deter anyone from reconnecting the supply while work is being carried out on equipment, a sign 'Danger – Electrician at Work' should be displayed on the isolator and the isolation 'secured' with a small padlock or the fuses removed so that no one can reconnect whilst work is being carried out on that piece of equipment. The Electricity at Work Regulations 1989 are very specific at Regulation 12(1) that we must ensure the disconnection and separation of electrical equipment from every source of supply and that this disconnection and separation is secure. Where a test instrument or voltage indicator is used to prove the supply dead, Regulation 4(3) of the Electricity at Work Regulations 1989 recommends that the following procedure is adopted.

Definition

Electrical isolation: We must ensure the disconnection and separation of electrical equipment from every source of supply and that this disconnection and separation is secure.

1. First connect the test device such as that shown in Fig. 2.3 to the supply which is to be isolated. The test device should indicate mains voltage.
2. Next, isolate the supply and observe that the test device now reads zero volts.
3. Then connect the same test device to a known live supply or proving unit such as that shown in Fig. 2.4 to 'prove' that the tester is still working correctly.
4. Finally secure the isolation and place warning signs; only then should work commence.

The test device being used by the electrician must incorporate safe test leads which comply with the Health and Safety Executive (HSE) Guidance Note 38 on electrical test equipment. These leads should incorporate barriers to prevent the user touching live terminals when testing and incorporating a protective fuse and be well insulated and robust, such as those shown in Fig. 2.5.

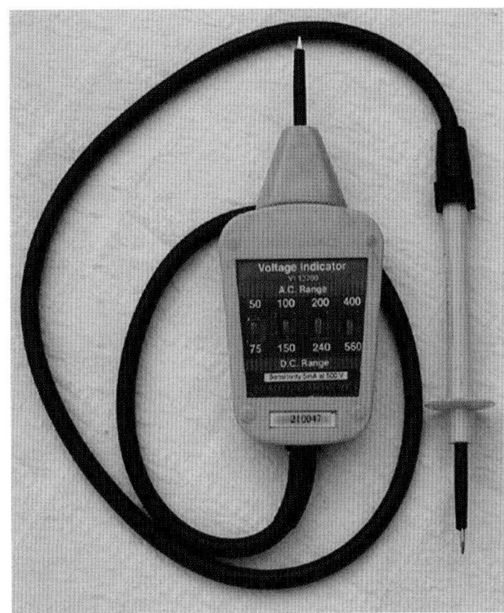

FIGURE 2.3
Typical voltage indicator.

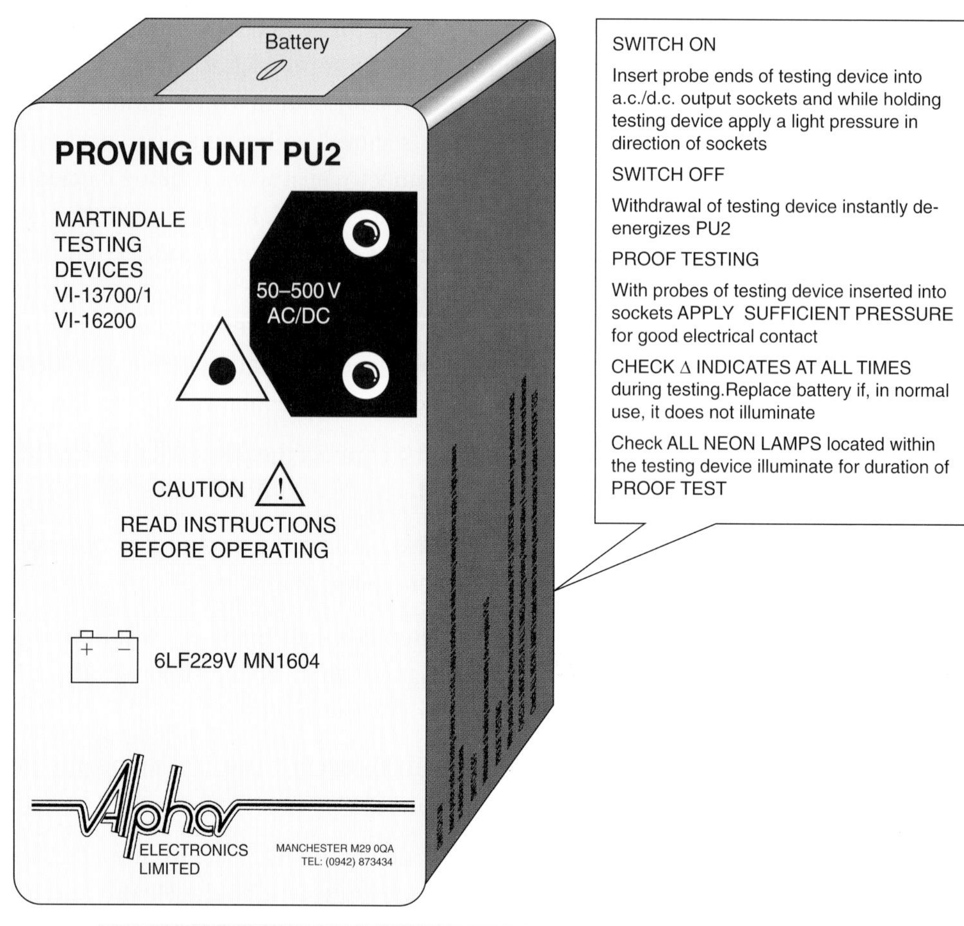

FIGURE 2.4
Voltage proving unit.

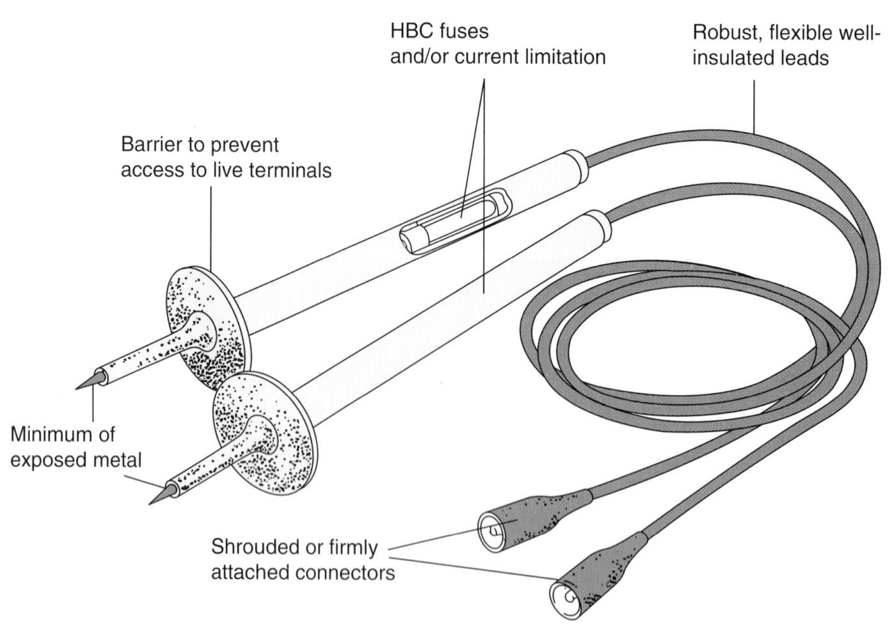

FIGURE 2.5
Recommended type of test probe and leads.

To isolate a piece of equipment or individual circuit successfully, competently, safely and in accordance with all the relevant regulations, we must follow a procedure such as that given by the flow diagram in Fig. 2.6. Start at the top and work down the flow diagram. When the heavy outlined amber boxes are reached, pause and ask yourself whether everything is satisfactory up to this point. If the answer is 'yes', move on. If the answer is 'no', go back as indicated by the diagram.

LIVE TESTING

The Electricity at Work Regulations 1989 at Regulation 4(3) tells us that it is preferable that supplies be made dead before work commences. However, it does acknowledge that some work, such as fault finding and testing, may require the electrical equipment to remain energized. Therefore, if the fault finding and testing can only be successfully carried out live then the person carrying out the fault diagnosis must:

- be trained so that they understand the equipment and the potential hazards of working live and can, therefore, be deemed 'competent' to carry out that activity;
- only use approved test equipment;
- set up appropriate warning notices and barriers so that the work activity does not create a situation dangerous to others.

While live testing may be required by workers in the electrotechnical industries in order to find the fault, live repair work must not be carried out. The individual circuit or piece of equipment must first be isolated before work commences in order to comply with the Electricity at Work Regulations 1989.

Permit-to-work system

Definition

The *permit-to-work procedure* is a type of 'safe system to work' procedure used in specialized and potentially dangerous plant process situations.

The **permit-to-work procedure** is a type of 'safe system to work' procedure used in specialized and potentially dangerous plant process situations. The procedure was developed for the chemical industry, but the principle is equally applicable to the management of complex risk in other industries or situations. For example:

- Working on part of an assembly line process where goods move through a complex, continuous process from one machine to another (e.g. the food industry).
- Repairs to railway tracks, tippers and conveyors.
- Working in confined spaces (e.g. vats and storage containers).
- Working on or near overhead crane tracks.
- Working underground or in deep trenches.
- Working on pipelines.
- Working near live equipment or unguarded machinery.
- Roof work.

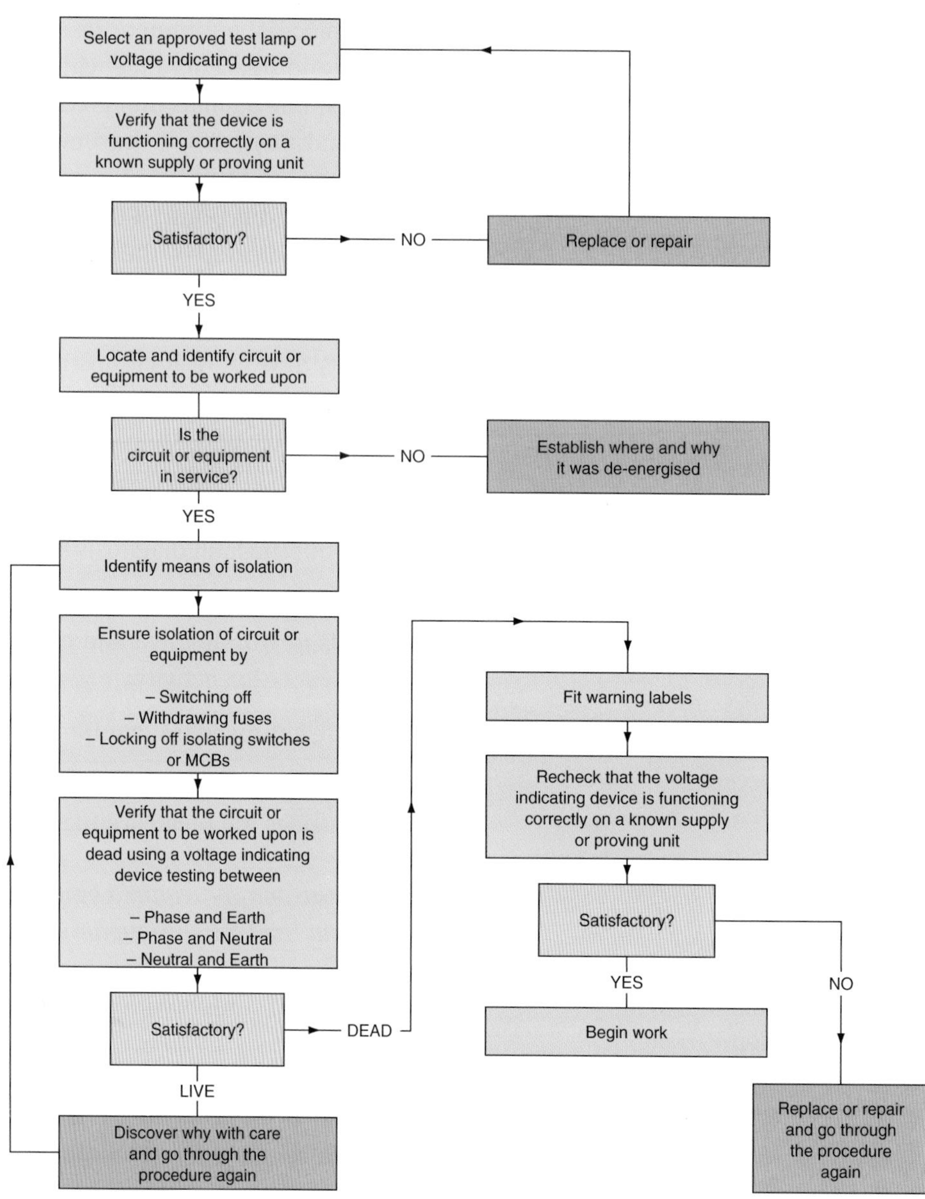

FIGURE 2.6
Flowchart for a secure isolation procedure.

- Working in hazardous atmospheres (e.g. the petroleum industry).
- Working near or with corrosive or toxic substances.

All the above situations are high-risk working situations that should be avoided unless you have received special training and will probably require the completion of a permit-to-work. Permits to work must adhere to the following eight principles:

1. Wherever possible the hazard should be eliminated so that the work can be done safely without a permit-to-work.
2. The Site Manager has overall responsibility for the permit-to-work even though he may delegate the responsibility for its issue.

3. The permit must be recognized as the master instruction, which, until it is cancelled, overrides all other instructions.
4. The permit applies to everyone on site, other trades and sub-contractors.
5. The permit must give detailed information, for example: (i) which piece of plant has been isolated and the steps by which this has been achieved (ii) what work is to be carried out (iii) the time at which the permit comes into effect.
6. The permit remains in force until the work is completed and is cancelled by the person who issued it.
7. No other work is authorized. If the planned work must be changed, the existing permit must be cancelled and a new one issued.
8. Responsibility for the plant must be clearly defined at all stages because the equipment that is taken out of service is released to those who are to carry out the work.

The people doing the work, the people to whom the permit is given, take on the responsibility of following and maintaining the safeguards set out in the permit, which will define what is to be done (no other work is permitted) and the time scale in which it is to be carried out.

Safety First

Isolation

Never carry out live repair work.

- First – test to verify circuit is 'alive'.
- Second – isolate the supply.
- Third – test to verify circuit is 'dead'.
- Fourth – secure the isolation.
- Fifth – test the tester.

The permit-to-work system must help communication between everyone involved in the process or type of work. Employers must train staff in the use of such permits and ideally, training should be designed by the company issuing the permit, so that sufficient emphasis can be given to particular hazards present and the precautions which will be required to be taken. For further details see Permit to Work @ www.hse.gov.uk

Safe manual handling

Definition

Manual handling is lifting, transporting or supporting loads by hand or by bodily force.

Manual handling is lifting, transporting or supporting loads by hand or by bodily force. The load might be any heavy object, a printer, a VDU, a box of tools or a stepladder. Whatever the heavy object is, it must be moved thoughtfully and carefully, using appropriate lifting techniques if personal pain and injury are to be avoided. **Many people hurt their back, arms and feet, and over one third of all 3 day reported injuries submitted to the HSE each year are the result of manual handling**.

When lifting heavy loads, correct lifting procedures must be adopted to avoid back injuries. Figure 2.7 demonstrates the technique. Do not lift objects from the floor with the back bent and the legs straight as this causes excessive stress on the spine. Always lift with the back straight and the legs bent so that the powerful leg muscles do the lifting work. Bend at the hips and knees to get down to the level of the object being lifted, positioning the body as close to the object as possible. Grasp the object firmly and, keeping the back straight and the head erect, use the leg muscles to raise in a smooth movement. Carry the load close to the body. When putting the object down, keep the back straight and bend at the hips and

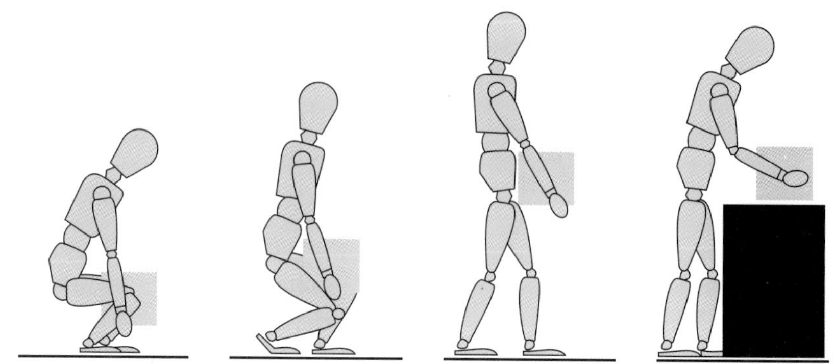

FIGURE 2.7
Correct manual lifting and carrying procedure.

knees, reversing the lifting procedure. A bad lifting technique will result in sprains, strains and pains. **There have been too many injuries over the years resulting from bad manual handling techniques. The problem has become so serious that the HSE has introduced new legislation** under the Health and Safety at Work Act 1974, the Manual Handling Operations Regulations 1992. Publications such as *Getting to Grips with Manual Handling* can be obtained from HSE Books; the address and Infoline are given in the Appendix.

Where a job involves considerable manual handling, employers must now train employees in the correct lifting procedures and provide the appropriate equipment necessary to promote the safe manual handling of loads.

Consider some 'good practice' when lifting loads:

- Do not lift the load manually if it is more appropriate to use a mechanical aid. Only lift or carry what you can easily manage.
- Always use a trolley, wheelbarrow or truck such as those shown in Fig. 2.8 when these are available.
- Plan ahead to avoid unnecessary or repeated movement of loads.
- Take account of the centre of gravity of the load when lifting – the weight acts through the centre of gravity.
- Never leave a suspended load unsupervised.
- Always lift and lower loads gently.
- Clear obstacles out of the lifting area.
- Use the manual lifting techniques described above and avoid sudden or jerky movements.
- Use gloves when manual handling to avoid injury from rough or sharp edges.
- Take special care when moving loads wrapped in grease or bubble-wrap.
- Never move a load over other people or walk under a suspended load.

Safety First

Lifting

- bend your legs when lifting from the floor
- keep your back straight
- use leg muscles to raise the weight in a smooth movement.

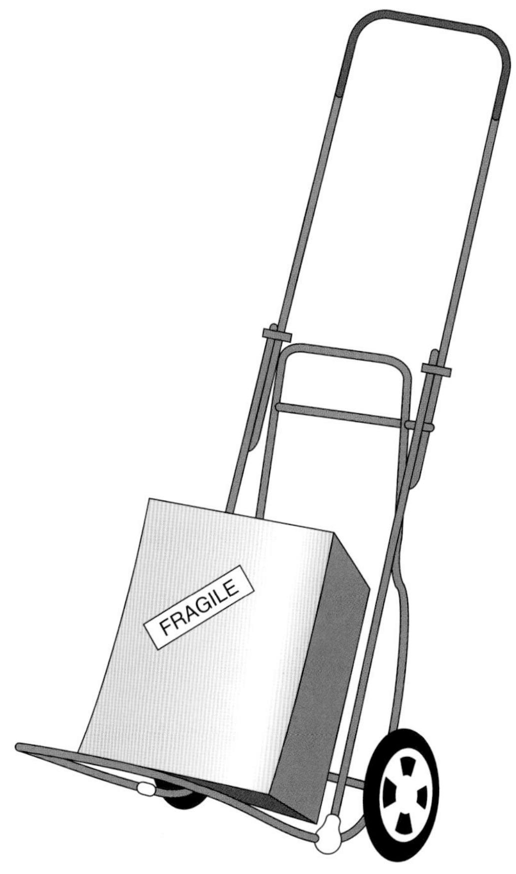

FIGURE 2.8
Always use a mechanical aid to transport a load when available.

Fire control

Definition

Fire is a chemical reaction which will continue if fuel, oxygen and heat are present.

Fire is a chemical reaction which will continue if fuel, oxygen and heat are present. To eliminate a fire *one* of these components must be removed. This is often expressed by means of the fire triangle shown in Fig. 2.9; all three corners of the triangle must be present for a fire to burn.

FUEL

Fuel is found in the construction industry in many forms: petrol and paraffin for portable generators and heaters; bottled gas for heating and soldering. Most solvents are flammable. Rubbish also represents a source of fuel: off-cuts of wood, roofing felt, rags, empty solvent cans and discarded packaging will all provide fuel for a fire.

To eliminate fuel as a source of fire, all flammable liquids and gases should be stored correctly, usually in an outside locked store. The working environment should be kept clean by placing rags in a metal bin with a lid. Combustible waste material should be removed from the work site or burned outside under controlled conditions by a competent person.

OXYGEN

Oxygen is all around us in the air we breathe, but can be eliminated from a small fire by smothering with a fire blanket, sand or foam. Closing doors

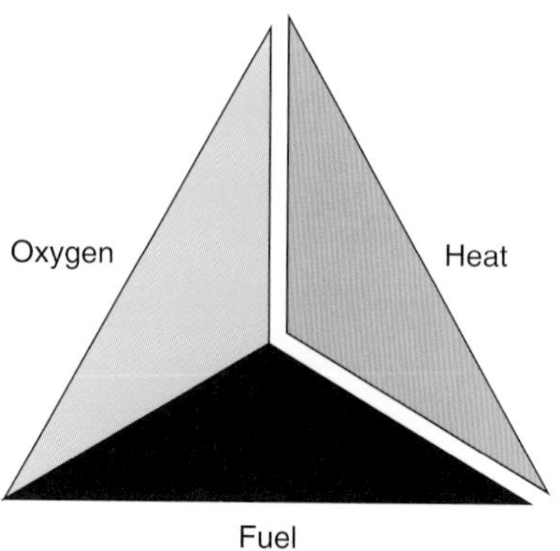

FIGURE 2.9
The fire triangle.

and windows but not locking them will limit the amount of oxygen available to a fire in a building and help to prevent it spreading.

Most substances will burn if they are at a high enough temperature and have a supply of oxygen. The minimum temperature at which a substance will burn is called the 'minimum ignition temperature' and for most materials this is considerably higher than the surrounding temperature. However, a danger does exist from portable heaters, blow torches and hot air guns which provide heat and can cause a fire by raising the temperature of materials placed in their path above the minimum ignition temperature. A safe distance must be maintained between heat sources and all flammable materials.

Safety First

Fire

If you discover a fire

- raise the alarm
- attack small fires with an extinguisher
- BUT only if you can do so without risk to your own safety.

HEAT

Heat can be removed from a fire by dousing with water, but water must not be used on burning liquids since the water will spread the liquid and the fire. Some fire extinguishers have a cooling action which removes heat from the fire.

Fires in industry damage property and materials, injure people and sometimes cause loss of life. Everyone should make an effort to prevent fires, but those which do break out should be extinguished as quickly as possible.

In the event of fire you should:

- raise the alarm;
- turn off machinery, gas and electricity supplies in the area of the fire;
- close doors and windows but without locking or bolting them;
- remove combustible materials and fuels away from the path of the fire, if the fire is small, and if this can be done safely;
- attack small fires with the correct extinguisher.

Only attack the fire if you can do so without endangering your own safety in any way. Always leave your own exit from the danger zone clear. Those not involved in fighting the fire should walk to a safe area or assembly point.

Fires are divided into four classes or categories:

- Class A – wood, paper and textile fires.
- Class B – liquid fires such as paint, petrol and oil.
- Class C – fires involving gas or spilled liquefied gas.
- Class D – very special types of fire involving burning metal.

Electrical fires do not have a special category because, once started, they can be identified as one of the four above types.

Fire extinguishers are for dealing with small fires, and different types of fire must be attacked with a different type of extinguisher. Using the wrong type of extinguisher could make matters worse. For example, water must not be used on a liquid or electrical fire. The normal procedure when dealing with electrical fires is to cut off the electrical supply and use an extinguisher which is appropriate to whatever is burning. Figure 2.10 shows the correct type of extinguisher to be used on the various categories of fire. The colour coding shown is in accordance with BS EN3: 1996.

To prevent a fire that has already started from spreading, you must remove one or more of the three elements of fuel, oxygen and heat from the fire.

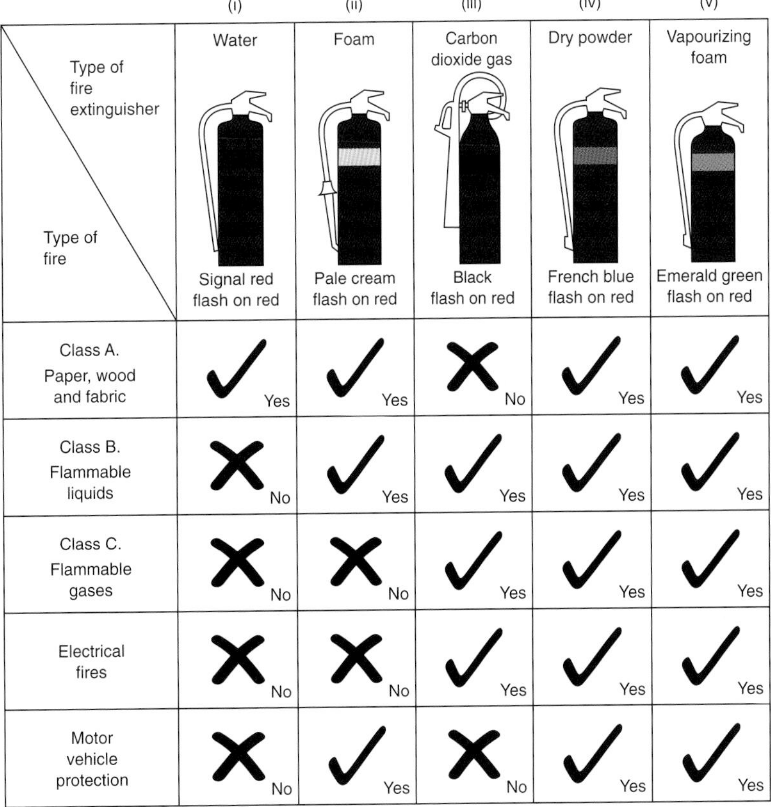

Type of fire extinguisher / Type of fire	(i) Water – Signal red flash on red	(ii) Foam – Pale cream flash on red	(iii) Carbon dioxide gas – Black flash on red	(iv) Dry powder – French blue flash on red	(v) Vapourizing foam – Emerald green flash on red
Class A. Paper, wood and fabric	✓ Yes	✓ Yes	✗ No	✓ Yes	✓ Yes
Class B. Flammable liquids	✗ No	✓ Yes	✓ Yes	✓ Yes	✓ Yes
Class C. Flammable gases	✗ No	✗ No	✓ Yes	✓ Yes	✓ Yes
Electrical fires	✗ No	✗ No	✓ Yes	✓ Yes	✓ Yes
Motor vehicle protection	✗ No	✓ Yes	✗ No	✓ Yes	✓ Yes

FIGURE 2.10
Fire extinguishers and their applications (colour codes to BS EN3:1996). The base colour of all fire extinguishers is red, with a different coloured flash to indicate the type.

Even small fires, once started, can generate sufficient heat to spread to any surrounding combustible material so this is another good reason for keeping work areas clean and tidy. Remove rubbish regularly. Off-cuts of wood, rags, empty solvent cans and discarded packaging will all provide fuel for a fire.

To remove oxygen from the fire, close doors and windows.

Definition

Fire extinguishers remove heat from a fire and are a first response for small fires.

Fire extinguishers remove heat from a fire and are a first response for small fires. Only attack a small fire with an extinguisher if you can do so without putting your own safety at risk and you feel confident to do so.

Evacuation procedures

Definition

Exit routes are usually indicated by a green and white 'running man' symbol. Evacuation should be orderly, do not run but walk purposefully to your designated assembly point.

When the fire alarm sounds you must leave the building immediately by any one of the escape routes indicated. **Exit routes** are usually indicated by a green and white 'running man' symbol. Evacuation should be orderly, do not run but walk purposefully to your designated assembly point.

Definition

The purpose of an *assembly point* is to get you away from danger to a place of safety where you will not be in the way of the emergency services.

The purpose of an **assembly point** is to get you away from danger to a place of safety where you will not be in the way of the emergency services. It also allows for people to be accounted for and to make sure that no one is left in the building. You must not re-enter the building until a person in authority gives permission to do so.

An evacuation in a real emergency can be a frightening experience, especially if you do not really know what to do, so take time to familiarize yourself with the fire safety procedures where you are working before an emergency occurs.

First aid

Despite all the safety precautions taken in the workplace to prevent injury to the workforce, accidents do happen and *you* may be the only other person able to take action to assist a workmate. If you are not a qualified first aider limit your help to obvious common sense assistance and call for help *but* do remember that if a workmate's heart or breathing has stopped as a result of an accident he has only minutes to live unless you act quickly. The Health and Safety (First Aid) Regulations 1981 and relevant approved codes of practice and guidance notes place a duty of care on all employers to provide *adequate* first aid facilities appropriate to the type of work being undertaken. Adequate facilities will relate to a number of factors such as:

- How many employees are employed?
- What type of work is being carried out?
- Are there any special or unusual hazards?
- Are employees working in scattered and/or isolated locations?
- Is there shift work or 'out of hours' work being undertaken?
- Is the workplace remote from emergency medical services?
- Are there inexperienced workers on site?
- What were the risks of injury and ill health identified by the company's Hazard Risk Assessment?

The Regulations state that:

> Employers are under a duty to provide such numbers of suitable persons as is *adequate and appropriate in the circumstances* for rendering first aid to his employees if they are injured or become ill at work. For this purpose a person shall not be suitable unless he or she has undergone such training and has such qualifications as the Health and Safety Executive may approve.

This is typical of the way in which the Health and Safety Regulations are written. The Regulations and codes of practice do not specify numbers, but set out guidelines in respect of the number of first aiders needed, dependent upon the type of company, the hazards present and the number of people employed.

Let us now consider the questions 'what is first aid?' and 'who might become a first aider?' The Regulations give the following definitions of first aid. '*First aid* is the treatment of minor injuries which would otherwise receive no treatment or do not need treatment by a doctor or nurse' *or* 'in cases where a person will require help from a doctor or nurse, first aid is treatment for the purpose of preserving life and minimizing the consequences of an injury or illness until such help is obtained'. A more generally accepted definition of first aid might be as follows: first aid is the initial assistance or treatment given to a casualty for any injury or sudden illness before the arrival of an ambulance, doctor or other medically qualified person.

Definition

First aid is the initial assistance or treatment given to a casualty for any injury or sudden illness before the arrival of an ambulance, doctor or other medically qualified person.

Now having defined **first aid**, who might become a first aider? A first aider is someone who has undergone a training course to administer first aid at work and holds a current first aid certificate. The training course and certification must be approved by the HSE. The aims of a first aider are to preserve life, to limit the worsening of the injury or illness and to promote recovery.

Definition

A *first aider* is someone who has undergone a training course to administer first aid at work and holds a current first aid certificate.

A **first aider** may also undertake the duties of an *appointed person*. An **appointed person** is someone who is nominated to take charge when someone is injured or becomes ill, including calling an ambulance if required. The appointed person will also look after the first aid equipment, including re-stocking the first aid box.

Definition

An *appointed person* is someone who is nominated to take charge when someone is injured or becomes ill, including calling an ambulance if required. The appointed person will also look after the first aid equipment, including re-stocking the first aid box.

Appointed persons should not attempt to give first aid for which they have not been trained but should limit their help to obvious common sense assistance and summon professional assistance as required. Suggested numbers of first aid personnel are given in Table 2.1. The actual number of first aid personnel must take into account any special circumstances such as remoteness from medical services, the use of several separate buildings and the company's hazard risk assessment. First aid personnel must be available at all times when people are at work, taking into account shift working patterns and providing cover for sickness absences.

Every company must have at least one first aid kit under the Regulations. The size and contents of the kit will depend upon the nature of the risks involved in the particular working environment and the number of employees. Table 2.2 gives a list of the contents of any first aid box to comply with the HSE Regulations.

Table 2.1 Suggested Numbers of First Aid Personnel

Category of risk	Numbers employed at any location	Suggested number of first aid personnel
Lower risk For example, shops and offices, libraries	Fewer than 50 50–100 More than 100	At least one appointed person At least one first aider One additional first aider for every 100 employed
Medium risk For example, light engineering and assembly work, food processing, warehousing	Fewer than 20 20–100 More than 100	At least one appointed person At least one first aider for every 50 employed (or part thereof) One additional first aider for every 100 employed
Higher risk For example, most construction, slaughterhouses, chemical manufacture, extensive work with dangerous machinery or sharp instruments	Fewer than five 5–50 More than 50	At least one appointed person At least one first aider One additional first aider for every 50 employed

There now follows a description of some first aid procedures which should be practised under expert guidance before they are required in an emergency.

Bleeding

If the wound is dirty, rinse it under clean running water. Clean the skin around the wound and apply a plaster, pulling the skin together.

If the bleeding is severe apply direct pressure to reduce the bleeding and raise the limb if possible. Apply a sterile dressing or pad and bandage firmly before obtaining professional advice.

To avoid possible contact with hepatitis or the AIDS virus, when dealing with open wounds, first aiders should avoid contact with fresh blood by wearing plastic or rubber protective gloves, or by allowing the casualty to apply pressure to the bleeding wound.

Burns

Remove heat from the burn to relieve the pain by placing the injured part under clean cold water. Do not remove burnt clothing sticking to the skin. Do not apply lotions or ointments. Do not break blisters or attempt to remove loose skin. Cover the injured area with a clean dry dressing.

Broken bones

Make the casualty as comfortable as possible by supporting the broken limb either by hand or with padding. Do not move the casualty unless by

Table 2.2 Contents of First Aid Boxes

Item	Number of employees				
	1–5	6–10	11–50	51–100	101–150
Guidance card on general first aid	1	1	1	1	1
Individually wrapped sterile adhesive dressings	10	20	40	40	40
Sterile eye pads, with attachment (Standard Dressing No. 16 BPC)	1	2	4	6	8
Triangular bandages	1	2	4	6	8
Sterile covering for serious wounds (where applicable)	1	2	4	6	8
Safety pins	6	6	12	12	12
Medium sized sterile unmedicated dressings (Standard Dressings No. 9 and No. 14 and the Ambulance Dressing No. 1)	3	6	8	10	12
Large sterile unmedicated dressings (Standard Dressings No. 9 and No. 14 and the Ambulance Dressing No. 1)	1	2	4	6	10
Extra large sterile unmedicated dressings (Ambulance Dressing No. 3)	1	2	4	6	8

Where tap water is not available, sterile water or sterile normal saline in disposable containers (each holding a minimum of 300 ml) must be kept near the first aid box. The following minimum quantities should be kept:

Number of employees	1–10	11–50	51–100	101–150
Quantity of sterile water	1 × 300 ml	3 × 300 ml	6 × 300 ml	6 × 300 ml

remaining in that position he is likely to suffer further injury. Obtain professional help as soon as possible.

Contact with chemicals

Wash the affected area very thoroughly with clean cold water. Remove any contaminated clothing. Cover the affected area with a clean sterile dressing and seek expert advice. It is a wise precaution to treat all chemical substances as possibly harmful; even commonly used substances can be dangerous if contamination is from concentrated solutions. When handling dangerous substances it is also good practice to have a neutralizing agent to hand.

Disposal of dangerous substances must not be into the main drains since this can give rise to an environmental hazard, but should be undertaken in accordance with Local Authority Regulations.

Exposure to toxic fumes

Get the casualty into fresh air quickly and encourage deep breathing if conscious. Resuscitate if breathing has stopped. Obtain expert medical advice as fumes may cause irritation of the lungs.

Definition

Asphyxiation is a condition caused by lack of air in the lungs leading to suffocation. Suffocation may cause discomfort by making breathing difficult or it may kill by stopping the breathing.

Asphyxiation

Asphyxiation is a condition caused by lack of air in the lungs leading to suffocation. Suffocation may cause discomfort by making breathing difficult or it may kill by stopping the breathing. There is a risk of asphyxiation to workers when:

- working in confined spaces,
- working in poorly ventilated spaces,
- working in paint stores and spray booths,
- working in the petro-chemical industry,
- working in any environment in which toxic fumes and gases are present.

Under the Management of Health and Safety at Work Regulations a risk assessment must be made if the environment may be considered hazardous to health. Safety procedures, including respiratory protective equipment, must be in place before work commences.

The treatment for fume inhalation or asphyxia is to get the patient into fresh air but only if you can do this without putting yourself at risk. If the patient is unconscious proceed with resuscitation as described below.

Sprains and bruising

A cold compress can help to relieve swelling and pain. Soak a towel or cloth in cold water, squeeze it out and place it on the injured part. Renew the compress every few minutes.

Breathing stopped – Resuscitation

Remove any restrictions from the face and any vomit, loose or false teeth from the mouth. Loosen tight clothing around the neck, chest and waist. To ensure a good airway, lay the casualty on his back and support the shoulders on some padding. Tilt the head backwards and open the mouth. If the casualty is faintly breathing, lifting the tongue clear of the airway may be all that is necessary to restore normal breathing. However, if the casualty does not begin to breathe, open your mouth wide and take a deep breath, close the casualty's nose by pinching with your fingers, and, sealing your lips around his mouth, blow into his lungs until the chest rises. Remove your mouth and watch the casualty's chest fall. Continue this procedure at your natural breathing rate. If the mouth is damaged or you have difficulty making a seal around the casualty's mouth, close his mouth and inflate the lungs through his nostrils. Give artificial respiration until natural breathing is restored or until professional help arrives.

Heart stopped beating – chest compressions

This sometimes happens following a severe electric shock. If the casualty's lips are blue, the pupils of his eyes widely dilated and the pulse in his neck cannot be felt, then he may have gone into cardiac arrest. Act quickly and lay the casualty on his back. Kneel down beside him and place the heel of one hand in the centre of his chest. Cover this hand with your other hand and interlace the fingers. Straighten your arms and press down on his chest sharply with the heel of your hands and then release the pressure. Continue to do this 15 times at the rate of one push per second. Check the casualty's pulse. If none is felt, give two breaths of artificial respiration and then a further 15 chest compressions. Continue this procedure until the heartbeat is restored and the artificial respiration until normal breathing returns. Pay close attention to the condition of the casualty while giving heart massage. When a pulse is restored the blueness around the mouth will quickly go away and you should stop the heart massage. Look carefully at the rate of breathing. When this is also normal, stop giving artificial respiration. Treat the casualty for shock, place him in the recovery position and obtain professional help.

Shock

Everyone suffers from shock following an accident. The severity of the shock depends upon the nature and extent of the injury. In cases of severe shock the casualty will become pale and his skin become clammy from sweating. He may feel faint, have blurred vision, feel sick and complain of thirst. Reassure the casualty that everything that needs to be done is being done. Loosen tight clothing and keep him warm and dry until help arrives. *Do not* move him unnecessarily or give him anything to drink.

Every accident must be reported to an employer and minor accidents reported to a supervisor, safety officer or first aider and the details of the accident and treatment given suitably documented as described in Chapter 1 of this book under the sub-heading 'Accident reports'.

If the accident results in death, serious injury or an injury that leads to an absence from work of more than 3 days, then your employer must report the accident to the local office of the HSE.

Emergency procedures – electric shock

Electric shock occurs when a person becomes part of the electrical circuit, as shown in Fig. 2.11. The level or intensity of the shock will depend upon many factors, such as age, fitness and the circumstances in which the shock is received. The lethal level is approximately 50 mA, above which muscles contract, the heart flutters and breathing stops. A shock above the 50 mA level is therefore fatal unless the person is quickly separated from the supply. Below 50 mA only an unpleasant tingling sensation may be experienced or you may be thrown across a room, roof or ladder, but the resulting fall may lead to serious injury.

To prevent people receiving an electric shock accidentally, all circuits contain protective devices. All exposed metal is earthed, fuses and MCBs are designed to trip under fault conditions.

Construction workers and particularly electricians do receive electric shocks, usually as a result of carelessness or unforeseen circumstances.

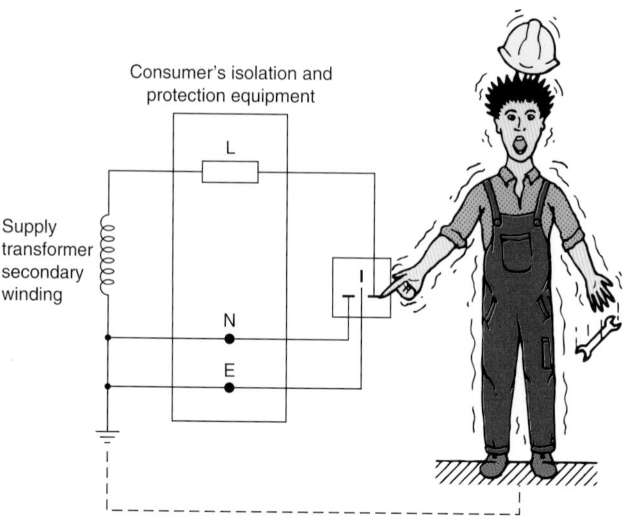

FIGURE 2.11

Touching live and earth or live and neutral makes a person part of the electrical circuit and can lead to an electric shock.

When this happens it is necessary to act quickly to prevent the electric shock becoming fatal. Actions to be taken upon finding a workmate receiving an electric shock are as follows:

- Switch off the supply if possible.
- Alternatively, remove the person from the supply *without touching him*, for example, push him off with a piece of wood, pull him off with a scarf, dry towel or coat.
- If breathing or heart has stopped, immediately call professional help by dialling 999 or 112 and asking for the ambulance service. Give precise directions to the scene of the accident. The casualty stands the best chance of survival if the emergency services can get a rapid-response paramedic team quickly to the scene. They have extensive training and will have specialist equipment with them.
- Only then should you apply resuscitation or cardiac massage until the patient recovers, or help arrives.
- Treat for shock.

To reduce the risk of an electric shock at work we should:

- Avoid contact with live parts by insulating all live parts and placing them out of reach by using barriers or temporary barriers.
- Check and inspect all cables and equipment for damage before using them.
- PAT test all portable equipment.
- Use only low voltage or battery tools.
- Use a secure electrical isolation procedure before beginning work as described earlier in this chapter.

Dangerous occurrences and hazardous malfunctions

Definition

Dangerous occurrence – is a 'near miss' that could easily have led to serious injury or loss of life. Near miss accidents occur much more frequently than injury accidents and are, therefore, a good indicator of hazard, which is why the HSE collects this data.

Dangerous occurrence – is a 'near miss' that could easily have led to serious injury or loss of life. Dangerous occurrences are defined in the Reporting of Injuries, Diseases and Dangerous Occurrences Regulations (RIDDOR) 1995. Near miss accidents occur much more frequently than injury accidents and are, therefore, a good indicator of hazard, which is why the HSE collects this data. As I write this in January 2008 a BA passenger aeroplane lost power to both engines as it prepared to land at Heathrow airport. The pilots glided the plane into a crash landing on the grass just short of the runway. This is one example of a dangerous occurrence which could so easily have been a disaster.

Consider another example – On a wet and windy night a large section of scaffold around a town centre building collapses. Fortunately this happens about midnight when no one was around because of the time and the bad weather. However, if it had occurred at midday, workers would have been using the scaffold and the streets would have been crowded with shoppers. This would be classified as a dangerous occurrence and must be reported to the HSE, who will investigate the cause and, using their wide range of powers, would either:

- stop all work,
- demand the dismantling of the structure,
- issue an Improvement Notice,
- issue a Prohibition Notice,
- prosecute those who have failed in their health and safety duties.

Other reportable dangerous occurrences are:

- the collapse of a lift,
- plant coming into contact with overhead power lines,
- any unexpected collapse which projects material beyond the site boundary,
- the overturning of a road tanker,
- a collision between a car and a train.

Definition

Hazardous malfunction – if a piece of equipment was to fail in its function, that is fail to do what it is supposed to do and, as a result of this failure have the potential to cause harm, then this would be defined as a hazardous malfunction.

Hazardous malfunction – if a piece of equipment was to fail in its function, that is fail to do what it is supposed to do and, as a result of this failure have the potential to cause harm, then this would be defined as a hazardous malfunction. Consider an example – if a 'materials lift' on a construction site was to collapse when the supply to its motor failed, this would be a hazardous malfunction. All the Regulations concerning work equipment state that it must be:

- suitable for its intended use;
- safe in use;
- maintained in a safe condition;
- used only by instructed persons;
- provided with suitable safety measures, protective devices and warning signs.

Check your Understanding

When you have completed these questions check out the answers at the back of the book.

Note: more than one multiple choice answer may be correct.

1. All equipment designed to be worn or held to protect against a risk to health and safety is one definition of:
   a. hazard
   b. risk
   c. PPE
   d. IEE.
2. Identify from the list below the potentially most dangerous work activity:
   a. isolating a live circuit
   b. working inside a grain silo
   c. fixing a socket outlet
   d. fixing a luminaire.
3. Some work situations are so potentially hazardous that you **must never:**
   a. work live
   b. work in isolation
   c. work above ground level
   d. work in the dark.
4. A type of safe system to work procedure used in potentially dangerous plant process situations is one definition of:
   a. PPE
   b. safe isolation
   c. permit-to-work
   d. working alone.
5. To avoid back injury when manually lifting heavy weights from ground level workers should:
   a. bend both legs and back
   b. bend legs but keep back straight
   c. keep legs straight but bend back
   d. keep both legs and back straight.
6. For any fire to continue to burn, three components must be present. These are:
   a. fuel, wood, cardboard
   b. petrol, oxygen, bottled gas
   c. flames, fuel, heat
   d. fuel, oxygen, heat.

7. The initial assistance given to a casualty for any sudden injury or illness is one definition of:
   a. assembly point
   b. nominated person
   c. first aider
   d. first aid.

8. Someone who has undertaken a training course in basic medical procedures is one definition of:
   a. assembly point
   b. nominated person
   c. first aider
   d. first aid.

9. Someone who will take charge when someone becomes ill or is injured at work is one definition of:
   a. assembly point
   b. nominated person
   c. first aider
   d. first aid.

10. A place where people come together following a fire alarm sounding is:
   a. assembly point
   b. nominated person
   c. first aider
   d. first aid.

11. A condition caused by a lack of air in the lungs is called:
   a. asphyxiation
   b. bleeding
   c. resuscitation
   d. winded.

12. A 'near miss' accident that would easily have led to serious injury or loss of life is called a:
   a. collapse
   b. disaster
   c. dangerous occurrence
   d. hazardous malfunction.

13. When working alone in a confined space there is a risk of:
    a. boredom
    b. becoming trapped
    c. asphyxiation
    d. hazardous malfunction.
14. Consider three types of PPE which you have used at work. For each one, sketch the symbol, state what it is called and what it is used for.
15. State six situations when it is more hazardous to work alone.
16. List four things that a worker can do to reduce the hazard of working at height as recommended by the Regulations.
17. List three of the potential risks when working in confined spaces.
18. Use bullet points to describe a secure system of electrical isolation.
19. Briefly explain what a 'permit-to-work' is and where it would be used.
20. Briefly explain a safe manual handling procedure. Perhaps use bullet points to identify the most important points.
21. Why does any fire continue to burn? How would you put out a fire at work. Highlight the most important personal safety considerations.
22. Why do workers need to go to the assembly point following the sounding of a fire alarm?
23. Briefly state the first aid provision provided by your employer for his workers.
24. List the procedure to be followed if you were to find a work colleague receiving an electric shock.
25. List five things that we can do to reduce the risk of an electric shock at work.
26. Very briefly describe the meaning of a 'dangerous occurrence' and a 'hazardous malfunction'.

Effective working practices

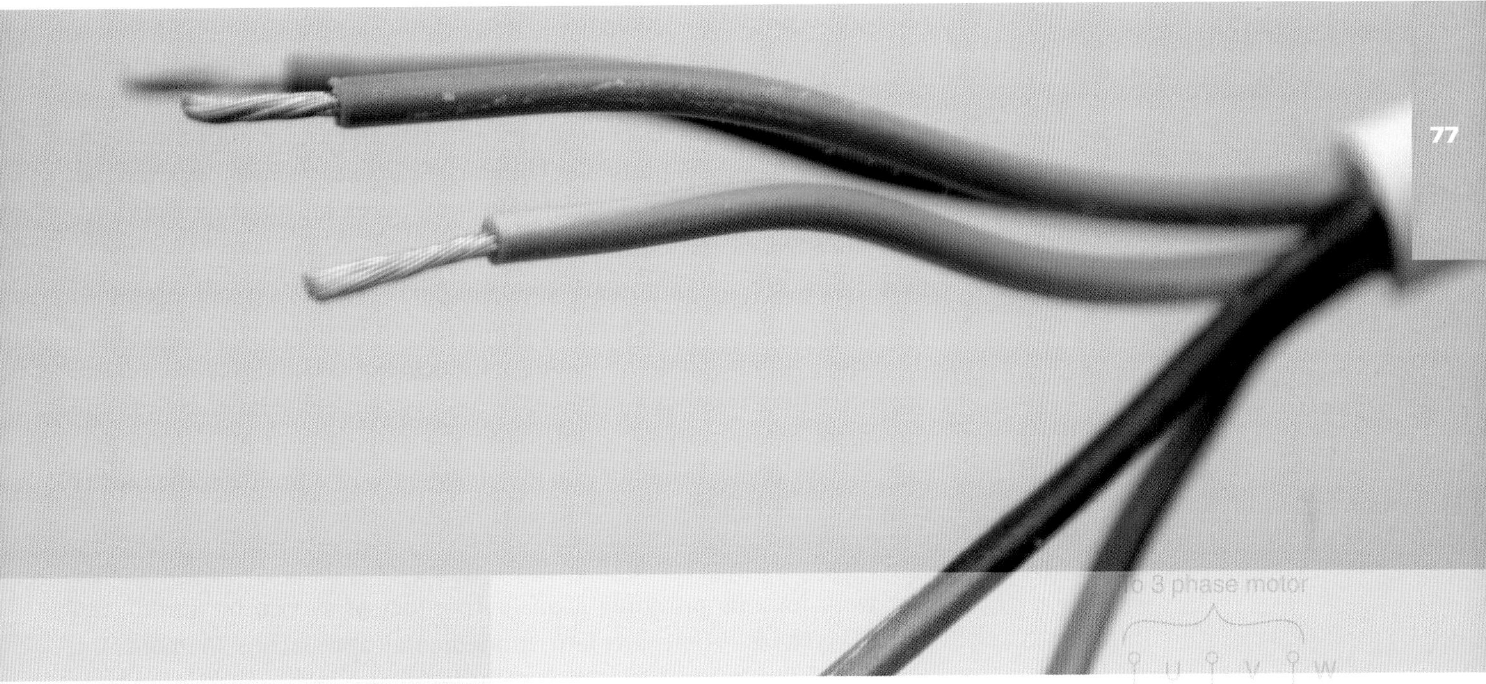

Unit 1 - Application of health and safety and electrical principles – Outcome 3

Underpinning knowledge: when you have completed this chapter you should be able to:

- state the meaning of team working
- explain employment legislation in terms of rights and responsibilities
- describe how to carry out electrotechnical activities safely and efficiently
- describe the quality standards of ISO 9000 and IiP
- describe the benefits of improving working practice

Effective working practices

Quality systems

When purchasing goods and services these days the customer is increasingly looking for good performance and reliability. Good performance means that a product will do what the customer wants it to do and reliability means that it will perform well for an acceptable period of time. Poor product reliability has been identified as one of the chief causes of customer dissatisfaction. Customers also look for durability and quality. Durability is closely linked to reliability and is a measure of the amount of use a customer receives from the product before it deteriorates.

Definition

Quality generally refers to the level of excellence, but in the business sense it means meeting the customer's expectations regarding performance, reliability and durability.

Quality generally refers to the level of excellence, but in the business sense it means meeting the customer's expectations regarding performance, reliability and durability. Quality is also a customer's subjective impression of a product or service which has been formed by images, advertisements, brand names or reputation. It is inferred from various tangible and intangible aspects of the product or service and may, in part, be due to the reputation built up by the particular company. Marks & Spencer, for example, have built up a formidable reputation from providing good-quality products and services.

In the early 1950s a motorcycle made in Japan was considered inferior to one made in Britain. Today the opposite is true. Japanese companies have used quality to become the leading producers of cars, televisions, photocopiers, radios, watches and cameras. After watching the Japanese capture the major share of these world markets, European and American companies have finally responded to the challenge and introduced the quality standards used so successfully by Japanese industry.

The customer's impression of quality is difficult to pin down but companies can work towards providing a quality product or service by introducing quality systems. There are four fundamental approaches to managing quality: quality control, quality assurance, total quality control and total quality management.

QUALITY CONTROL

Post-production inspection is the traditional form of quality control. It was introduced in the 1920s to improve the quality of mass-produced goods. Statistical sampling of the finished product took place, where, for example, one part in every hundred was tested. If the sample was found to be faulty then all 100 parts would be scrapped. If the sample was found to be acceptable it was assumed that all 100 parts were satisfactory. The problem with quality control is that the focus of attention is on the finished product rather than on the manufacturing process. Quality control never deals with the cause of the problem and, as a result, many defective products roll off the assembly line. Also, any scrapped products become built-in costs which reduce company profits and increase the product price in the shops. Defects and malfunctions have become acceptable within certain tolerance limits, but how often these days do we buy a faulty video, television or camera? Hardly ever, because Japanese industry has moved to a 'zero defects' quality management system.

QUALITY ASSURANCE

Unlike quality control, which focuses upon post-production inspection, quality assurance emphasizes defect prevention through statistical quality control and by monitoring processes to eliminate the production of bad parts.

Each part of a process has procedures written down which have been found to be the most effective. The procedures and standard forms of documentation are followed implicitly to ensure product conformity. These written procedures, used in conjunction with one of the recognized quality standards such as BS 5750 or ISO 9000, have become synonymous with quality assurance.

TOTAL QUALITY CONTROL

Total quality control attempts to expand the quality assurance philosophy to encompass all company activities. It focuses upon the elimination of waste and views the continuous improvement of systems and procedures as essential to an organization's survival. It was slow to be adopted by Western companies because it did not easily fit the organizational structures. Typically, European companies had strong vertical management structures with little opportunity for the workers' voice to be heard. Also, managers themselves tended to work independently of each other and, as a result, efforts to address company-wide issues such as quality were often met with indifference or resistance by the individuals involved.

This attitude is in sharp contrast to that of the Japanese people. They embraced the word 'total' and introduced quality assurance throughout their organizations. They have also introduced a new term 'company-wide quality control' which seeks to achieve continuous quality improvements throughout the entire organization. In the West, this company-wide quality management philosophy is known as 'total quality management'.

TOTAL QUALITY MANAGEMENT

Total quality management makes quality a way of life. It is no longer 'inspected in', 'built in' or even 'organized in': quality is 'managed in' at all levels. It is based upon four principles: meeting the customer's requirements; striving to do error-free work; managing by prevention and measuring the cost of non-quality.

Meeting the customer's requirements is the simple driving force behind total quality management. Many companies focus on meeting the needs of external customers, that is, those who buy the product or service, but this system accords equal importance to internal customers: other workers, supervisors, salesmen and managers all depend upon each other to provide a quality product or service.

Striving to do error-free work means providing a quality product or service first time, every time. A total quality management company strives to create an environment which seeks perfection at all levels of the operation, a corporate attitude which encourages the workforce to ask why an error has occurred, to track down the root cause and then take action to prevent it from happening again.

Managing by prevention means that workers at all levels must be encouraged to anticipate problems and be given the power to make permanent changes to procedures to prevent future errors. As the emphasis on preventing errors grows, the ability to meet a customer's requirements first time, every time, increases.

The cost of non-quality is the money a company would otherwise spend on detecting, correcting and preventing errors. The real benefits of a total quality management system are to be found in the education and training of the individuals, the improvement in contentment expressed by the workforce and the quality of the finished product or service.

BRITISH STANDARD QUALITY

British Standard (BS) 5750 (published in 1979) and the ISO 9000 series, the world standard for quality assurance (published in 1987), have become synonymous with quality assurance and are at the heart of most quality management systems in Europe. They specify the organizational framework for the quality management of systems, for product design, development, production, installation and servicing.

Definition

A *BS 5750/ISO 9000 certificate* provides a framework for a company to establish quality procedures and identify ways of improving its particular product or service. An essential part of any quality system is accurate record-keeping and detailed documentation which ensures that procedures are being followed and producing the desired results.

A **BS 5750/ISO 9000 certificate** provides a framework for a company to establish quality procedures and identify ways of improving its particular product or service. An essential part of any quality system is accurate record-keeping and detailed documentation which ensures procedures are being followed and producing the desired results.

Many electrotechnical companies are now accredited to ISO 9001: 2000 which means all of the company's standard systems and procedures have been documented into an approved quality management system. All procedures are internally audited throughout the year on a rolling programme to make sure that they are working effectively. Once a year an external audit of the systems takes place by an inspector nominated by ISO 9001. If the inspector is assured that the system is being operated effectively, then the company continues to use the quality system for a further year and is entitled to display the Quality Management ISO 9001 logo on vehicles and stationery. This says to potential customers 'we are a serious professional company working to the best standards of our industry and providing a quality service'.

Another quality system dedicated to improving a company's performance through the development of its employees is 'Investors in People'.

Investors in people

Most people would agree that the people an organization employs are the most valuable asset of the business. Conscientious workers are hard to find and difficult to keep. For any business to succeed, everyone must perform to the best of their ability from the youngest trainee to the Managing Director.

Definition

Investors in People is a National Quality Standard that focuses on the needs of the people working within an organization.

It recognizes that a company or business is investing some of its profits in its workforce in order to improve the efficiency and performance of the organization.

Investors in People (IiP) is a National Quality Standard that focuses on the needs of the people working within an organization.

It recognizes that a company or business is investing some of its profits in its workforce in order to improve the efficiency and performance of the organization. The objective is to create an environment where what people can do and are motivated to do, matches what the company needs them to do to improve.

The IiP standard lays down a set of 'principles' and 'indicators' of good practice which the participating organization must meet. The IiP standards are the same for all types of organization, large and small, and recognizes that each company must find its own way of achieving success through the development of its employees.

The IiP was started in 1990 and was driven by a partnership between leading businesses and organizations such as the Confederation of British Industry (CBI), Trade Union Councils (TUC), the Institute of Personnel and Development and the National Training Task Force. It is now a nationally recognized quality standard with over 36,000 qualifying organizations able to display the coveted 'IiP' UK Charter mark on their vehicles and stationery.

For more information on IiP go to www.investorsinpeople.co.uk

The benefits of improving working practices and procedures are:

- improved customer satisfaction,
- improved productivity,
- more efficient use of resources,
- increased profitability for the electrotechnical organization.

Try This

Quality Mark

- Does your Company hold ISO 9000 accreditation or IiP?
- Do they display the logo on company vehicles?
- How does this accreditation affect your work?

TEAM WORKING

Definition

Team working is about working with other people.

Team working is about working with other people, probably with other employees from the company you work for. Working together, helping each other, sharing the load in order to get the job done to a good standard of workmanship in the time allowed. All the separate parts of the job have to be finished and eventually brought together at completion. The team can also be much larger than just those people who work for the same company. We are often dependent on other trades completing their work before we can start ours. The ceiling fitters must install the suspended ceiling before we can drop the recessed modular fluorescent fittings in place and connect them. So, in this case, two different trades are interdependent, working as a team to complete a suspended ceiling job.

A lot of research has been done over the years about what makes a good team and how the relationship of individual members of a team develop over time. One such model is called 'Forming, Storming, Norming and Performing'.

Forming is the first stage of the developing team where the separate individuals come together. They behave as individuals, their responsibility to the team is unclear and they feel confused about what they should be doing. At this stage the team leader will be telling everyone what to do for the collective good because they are all acting as individuals.

Storming is the stage where people begin to see a role for themselves within the team. They will challenge other team members and the team leader, who must become less dominant and more encouraging.

Norming is the stage where team members have generally reached an agreement upon their individual roles and responsibilities to the group. They discuss together and reach agreement upon the best way to perform a task together. The team may share the leadership role.

Performing is the final stage of the development of the team. Everyone knows what they are doing and how their input fits into everyone else's work in the team. The team leaders role is to oversee the project. There is no requirement for instruction or assistance because everyone knows what they have to do to be successful. The individual members support each other, jollying each other along if necessary, giving help when required and generally looking after each other. They have a shared vision and goal.

To work safely in a busy electrotechnical environment alongside other workers we must, in the words of the Health and Safety Act:

- not interfere with or misuse anything provided to protect our health and safety and
- take reasonable care to avoid injury to ourselves and other people as a result of our work activities.

To complete an electrotechnical project successfully, safely and efficiently it will require careful planning. The planning process should involve the following activities:

- Check the drawings, instructions and specifications to make sure that all the specified work is carried out but not more than that which was specified.
- Check that the work area is suitable and safe at all times.
- Create a logical sequence for all work activities and identify tasks remaining to be done before completion of the project.

- Make a list of tools, materials and equipment required to complete the project. Will the project require specialist equipment that needs to be hired from another contractor or specially purchased. Make sure everything is available on site when it is required.
- Make an assessment of the skill that will be required of the electrotechnical staff. Do some work tasks require a specialist's knowledge or skill or someone with greater experience?
- Co-ordinate the work tasks with other trades on site when necessary. For example, a suspended ceiling must be finished before the recessed modular fittings can be installed.
- Make sure that everyone in the electrotechnical team observes all safety procedures and practices.
- When the project is completed make sure that the customer's specified requirements have been met and those of BS 7671 through the process of commissioning, inspection, testing and certification.

Laws protecting people

In Chapter 1 of this book we looked at some of the major pieces of legislation that affect our working environment and some of the main pieces of environmental law. Let us now look at some of the laws and regulations that protect and affect us as individuals, and our human rights and responsibilities.

Employment Rights Act 1996

If you work for a company you are an employee and you will have a number of legal rights under the Employment Rights Act 1996.

As a trainee in the electrotechnical industry you are probably employed by a company and, therefore, are an employee. There are strict guidelines regarding those who are employed and those who are self-employed. Indicators of being employed are listed below:

- You work wholly or mainly for one company and work is centred upon the premises of the company.
- You do not risk your own money.
- You have no business organization such as a storage facility or stock in trade.
- You do not employ anyone.
- You work a set number of hours in a given period and are paid by the hour and receive a weekly or monthly wage or salary.
- Someone else has the right to control what you do at work even if such control is rarely practised.

Indicators of being self-employed are as follows:

- You supply the materials, plant and equipment necessary to do the job.
- You give a price for doing a job and will bear the consequences if your price is too low or something goes wrong.
- You have the right to hire other people who will answer to you and are paid by you to do a job.
- You may be paid an agreed amount for a job regardless of how long it takes or be paid according to some formula, for example, a fee to 'first fix' a row of houses.
- Within an overall deadline, you have the right to decide how and when the work will be done.

The titles 'employed' or 'self-employed' are not defined by statute but have emerged through cases coming before the courts. The above points will help in deciding the precise nature of the working relationship.

Home working is a growing trend which prompts the question as to whether home workers are employed or self-employed. As in any circumstance, it will depend upon the specific conditions of employment, and the points mentioned above may help to decide the question.

The Inland Revenue look with concern at those people who claim to be self-employed but do all or most of their work for one company. There is a free leaflet available from local Inland Revenue Offices, IR 56 – titled 'Employed or Self-Employed' – which will give further guidance if required.

If you are an employee you have a special relationship in law with your employer which entitles you to the following benefits:

- A written statement of the particulars of your employment. It is clearly in the interests of both parties to understand at the outset of their relationship the terms and conditions of employment. The legal relationship between employer and employee is one of contract. Both parties are bound by the agreed terms but the contract need not necessarily be in writing, although contracts of apprenticeship must be in writing.
- The date your employment started.
- The continuity of service, that is, whether employment with a previous employer is to count as part of an employee's continuous service. Continuous service is normally with one employer but there are exceptions, for example, if a business is transferred or taken over or there is a change of partners or trustees. This is important because many employees rights depend on the need to show that he or she has worked for the 'appropriate period' and this is known as '**continuous service**'.
- The job title.
- The normal place from which you work.
- A brief description of your work.

- The hours to be worked.
- Holiday entitlement and holiday pay.
- Sick pay entitlement.
- Pension scheme arrangements.
- The length of notice which an employee is obliged to give and is entitled to receive to terminate his contract of employment.
- Where the employment is not intended to be permanent, the period for which it is expected to continue and the date when it is to end.
- Disciplinary and grievance procedures.
- The rate of pay and frequency, weekly or monthly.
- An itemized pay statement showing
  - the gross amount of the wage or salary;
  - the amounts of any deductions and the purpose for which they have been made. This will normally be tax and National Insurance contributions, but may also include payments to professional bodies or Trade Unions;
  - the net amount of salary being paid.

An **employer** has responsibilities to all employees. Even if the responsibilities are not written down in the contract of employment, they are implied by law. Case histories speak of a relationship of trust, confidence and respect. These responsibilities include:

- The obligation to pay an employee for work done.
- The obligation to treat an employee fairly.
- The obligation to take reasonable care of an employee's health and safety.
- An obligation to provide equal treatment both for men and women.

An **employee** also has responsibilities to his employer. These include:

- Carrying out the tasks for which you are employed with all reasonable skill and care.
- Conducting yourself in such a way as would best serve your employer's interests.
- Carrying out all reasonable orders.

An employee is not expected to carry out any order that is plainly illegal or unreasonable. 'Illegal' is quite easy to define – anything which is against the law, for example, driving a vehicle for which you do not hold a licence or falsifying documents or accounts. 'Unreasonable' is more difficult to define, what is reasonable to one person may be quite unreasonable to another person.

Finally, employees are under a general duty not to disclose confidential information relating to their employer's affairs that they might obtain in the course of their work. Employees are also under a general duty not to assist a competitor of their employer. This is one aspect of the employee's duty to ensure that the relationship between employer and employee is one of trust. Even when an employee has left an employer, confidential information is not to be disclosed.

Health and Safety (First Aid) Regulations 1981

People can suffer an injury or become ill whilst at work. It does not matter whether the injury or illness is caused by the work they do or not, what is important is that they are able to receive immediate attention by a competent person or that an ambulance is called in serious cases. First aid at work covers the arrangements that an employer must make to ensure that this happens. It can save lives and prevent a minor incident becoming a major one.

The Health and Safety (First Aid) Regulations 1981 requires employers to provide '**adequate**' and '**appropriate**' equipment, facilities and personnel to enable first aid to be given to employees if they are injured or become ill at work. What is adequate and appropriate will depend upon the type of work being carried out by the employer. The minimum provision is a suitably stocked first aid box and a competent person to take charge of first aid arrangements.

Employers must consider:

- How many people are employed and, therefore, how many first aid boxes will be required?
- What is the pattern of working hours, shift work, night work, is a 'first aider' available for everyone at all times?
- How many trained 'first aiders' will be required?
- Where will first aid boxes be made available?
- Do employees travel frequently or work alone?
- Will it be necessary to issue personal first aid boxes if employees travel or work away from the company's main premises?
- How hazardous is the work being done – what are the risks?
- Are different employees at different levels of risk?
- What has been the accident or sickness record of staff in the past?

Although there is no legal responsibility for employers to make provision for non-employees, the HSE strongly recommends that they are included in any first aid provision.

We looked at first aid provision at work in the last chapter of this book, Chapter 2.

Data Protection Act 1998

The right to privacy is a fundamental human right and one that many of us take for granted. Most of us, for instance, would not want our medical records freely circulated, and many people are sensitive about revealing their age, religious beliefs, family circumstances or academic qualifications. In the United Kingdom, even the use of name and address files for mail shots is often felt to be an invasion of privacy.

With the advent of large computerized databases it is now possible for sensitive personal information to be stored without the individual's knowledge and accessed by, say, a prospective employer, credit card company or insurance company in order to assess somebody's suitability for employment, credit or insurance.

The Data Protection Act 1984 grew out of public concern about personal privacy in the face of rapidly developing computer technology.

The act covers 'personal data' which is 'automatically processed'. It works in two ways, giving individuals certain rights whilst requiring those who record and use personal information on computer, to be open about that use and to follow proper practices.

The Data Protection Act 1998 was passed in order to implement a European Data Protection Directive. This Directive sets a standard for data protection throughout all the countries of the European Union, and the new act was brought into force in March 2000. The act gives the following useful definitions:

- *Data subjects*: the individuals to whom the personal data relate – we are all data subjects.
- *Data users*: those who control the contents and use a collection of personal data. They can be any type of company or organization, large or small, within the public or private sector.
- *Personal data*: information about living, identifiable individuals. Personal data does not have to be particularly sensitive information and can be as little as a name and address.
- *Automatically processed*: processed by computer or other technology such as document image processing systems. The act does not currently cover information which is held on manual records, for example, in ordinary paper files.

Registered data users must comply with the eight Data Protection principles of good information handling practice contained in the act. Broadly these state that data must be:

1. obtained and processed fairly and lawfully;
2. held for the lawful purposes described in the data users' register entry;
3. used for the purposes and disclosed only to those people described in the register entry;
4. adequate, relevant and not excessive in relation to the purposes for which they are held;
5. accurate and, where necessary, kept up to date;

6. held no longer than is necessary for the registered purpose;
7. accessible to the individual concerned who, where appropriate, has the right to have information about themselves corrected or erased;
8. surrounded by proper security.

EXEMPTIONS FROM THE ACT

- The act does not apply to payroll, pensions and accounts data, nor to names and addresses held for distribution purposes.
- Registration may not be necessary if the data is for personal, family, household or recreational use.
- Data subjects do not have a right to access data if the sole aim of collecting it is for statistical or research purposes.
- Data can be disclosed to the data subject's agent (e.g. lawyer or accountant), to persons working for the data user, and in response to urgent need to prevent injury or damage to health.

Additionally, there are exemptions for special categories, including data held:

- in connection with national security,
- for prevention of crime,
- for the collection of tax or duty.

THE RIGHTS OF DATA SUBJECTS

The Data Protection Act allows individuals to have access to information held about themselves on computer and where appropriate to have it corrected or deleted.

As an individual you are entitled, on making a written request to a data user, to be supplied with a copy of any personal data held about yourself. The data user may charge a fee of up to £10 for each register entry for supplying this information but in some cases it is supplied free.

Usually the request must be responded to within 40 days. If not, you are entitled to complain to the Registrar or apply to the courts for correction or deletion of the data.

Apart from the right to complain to the Registrar, data subjects also have a range of rights which they may exercise in the civil courts. These are:

- Right to compensation for unauthorized disclosure of data.
- Right to compensation for inaccurate data.
- Right of access to data and to apply for rectification or erasure where data is inaccurate.
- Right to compensation for unauthorized access, loss or destruction of data.

For more information see www.dataprotection.gov.uk

Prejudice and discrimination

It is because we are all different to each other that life is so interesting and varied. Our culture is about the way of life that we have, the customs, ideas and experiences that we share and the things that we find acceptable and unacceptable. Different groups of people have different cultures. When people have a certain attitude towards you, or the group of people to which you belong, or a belief about you that is based upon lack of knowledge, understanding or myth, this is prejudice.

When prejudice takes form or action it becomes discrimination and this often results in unfair treatment of people. Regardless of our age, ability, sex, religion, race or sexuality we should all be treated equally and with respect. If we are treated differently because of our differences, we are being discriminated against.

If you are being discriminated against or you see it happening to someone else, you do not have to put up with it. Stay calm and do not retaliate but report it to someone, whoever is the most appropriate person, your supervisor, trainer or manager. If you are a member of a Trade Union you may be able to get help from them if it is an employment related matter.

There are three areas covered by legislation at the moment, these are race, sex and disability. In the next few years the law will change to make it unlawful to discriminate in the training or workplace on the grounds of sexual orientation, religious belief and age.

The Race Relations Act 1976 and Amendment Act 2000

The 1976 Race Relations Act (RRA) made employers liable for acts of racial discrimination committed by their employees in the course of their employment. However, police officers are office holders, not employees, and, therefore, Chief Officers of the police were not liable under the 1976 Act for acts of racial discrimination. The Commission for Racial Equality proposed that the act be extended to include all public services and the amendment came into force in 2000.

It is illegal to discriminate against someone because of their race, colour, nationality, citizenship or ethnic origin.

Institutional racism is when the policies or practices of an organization or institution results in its failure to provide an appropriate service to people because of their colour, culture or ethnic origin. It may mean that the organization or institution does, or does not do something, or that someone is treated less favourably. This includes public services as well as educational institutions.

There are some exceptions in the RRA. It does not apply to certain jobs where people from a certain ethnic or racial background are required for authenticity. These are known as 'genuine occupational qualifications' and might apply to actors and restaurants.

The Commission for Racial Equality website can be found at www.cre.gov.uk

Sex Discrimination Act 1975

'Sexism' takes place every time a person, usually a woman, is discriminated against because of their sex. The Sex Discrimination Act of 1975 makes it unlawful to discriminate against people on sexual grounds in areas relating to recruitment, promotion or training. Job advertisements must not discriminate in their language but they can make it clear that they are looking for people of a particular sex. If, though, a person of either sex applies then they must be treated equally and fairly.

There are some exceptions in the Sex Discrimination Act (SDA) known as '**genuine occupational qualifications**' that might apply to artists, models, actors and some parts of the priesthood in the church. Some exceptions can also apply when appointing people to occupations where 'decency' is required, for example, in changing room attendants in swimming pools, gymnasiums, etc., and women are not allowed to work underground.

Sex discrimination is when someone is treated less favourably because of sex or marital status. It includes sexual harassment and unfavourable treatment because a woman is pregnant. Employers fear a high level of absenteeism, often unjustified, from a mother who is trying to juggle the conflicting demands of work and motherhood. This is known as 'direct sex discrimination'.

'Indirect sex discrimination' occurs when a condition of the job is applied to both sexes but excludes or disadvantages a larger proportion of one sex and is not justifiable. For example, an unnecessary height requirement of 180 cm (5′ 10″) would discriminate against women because less women would be able to meet this requirement.

The Equal Opportunities Commission has published a Code of Practice that gives guidance on best practice in the promotion of equality of opportunity in employment. Further information can be found on the SDA website at www.eoc.org.uk

Disability Discrimination Act 1995

There are more than 8.5 million disabled people in the United Kingdom. The Disability Discrimination Act (DDA) makes it unlawful to discriminate against a disabled person in the areas of employment, access to goods and services and buying or renting land or property.

It is now unlawful for employers of more than 15 people to discriminate against employees or job applicants on the grounds of disability. Reasonable adjustments must be made for people with disabilities and employers must ensure that discrimination does not occur in the workplace.

Under Part 111 of the DDA, from 1 October 2004, service providers will have to take reasonable steps to remove, alter or provide reasonable means of avoiding physical features that make it impossible or unreasonably difficult for disabled people to use their services. The duty requires service providers to make '**reasonable**' adjustments to their premises so that disabled people can use the service and are not restricted by physical barriers. If this is not possible then the service should be provided by means of a reasonable alternative such as bringing goods to the disabled person or helping the person to find items.

All organizations which provide goods, facilities or services to the public are covered by the DDA including shops, offices, public houses, leisure facilities, libraries, museums, banks, cinemas, churches and many more, in fact there are few exemptions.

Some service providers will need to incur significant capital expenditure in order to comply with the DDA. What is '**reasonable**' will depend upon the state and condition of the service provider's premises. A subjective standard will apply when determining what is reasonable under the circumstances at a given location. Whether or not an adjustment is reasonable will ultimately be a question of fact for the courts.

Further information can be found on the DDA website at www.disability.gov.uk

The Human Rights Act 1998

The Human Rights Act (HRA) 1998 came into force on 2 October 2000 bringing the European Convention on Human Rights into UK law. It means that if you think your human rights have been violated, you can take action through the British court system, rather than taking it to the European Court of Human Rights. The act makes it unlawful for a '**Public Authority**' to act in a way that goes against any of the rights laid down in the convention unless an Act of Parliament meant that it could not have acted differently. The basic human rights in the Human Rights Act are:

- the right to life,
- the right to a fair trial,
- the right to respect for your private and family life,
- the right to marry,
- the right to liberty and security,
- prohibition of torture,
- prohibition of slavery and forced labour,
- prohibition of discrimination,
- prohibition of the abuse of rights,
- freedom of thought, conscience and religion,
- freedom of expression,
- freedom of assembly and association,
- no punishment without law.

If you feel that your human rights have been violated, you should seek advice from a solicitor. Rights under the act can only be used against a public authority such as the police or a local authority. They cannot be used against a private company. For more information see www.humanrights.gov.uk

Check your Understanding

When you have completed these questions check out the answers at the back of the book.

Note: more than one multiple choice answer may be correct.

1. Meeting the customer's expectations regarding performance, reliability and durability is one definition of:
   a. team working
   b. quality
   c. ISO 9000 system
   d. IiP standard.
2. A system which provides a framework for a company to establish quality procedures that improves its product and services is one definition of:
   a. team working
   b. quality
   c. ISO 9000 system
   d. IiP standard.
3. A system which focuses upon the need of the people working within an organization is one definition of:
   a. team working
   b. quality
   c. ISO 9000 system
   d. IiP standard.
4. Working together with other people to get the job done to the best standards is one definition of:
   a. team working
   b. quality
   c. ISO 9000 system
   d. IiP standard.
5. In five short bullet point statements describe an ISO 9000 quality system.
6. In four short bullet point statements describe the IiP standard.
7. In four bullet point statements describe the benefits of improving working practices and procedures.
8. Give a very brief description or definition of the following words – team working, forming, storming, norming and performing.
9. State eight activities which lead to the successful planning and completion of an electrotechnical project.

10. Make a list of the legal rights which we all have as individuals under the various laws protecting people.

11. Make a list of the responsibilities which we as individuals have in order to ensure other people's rights.

CH 4

Electrical systems and components

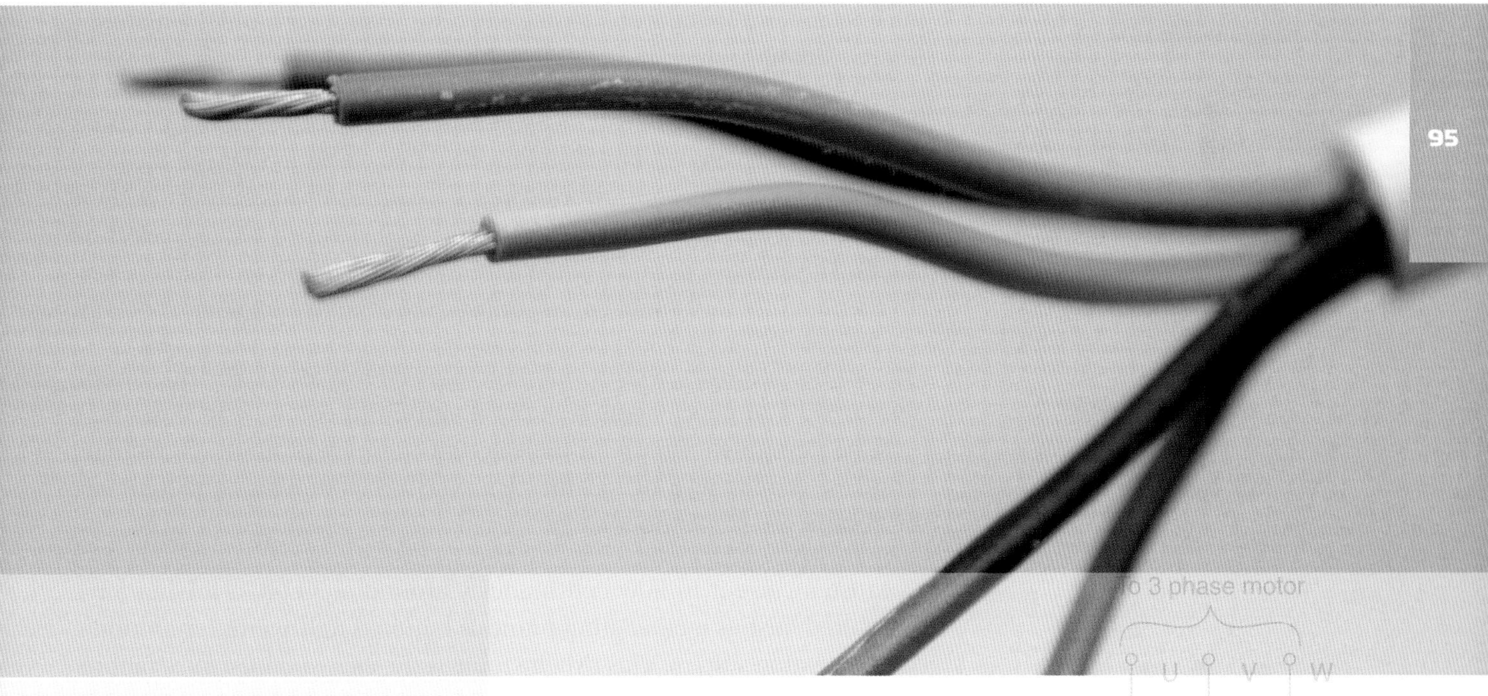

Unit 1 – Application of health and safety and electrical principles – Outcome 4

Underpinning knowledge: when you have completed this chapter you should be able to:

- describe resistance and Ohm's Law
- describe magnetism and magnetic circuits
- describe inductance and inductive components
- describe capacitors and their construction
- state the effect of resistance, inductance, capacitance and impedance in a.c. circuits
- describe basic electronic circuits
- describe luminaires and calculate lighting levels

Electrical science

All matter is made up of atoms which arrange themselves in a regular framework within the material. The atom is made up of a central, positively charged nucleus, surrounded by negatively charged electrons. The electrical properties of a material depend largely upon how tightly these electrons are bound to the central nucleus.

Definition

A *conductor* is a material in which the electrons are loosely bound to the central nucleus and are, therefore, free to drift around the material at random from one atom to another.

A **conductor** is a material in which the electrons are loosely bound to the central nucleus and are, therefore, free to drift around the material at random from one atom to another, as shown in Fig. 4.1(a). Materials which are good conductors include copper, brass, aluminium and silver.

Definition

An *insulator* is a material in which the outer electrons are tightly bound to the nucleus and so there are no free electrons to move around the material.

An **insulator** is a material in which the outer electrons are tightly bound to the nucleus and so there are no free electrons to move around the material. Good insulating materials are PVC, rubber, glass and wood.

If a battery is attached to a conductor as shown in Fig. 4.1(b), the free electrons drift purposefully in one direction only. The free electrons close to the positive plate of the battery are attracted to it since unlike charges attract, and the free electrons near the negative plate will be repelled from it. For each electron entering the positive terminal of the battery, one will be ejected from the negative terminal, so the number of electrons in the conductor remains constant.

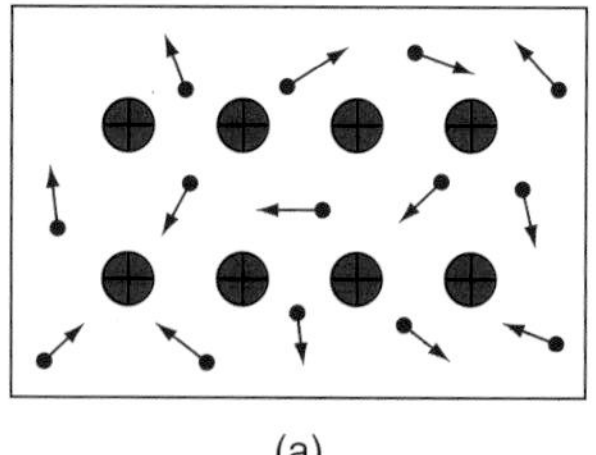

(a)

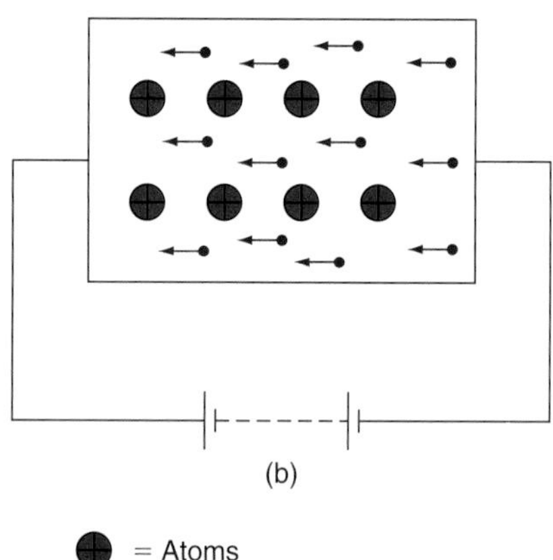

(b)

FIGURE 4.1
Atoms and electrons in a material.

This drift of electrons within a conductor is known as an electric *current*, measured in amperes (symbol *I*). For a current to continue to flow, there must be a complete circuit for the electrons to move around. If the circuit is broken by opening a switch, for example, the electron flow and therefore the current will stop immediately.

To cause a current to flow continuously around a circuit, a driving force is required, just as a circulating pump is required to drive water around a central heating system. This driving force is the *electromotive force* (abbreviated as emf). Each time an electron passes through the source of emf, more energy is provided to send it on its way around the circuit.

An emf is always associated with energy conversion, such as chemical to electrical in batteries and mechanical to electrical in generators. The energy introduced into the circuit by the emf is transferred to the load terminals by the circuit conductors. The *potential difference* (abbreviated as p.d.) is the change in energy levels measured across the load terminals. This is also called the volt drop or terminal voltage, since emf and p.d. are both measured in volts. Every circuit offers some opposition to current flow, which we call the circuit *resistance*, measured in ohms (symbol Ω), to commemorate the famous German physicist George Simon Ohm, who was responsible for the analysis of electrical circuits.

Ohm's law

In 1826, George Ohm published details of an experiment he had done to investigate the relationship between the current passing through and the potential difference between the ends of a wire. As a result of this experiment, he arrived at a law, now known as **Ohm's law**, which says that the current passing through a conductor under constant temperature conditions is proportional to the potential difference across the conductor. This may be expressed mathematically as:

Definition

Ohm's law, which says that the current passing through a conductor under constant temperature conditions is proportional to the potential difference across the conductor.

$$V = I \times R \text{ (V)}$$

Transposing this formula, we also have:

$$I = \frac{V}{R} \text{ (A)} \quad \text{and} \quad R = \frac{V}{I} \text{ (}\Omega\text{)}$$

Example 1

An electric heater, when connected to a 230 V supply, was found to take a current of 4 A. Calculate the element resistance.

$$R = \frac{V}{I}$$

$$\therefore R = \frac{230 \text{ V}}{4 \text{ A}} = 57.5 \, \Omega$$

Example 2

The insulation resistance measured between phase conductors on a 400 V supply was found to be 2 MΩ. Calculate the leakage current.

$$I = \frac{V}{R}$$

$$\therefore I = \frac{400 \text{ V}}{2 \times 10^6 \, \Omega} = 200 \times 10^{-6} \text{ A} = 200 \, \mu\text{A}$$

Example 3

When a 4 Ω resistor was connected across the terminals of an unknown d.c. supply, a current of 3 A flowed. Calculate the supply voltage.

$$V = I \times R$$

$$\therefore V = 3 \text{ A} \times 4 \, \Omega = 12 \text{ V}$$

Resistivity

The resistance or opposition to current flow varies for different materials, each having a particular constant value. If we know the resistance of, say, 1 m of a material, then the resistance of 5 m will be five times the resistance of 1 m.

Definition

The *resistivity* (symbol ρ – the Greek letter 'rho') of a material is defined as the resistance of a sample of unit length and unit cross-section.

The **resistivity** (symbol ρ – the Greek letter 'rho') of a material is defined as the resistance of a sample of unit length and unit cross-section. Typical values are given in Table 4.1. Using the constants for a particular material we can calculate the resistance of any length and thickness of that material from the equation.

$$R = \frac{\rho l}{a} \ (\Omega)$$

where

Key Fact

Resistance

- If the length is doubled, resistance is doubled.
- If the thickness is doubled, resistance is halved.

ρ = the resistivity constant for the material (Ω m)

l = the length of the material (m)

a = the cross-sectional area of the material (m^2).

Table 4.1 gives the resistivity of silver as $16.4 \times 10^{-9}\,\Omega$m, which means that a sample of silver 1 m long and 1 m in cross-section will have a resistance of $16.4 \times 10^{-9}\,\Omega$.

Example 1

Calculate the resistance of 100 m of copper cable of 1.5 mm^2 cross-sectional area if the resistivity of copper is taken as $17.5 \times 10^{-9}\,\Omega$m.

$$R = \frac{\rho l}{a} \ (\Omega)$$

$$\therefore R = \frac{17.5 \times 10^{-9}\,\Omega \times 100\text{ m}}{1.5 \times 10^{-6}\text{m}^2} = 1.16\ \Omega$$

Example 2

Calculate the resistance of 100 m of aluminium cable of 1.5 mm^2 cross-sectional area if the resistivity of aluminium is taken as $28.5 \times 10^{-9}\,\Omega$m.

$$R = \frac{\rho l}{a} \ (\Omega)$$

$$\therefore R = \frac{28.5 \times 10^{-9}\,\Omega\text{m} \times 100\text{ m}}{1.5 \times 10^{-6}\text{m}^2} = 1.9\ \Omega$$

The above examples show that the resistance of an aluminium cable is some 60% greater than a copper conductor of the same length and cross-section.

Table 4.1 Resistivity Values

Material	Resistivity (Ω m)
Silver	16.4×10^{-9}
Copper	17.5×10^{-9}
Aluminium	28.5×10^{-9}
Brass	75.0×10^{-9}
Iron	100.0×10^{-9}

Therefore, if an aluminium cable is to replace a copper cable, the conductor size must be increased to carry the rated current as given by the tables in Appendix 4 of the *IEE Regulations* and Appendix 6 of the *On Site Guide.*

The other factor which affects the resistance of a material is the temperature, and we will consider this next.

Temperature coefficient

The resistance of most materials changes with temperature. In general, conductors increase their resistance as the temperature increases and insulators decrease their resistance with a temperature increase. Therefore, an increase in temperature has a bad effect on the electrical properties of a material.

Each material responds to temperature change in a different way, and scientists have calculated constants for each material which are called the *temperature coefficient of resistance* (symbol α – the Greek letter 'alpha'). Table 4.2 gives some typical values.

Using the constants for a particular material and substituting values into the following formulae the resistance of a material at different temperatures may be calculated. For a temperature increase from 0°C

$$R_t = R_0(1 + \alpha t) \; (\Omega)$$

Table 4.2 Temperature Coefficient Values

Material	Temperature coefficient (Ω/Ω°C)
Silver	0.004
Copper	0.004
Aluminium	0.004
Brass	0.001
Iron	0.006

where

R_t = the resistance at the new temperature t°C

R_0 = the resistance at 0°C

α = the temperature coefficient for the particular material.

For a temperature increase between two intermediate temperatures above 0°C:

$$\frac{R_1}{R_2} = \frac{(1+\alpha t_1)}{(1+\alpha t_2)}$$

where

R_1 = the resistance at the original temperature

R_2 = the resistance at the final temperature

α = the temperature coefficient for the particular material.

If we take a 1 Ω resistor of, say, copper, and raise its temperature by 1°C, the resistance will increase to 1.004 Ω. This increase of 0.004 Ω is the temperature coefficient of the material.

Example 1

The field winding of a d.c. motor has a resistance of 100 Ω at 0°C. Determine the resistance of the coil at 20°C if the temperature coefficient is 0.004 Ω/Ω°C.

$$R_t = R_0\,(1+\alpha t)\ (\Omega)$$
$$\therefore R_t = 100\ \Omega\ (1+0.004\ \Omega/\Omega°\text{C}\times 20°\text{C})$$
$$R_t = 100\ \Omega\ (1+0.08)$$
$$R_t = 108\ \Omega$$

Example 2

The field winding of a shunt generator has a resistance of 150 Ω at an ambient temperature of 20°C. After running for some time the mean temperature of the generator rises to 45°C. Calculate the resistance of the winding at the higher temperature if the temperature coefficient of resistance is 0.004 Ω/Ω°C.

$$\frac{R_1}{R_2} = \frac{(1+\alpha t_1)}{(1+\alpha t_2)}$$
$$\frac{150\ \Omega}{R_2} = \frac{1+0.004\ \Omega/\Omega°\text{C}\times 20°\text{C}}{1+0.004\ \Omega/\Omega°\text{C}\times 45°\text{C}}$$
$$\frac{150\ \Omega}{R_2} = \frac{1.08}{1.18}$$
$$\therefore R_2 = \frac{150\ \Omega\times 1.18}{1.08} = 164\ \Omega$$

It is clear from the last two sections that the resistance of a cable is affected by length, thickness, temperature and type of material. Since Ohm's law tells us that current is inversely proportional to resistance, these factors must also influence the current carrying capacity of a cable. The tables of current ratings in Appendix 4 of the IEE Regulations contain correction factors so that current ratings may be accurately determined under defined installation conditions. Cable selection is considered in Chapter 7.

Resistors

In an electrical circuit resistors may be connected in series, in parallel, or in various combinations of series and parallel connections.

Key Fact

Resistance

When the temperature increases:

- conductor resistance increases
- insulator resistance decreases.

SERIES-CONNECTED RESISTORS

In any series circuit a current I will flow through all parts of the circuit as a result of the potential difference supplied by a battery V_T. Therefore, we say that in a series circuit the current is common throughout that circuit.

When the current flows through each resistor in the circuit, R_1, R_2 and R_3 for example in Fig. 4.2, there will be a voltage drop across that resistor whose value will be determined by the values of I and R, since from Ohm's law $V = I \times R$. The sum of the individual voltage drops, V_1, V_2 and V_3 for example in Fig. 4.2, will be equal to the total voltage V_T.

We can summarize these statements as follows. For any series circuit, I is common throughout the circuit and,

$$V_T = V_1 + V_2 + V_3 \tag{4.1}$$

Let us call the total circuit resistance R_T. From Ohm's law we know that $V = I \times R$ and therefore

$$\begin{aligned}\text{Total voltage } V_T &= I \times R_T \\ \text{Voltage drop across } R_1 \text{ is } V_1 &= I \times R_1 \\ \text{Voltage drop across } R_2 \text{ is } V_2 &= I \times R_2 \\ \text{Voltage drop across } R_3 \text{ is } V_3 &= I \times R_3\end{aligned} \tag{4.2}$$

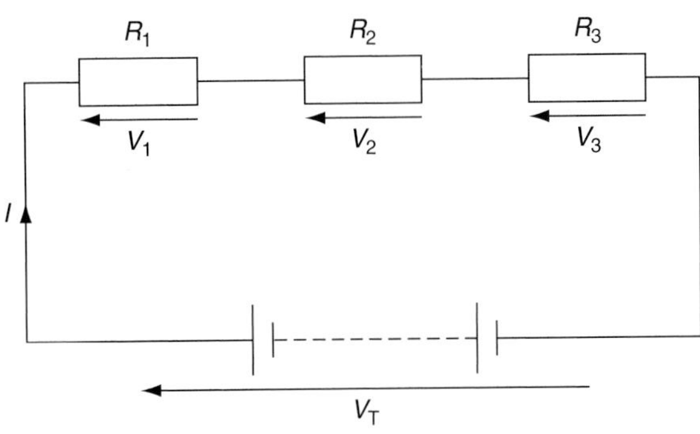

FIGURE 4.2
A series circuit.

We are looking for an expression for the total resistance in any series circuit and, if we substitute Equations (4.2) into Equation (4.1) we have:

$$V_T = V_1 + V_2 + V_3$$
$$\therefore\ I \times R_T = I \times R_1 + I \times R_2 + I \times R_3$$

Now, since I is common to all terms in the equation, we can divide both sides of the equation by I. This will cancel out I to leave us with an expression for the circuit resistance:

$$R_T = R_1 + R_2 + R_3$$

Note that the derivation of this formula is given for information only. Craft students need only state the expression $R_T = R_1 + R_2 + R_3$ for series connections.

PARALLEL-CONNECTED RESISTORS

In any parallel circuit, as shown in Fig. 4.3, the same voltage acts across all branches of the circuit. The total current will divide when it reaches a resistor junction, part of it flowing in each resistor. The sum of the individual currents, I_1, I_2 and I_3 for example in Fig. 4.3, will be equal to the total current I_T.

We can summarize these statements as follows. For any parallel circuit, V is common to all branches of the circuit and,

$$I_T = I_1 + I_2 + I_3 \qquad (4.3)$$

Let us call the total resistance R_T.

From Ohm's law we know, that $I = \frac{V}{R}$, and therefore

$$\begin{aligned} \text{the total current } I_T &= \frac{V}{R_T} \\ \text{the current through } R_1 \text{ is } I_1 &= \frac{V}{R_1} \\ \text{the current through } R_2 \text{ is } I_2 &= \frac{V}{R_2} \\ \text{the current through } R_3 \text{ is } I_3 &= \frac{V}{R_3} \end{aligned} \qquad (4.4)$$

Key Fact

Series resistors

The total resistance in a series circuit can be found from $R_T = R_1 + R_2$

We are looking for an expression for the equivalent resistance R_T in any *parallel* circuit and, if we substitute Equations (4.4) into Equation (4.3) we have:

$$I_T = I_1 + I_2 + I_3$$
$$\therefore\ \frac{V}{R_T} = \frac{V}{R_1} + \frac{V}{R_2} + \frac{V}{R_3}$$

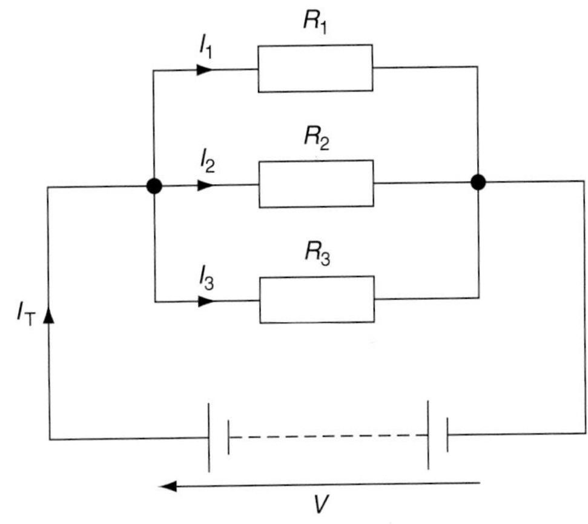

FIGURE 4.3
A parallel circuit.

Now, since V is common to all terms in the equation, we can divide both sides by V, leaving us with an expression for the circuit resistance:

$$\frac{1}{R_T} = \frac{1}{R_1} + \frac{1}{R_2} + \frac{1}{R_3}$$

Note that the derivation of this formula is given for information only. Craft students need only state the expression $1/R_T = 1/R_1 + 1/R_2 + 1/R_3$ for parallel connections.

Example

Three 6 Ω resistors are connected (a) in series (see Fig. 4.4) and (b) in parallel (see Fig. 4.5), across a 12 V battery. For each method of connection, find the total resistance and the values of all currents and voltages.

For any series connection:

$$R_T = R_1 + R_2 + R_3$$
$$\therefore R_1 = 6\ \Omega + 6\ \Omega + 6\ \Omega = 18\ \Omega$$

$$\text{Total current } I_T = \frac{V}{R_T}$$

$$\therefore I_T = \frac{12\ \text{V}}{18\ \Omega} = 0.67\ \text{A}$$

The voltage drop across R_1 is:

$$V_1 = I \times R_1$$
$$\therefore V_1 = 0.67\ \text{A} \times 6\ \Omega = 4\ \text{V}$$

The voltage drop across R_2 is:

$$V_2 = I \times R_2$$
$$\therefore V_2 = 0.67\ \text{A} \times 6\ \Omega = 4\ \text{V}$$

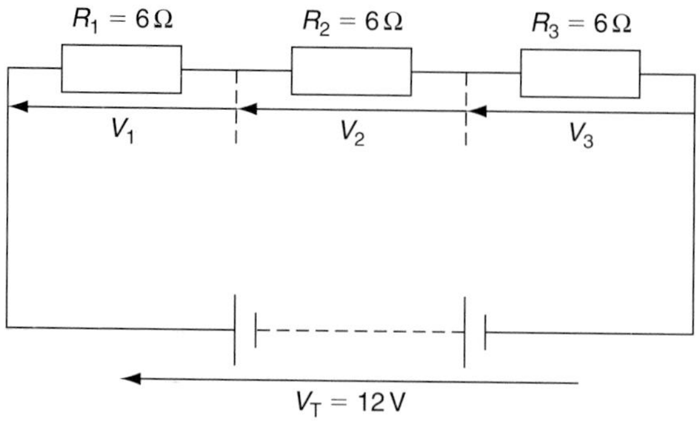

FIGURE 4.4
Resistors in series.

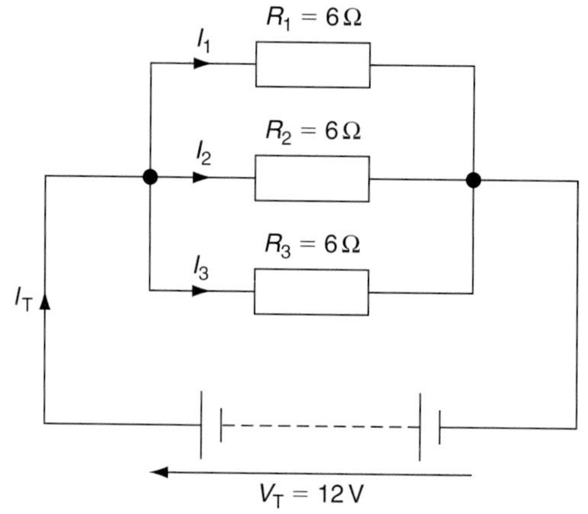

FIGURE 4.5
Resistors in parallel.

The voltage drop across R_3 is:

$$V_3 = I \times R_3$$

$$\therefore V_3 = 0.67\text{ A} \times 6\,\Omega = 4\text{ V}$$

For any parallel connection:

$$\frac{1}{R_T} = \frac{1}{R_1} + \frac{1}{R_2} + \frac{1}{R_3}$$

$$\therefore \frac{1}{R_T} = \frac{1}{6\,\Omega} + \frac{1}{6\,\Omega} + \frac{1}{6\,\Omega}$$

$$\frac{1}{R_T} = \frac{1+1+1}{6\,\Omega} = \frac{3}{6\,\Omega}$$

$$R_T = \frac{6\,\Omega}{3} = 2\,\Omega$$

$$\text{Total current } I_T = \frac{V}{R_T}$$

$$\therefore I_T = \frac{12\text{ V}}{2\,\Omega} = 6\text{ A}$$

The current flowing through R_1 is:

$$I_1 = \frac{V}{R_1}$$

$$\therefore I_1 = \frac{12\text{ V}}{6\ \Omega} = 2\text{ A}$$

The current flowing through R_2 is:

$$I_2 = \frac{V}{R_2}$$

$$\therefore I_2 = \frac{12\text{ V}}{6\ \Omega} = 2\text{ A}$$

The current flowing through R_3 is:

$$I_3 = \frac{V}{R_3}$$

$$\therefore I_3 = \frac{12\text{ V}}{6\ \Omega} = 2\text{ A}$$

SERIES AND PARALLEL COMBINATIONS

The most complex arrangement of series and parallel resistors can be simplified into a single equivalent resistor by combining the separate rules for series and parallel resistors.

Example 1

Resolve the circuit shown in Fig. 4.6 into a single resistor and calculate the potential difference across each resistor.

By inspection, the circuit contains a parallel group consisting of R_3, R_4 and R_5, and a series group consisting of R_1 and R_2 in series with the equivalent resistor for the parallel branch.

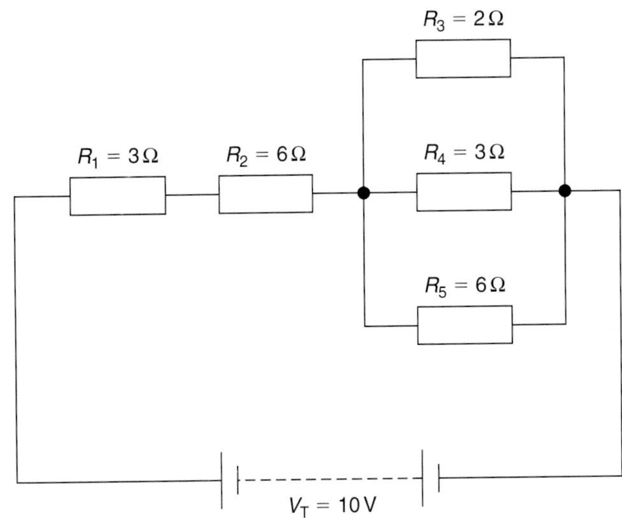

FIGURE 4.6
A series/parallel circuit.

Consider the parallel group. We will label this group R_P. Then,

$$\frac{1}{R_P} = \frac{1}{R_3} + \frac{1}{R_4} + \frac{1}{R_5}$$

$$\frac{1}{R_P} = \frac{1}{2\,\Omega} + \frac{1}{3\,\Omega} + \frac{1}{6\,\Omega}$$

$$\frac{1}{R_P} = \frac{3+2+1}{6\,\Omega} = \frac{6}{6\,\Omega}$$

$$R_P = \frac{6\,\Omega}{6} = 1\,\Omega$$

Figure 4.6 may now be represented by the more simple equivalent circuit is shown in Fig. 4.7

Since all resistors are now in series,

$$R_T = R_1 + R_2 + R_P$$
$$\therefore R_T = 3\,\Omega + 6\,\Omega + 1\,\Omega = 10\,\Omega$$

Thus, the circuit may be represented by a single equivalent resistor of value 10 Ω as shown in Fig. 4.8. The total current flowing in the circuit may be found by using Ohm's law:

$$I_T = \frac{V_T}{R_T} + \frac{10\text{ V}}{10\,\Omega} = 1\text{ A}$$

Key Fact

Parallel Resistors
The total resistance in a parallel circuit can be found from $\frac{1}{R_T} = \frac{1}{R_1} + \frac{1}{R_2}$

The potential differences across the individual resistors are:

$$V_1 = I \times R_1 = 1\text{A} \times 3\Omega = 3\text{V}$$
$$V_2 = I \times R_2 = 1\text{A} \times 6\Omega = 6\text{V}$$
$$V_P = I \times R_P = 1\text{ A} \times 1\Omega = 1\text{V}$$

Since the same voltage acts across all branches of a parallel circuit the same p.d. of 1 V will exist across each resistor in the parallel branch R_3, R_4 and R_5.

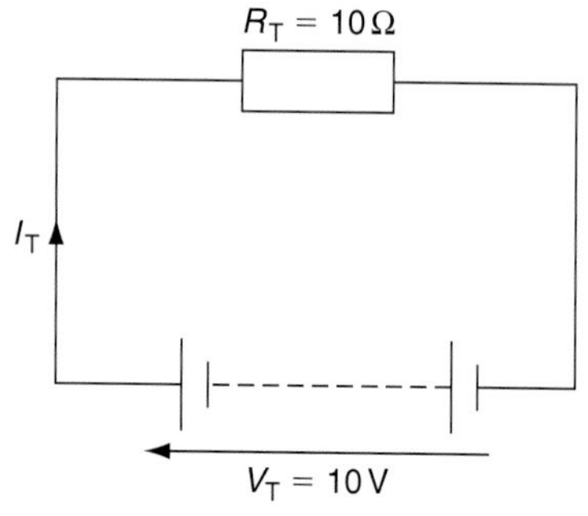

FIGURE 4.8
Single equivalent resistor for Fig. 4.6.

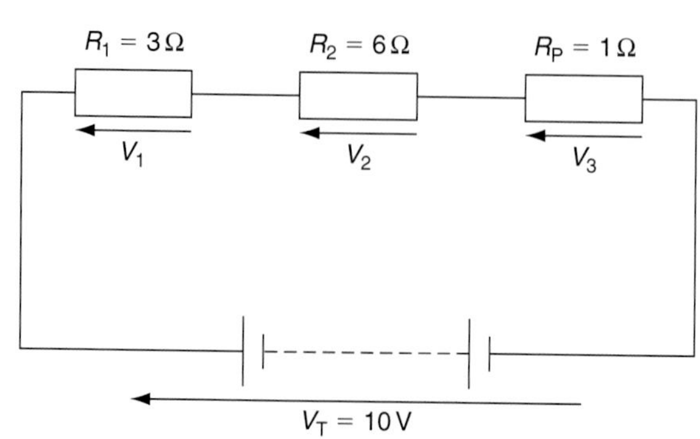

FIGURE 4.7
Equivalent series circuit.

Example 2

Determine the total resistance and the current flowing through each resistor for the circuit shown in Fig. 4.9.

By inspection, it can be seen that R_1 and R_2 are connected in series while R_3 is connected in parallel across R_1 and R_2. The circuit may be more easily understood if we redraw it as in Fig. 4.10.

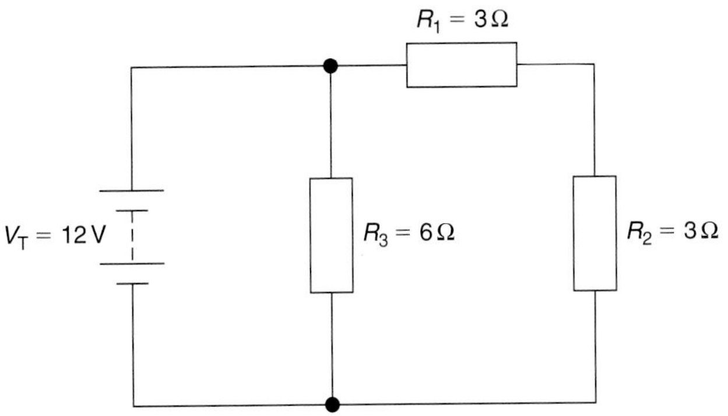

FIGURE 4.9
A series/parallel circuit for Example 2.

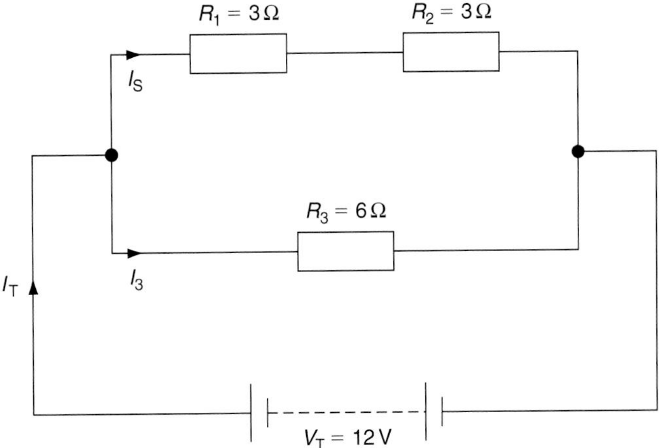

FIGURE 4.10
Equivalent circuit for Example 2.

For the series branch, the equivalent resistor can be found from:

$$R_S = R_1 + R_2$$
$$\therefore R_S = 3\,\Omega + 3\,\Omega = 6\,\Omega$$

Figure 4.10 may now be represented by a more simple equivalent circuit, as in Fig. 4.11.

Since the resistors are now in parallel, the equivalent resistance may be found from:

$$\frac{1}{R_T} = \frac{1}{R_S} + \frac{1}{R_3}$$

$$\therefore \frac{1}{R_T} = \frac{1}{6\,\Omega} + \frac{1}{6\,\Omega}$$

$$\frac{1}{R_T} = \frac{1+1}{6\,\Omega} = \frac{2}{6\,\Omega}$$

$$R_T = \frac{6\,\Omega}{2} = 3\,\Omega$$

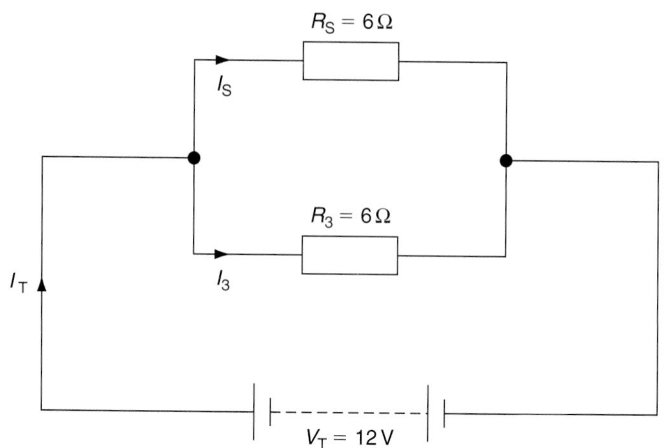

FIGURE 4.11
Simplified equivalent circuit for Example 2.

The total current is:

$$I_T = \frac{V}{R_T} = \frac{12\ \text{V}}{3\ \Omega} = 4\ \text{A}$$

Let us call the current flowing through resistor R_3 I_3:

$$\therefore I_3 = \frac{V}{R_3} = \frac{12\ \text{V}}{6\ \Omega} = 2\ \text{A}$$

Let us call the current flowing through both resistors R_1 and R_2, as shown in Fig. 4.10, I_S:

$$\therefore I_S = \frac{V}{R_S} = \frac{12\ \text{V}}{6\ \Omega} = 2\ \text{A}$$

Power and energy

POWER

Power is the rate of doing work and is measured in watts.

$$\text{Power} = \frac{\text{Work done}}{\text{Time taken}}\ (\text{W})$$

In an electrical circuit:

$$\text{Power} = \text{Voltage} \times \text{Current}\ (\text{W}) \qquad (4.5)$$

Now from Ohm's law:

$$\text{Voltage} = I \times R (\text{V}) \qquad (4.6)$$

$$\text{Current} = \frac{V}{R}\ (\text{A}) \qquad (4.7)$$

Substituting Equation (4.6) into Equation (4.5), we have:

$$\text{Power} = (I \times R) \times \text{Current} = I^2 \times R \text{ (W)}$$

and substituting Equation (4.7) into Equation (4.5), we have:

$$\text{Power} = \text{Voltage} \times \frac{V}{R} = \frac{V^2}{R} \text{ (W)}$$

We can find the power of a circuit by using any of the three formulae:

$$P = V \times I, \quad P = I^2 \times R, \quad P = \frac{V^2}{R}$$

ENERGY

Energy is a concept which engineers and scientists use to describe the ability to do work in a circuit or system.

$$\begin{aligned} \text{Energy} &= \text{Power} \times \text{Time} \\ \text{but, since Power} &= \text{Voltage} \times \text{Current} \\ \text{then Energy} &= \text{Voltage} \times \text{Current} \times \text{Time} \end{aligned}$$

The SI unit of energy is the joule, where time is measured in seconds. For practical electrical installation circuits this unit is very small and therefore the kilowatt-hour (kWh) is used for domestic and commercial installations. Electricity Board meters measure 'units' of electrical energy, where each 'unit' is 1 kWh. So,

$$\begin{aligned} \text{Energy in joules} &= \text{Voltage} \times \text{Current} \\ &\quad \times \text{Time in seconds} \\ \text{Energy in kWh} &= \text{kW} \times \text{Time in hours} \end{aligned}$$

Example 1

A domestic immersion heater is switched on for 40 minutes and takes 15 A from a 200 V supply. Calculate the energy used during this time.

$$\text{Power} = \text{Voltage} \times \text{Current}$$

$$\text{Power} = 200\text{ V} \times 15\text{ A} = 3000\text{ W or } 3\text{ kW}$$

$$\text{Energy} = \text{kW} \times \text{Time in hours}$$

$$\text{Energy} = 3\text{ kW} \times \frac{40\text{ min}}{60\text{ min/h}} = 2\text{ kWh}$$

This immersion heater uses 2 kWh in 40 minutes, or 2 'units' of electrical energy every 40 minutes.

Example 2

Two 50 Ω resistors may be connected to a 200V supply. Determine the power dissipated by the resistors when they are connected (a) in series, (b) each resistor separately connected and (c) in parallel.

For (a), the equivalent resistance when resistors are connected in series is given by:

$$R_T = R_1 + R_2$$

$$\therefore R_T = 50\ \Omega + 50\ \Omega = 100\ \Omega$$

$$\text{Power} = \frac{V^2}{R_T}\ (\text{W})$$

$$\therefore \text{Power} = \frac{200\ \text{V} \times 200\ \text{V}}{100\ \Omega} = 400\ \text{W}$$

For (b), each resistor separately connected has a resistance of 50 Ω.

$$\text{Power} = \frac{V^2}{R}\ (\text{W})$$

$$\therefore \text{Power} = \frac{200\ \text{V} \times 200\ \text{V}}{50\ \Omega} = 800\ \text{W}$$

For (c), the equivalent resistance when resistors are connected in parallel is given by:

$$\frac{1}{R_T} = \frac{1}{R_1} + \frac{1}{R_2}$$

$$\therefore \frac{1}{R_T} = \frac{1}{50\ \Omega} + \frac{1}{50\ \Omega}$$

$$\frac{1}{R_T} = \frac{1+1}{50\ \Omega} = \frac{2}{50\ \Omega}$$

$$R_T = \frac{50\ \Omega}{2} = 25\ \Omega$$

$$\text{Power} = \frac{V^2}{R_T}\ (\text{W})$$

$$\therefore \text{Power} = \frac{200\ \text{V} \times 200\ \text{V}}{25\ \Omega} = 1600\ \text{W}$$

This example shows that by connecting resistors together in different combinations of series and parallel connections, we can obtain various power outputs: in this example, 400, 800 and 1600 W. This theory finds a practical application in the three heat switch used to control a boiling ring or cooker hotplate.

Alternating waveforms

The supply which we obtain from a car battery is a unidirectional or d.c. supply, whereas the mains electricity supply is alternating or a.c. (see Fig. 4.12).

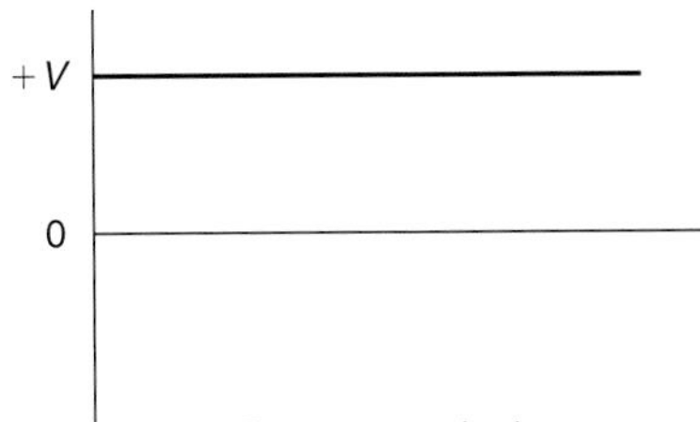

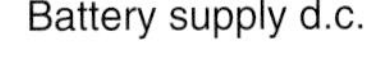

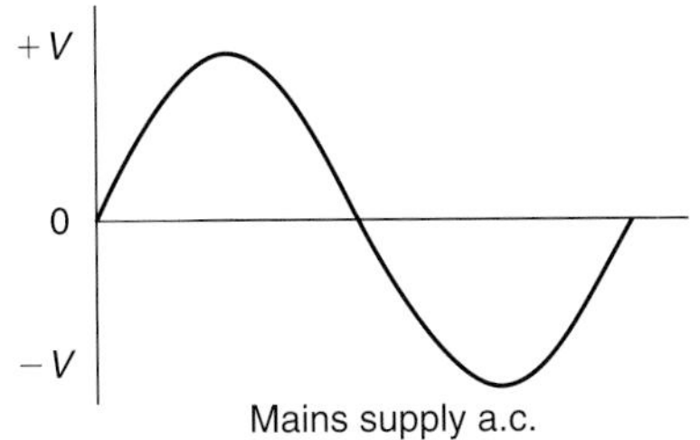

FIGURE 4.12
Unidirectional and alternating supply.

Most electrical equipment makes use of alternating current supplies, and for this reason a knowledge of alternating waveforms and their effect on resistive, capacitive and inductive loads is necessary for all practising electricians.

When a coil of wire is rotated inside a magnetic field a voltage is induced in the coil. The induced voltage follows a mathematical law known as the sinusoidal law and, therefore, we can say that a sine wave has been generated. Such a waveform has the characteristics displayed in Fig. 4.13.

In the United Kingdom we generate electricity at a frequency of 50 Hz and the time taken to complete each cycle is given by:

$$T = \frac{1}{f}$$

$$\therefore T = \frac{1}{50\,\text{Hz}} = 0.02\,\text{s}$$

An alternating waveform is constantly changing from zero to a maximum, first in one direction, then in the opposite direction, and so the instantaneous values of the generated voltage are always changing. A useful description of the electrical effects of an a.c. waveform can be given by the maximum, average and rms values of the waveform.

The maximum or peak value is the greatest instantaneous value reached by the generated waveform. Cable and equipment insulation levels must be equal to or greater than this value.

The average value is the average over one half-cycle of the instantaneous values as they change from zero to a maximum and can be found from the following formula applied to the sinusoidal waveform shown in Fig. 4.14:

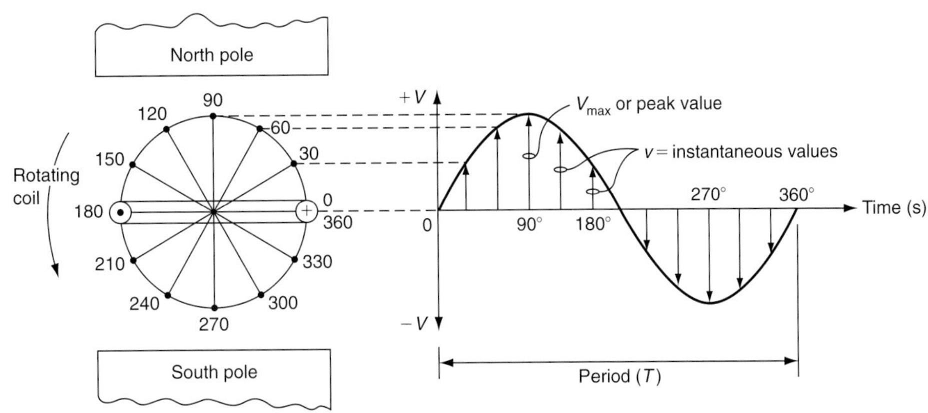

FIGURE 4.13
Characteristics of a sine wave.

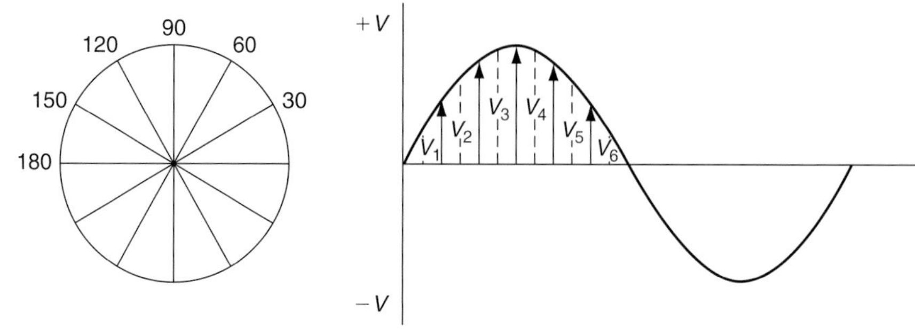

FIGURE 4.14
Sinusoidal waveform showing instantaneous values of voltage.

$$V_{av} = \frac{V_1 + V_2 + V_3 + V_4 + V_5 + V_6}{6}$$
$$= 0.637 V_{max}$$

For any sinusoidal waveform the average value is equal to 0.637 of the maximum value.

Key Fact

a.c. waveforms

- Maximum value = V_{max}
- Average value = $0.637V_{max}$
- RMS value = $0.7071 V_{max}$

The rms value is the square root of the mean of the individual squared values and is the value of an a.c. voltage which produces the same heating effect as a d.c. voltage. The value can be found from the following formula applied to the sinusoidal waveform shown in Fig. 4.14.

$$V_{rms} = \sqrt{\frac{V_1^2 + V_2^2 + V_3^2 + V_4^2 + V_5^2 + V_6^2}{6}}$$
$$= 0.7071\ V_{max}$$

For any sinusoidal waveform the rms value is equal to 0.7071 of the maximum value.

Example

The sinusoidal waveform applied to a particular circuit has a maximum value of 325.3V. Calculate the average and rms value of the waveform.

$$\text{Average value } V_{av} = 0.637 \times V_{max}$$
$$\therefore V_{av} = 0.637 \times 325.3 = 207.2\text{ V}$$
$$\text{rms value } V_{rms} = 0.7071 \times V_{max}$$
$$V_{rms} = 0.7071 \times 325.3 = 230\text{ V}$$

When we say that the main supply to a domestic property is 230V we really mean 230V rms. Such a waveform has an average value of about 207.2V and a maximum value of almost 325.3V but because the rms value gives the d.c. equivalent value we almost always give the rms value without identifying it as such.

Magnetism

The Greeks knew as early as 600 BC that a certain form of iron ore, now known as magnetite or lodestone, had the property of attracting small pieces of iron. Later, during the Middle Ages, navigational compasses were made using the magnetic properties of lodestone. Small pieces of lodestone attached to wooden splints floating in a bowl of water always came to rest pointing in a north–south direction. The word lodestone is derived from an old English word meaning 'the way', and the word magnetism is derived from Magnesia, the place where magnetic ore was first discovered.

Iron, nickel and cobalt are the only elements which are attracted strongly by a magnet. These materials are said to be *ferromagnetic*. Copper, brass, wood, PVC and glass are not attracted by a magnet and are, therefore, described as *non-magnetic*.

SOME BASIC RULES OF MAGNETISM

1. Lines of magnetic flux have no physical existence, but they were introduced by Michael Faraday (1791–1867) as a way of explaining the magnetic energy existing in space or in a material. They help us to visualize and explain the magnetic effects. The symbol used for magnetic flux is the Greek letter Φ (phi) and the unit of magnetic flux is the weber (symbol Wb), pronounced 'veber', to commemorate the work of the German physicist Wilhelm Weber (1804–1891).
2. Lines of magnetic flux always form closed loops.
3. Lines of magnetic flux behave like stretched elastic bands, always trying to shorten themselves.
4. Lines of magnetic flux never cross over each other.
5. Lines of magnetic flux travel along a magnetic material and always emerge out of the 'north pole' end of the magnet.
6. Lines of magnetic flux pass through space and non-magnetic materials undisturbed.
7. The region of space through which the influence of a magnet can be detected is called the *magnetic field* of that magnet.
8. The number of lines of magnetic flux within a magnetic field is a measure of the flux density. Strong magnetic fields have a high flux density. The symbol used for flux density is B, and the unit of flux density is the tesla (symbol T), to commemorate the work of the Croatian-born American physicist Nikola Tesla (1857–1943).
9. The places on a magnetic material where the lines of flux are concentrated are called the magnetic poles.
10. Like poles repel; unlike poles attract. These two statements are sometimes called the 'first laws of magnetism' and are shown in Fig. 4.16.

Example

The magnetizing coil of a radio speaker induces a magnetic flux of 360 μWb in an iron core of cross-sectional area 300 mm². Calculate the flux density in the core.

$$\text{Flux density } B = \frac{\Phi}{\text{area}} \text{ (tesla)}$$

$$B = \frac{360 \times 10^{-6} \text{ (Wb)}}{300 \times 10^{-6} \text{ (m}^2\text{)}}$$

$$B = 1.2 \text{ T}$$

MAGNETIC FIELDS

If a permanent magnet is placed on a surface and covered by a piece of paper, iron filings can be shaken on to the paper from a dispenser. Gently tapping the paper then causes the filings to take up the shape of the magnetic field surrounding the permanent magnet. The magnetic fields around a permanent magnet are shown in Figs. 4.15 and 4.16.

Electromagnetism

Electricity and magnetism have been inseparably connected since the experiments by Oersted and Faraday in the early nineteenth century. An electric current flowing in a conductor produces a magnetic field 'around' the conductor which is proportional to the current. Thus a small current produces a weak magnetic field, while a large current will produce a strong magnetic field. The magnetic field 'spirals' around the conductor, as shown in Fig. 4.17 and its direction can be determined by the 'dot' or 'cross' notation and the 'screw rule'. To do this, we think of the current as being represented by a dart or arrow inside the conductor. The dot represents current coming towards us when we would see the point of the arrow or dart inside the conductor. The cross represents current going away from us when we

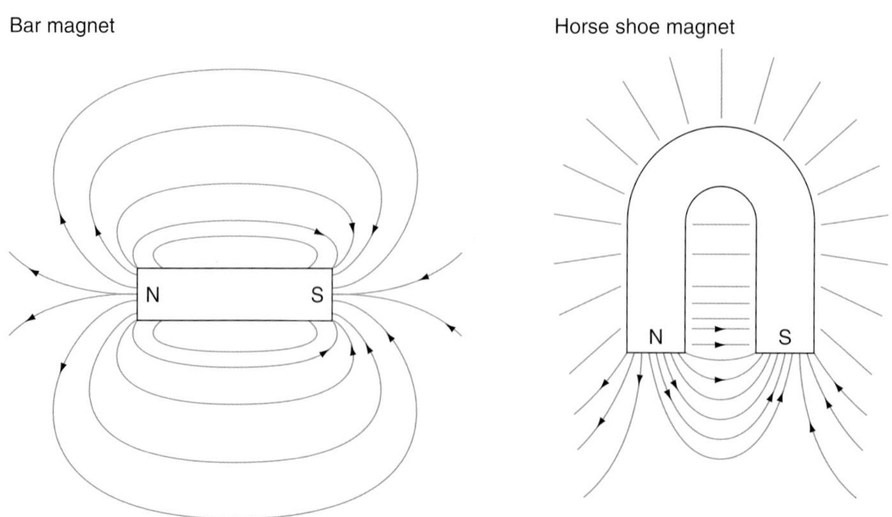

FIGURE 4.15
Magnetic fields around a permanent magnet.

would see the flights of the dart or arrow. Imagine a corkscrew or screw being turned so that it will move in the direction of the current. Therefore, if the current was coming out of the paper, as shown in Fig. 4.17(a), the magnetic field would be spiralling anticlockwise around the conductor. If the current was going into the paper, as shown in Fig. 4.17(b), the magnetic field would spiral clockwise around the conductor.

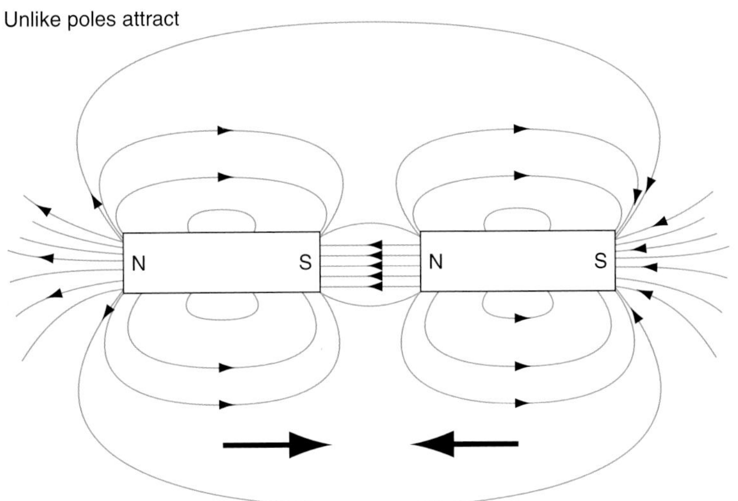

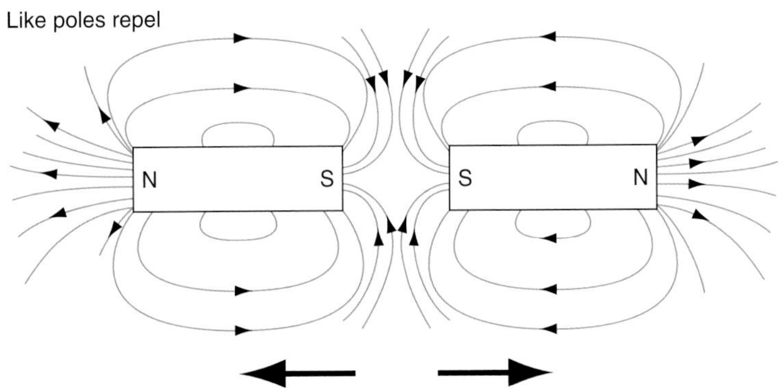

FIGURE 4.16
The first laws of magnetism.

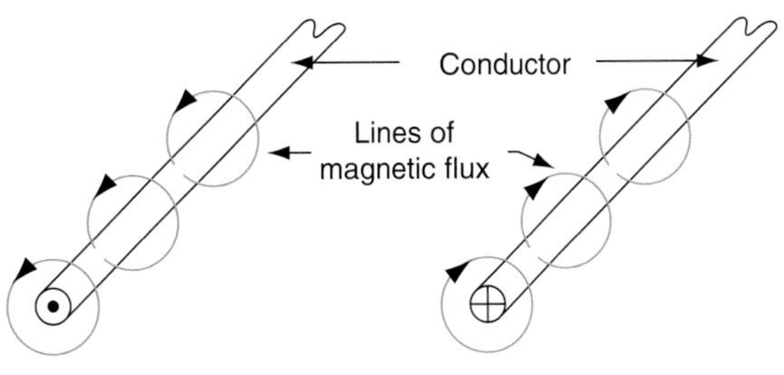

FIGURE 4.17
Magnetic fields around a current carrying conductor.

A current flowing in a *coil* of wire or solenoid establishes a magnetic field which is very similar to that of a bar magnet. Winding the coil around a soft iron core increases the flux density because the lines of magnetic flux concentrate on the magnetic material. The advantage of the electromagnet when compared with the permanent magnet is that the magnetism of the electromagnet can be switched on and off by a functional switch controlling the coil current. This effect is put to practical use in the electrical relay as used in a motor starter or alarm circuit. Fig. 4.18 shows the structure and one application of the solenoid.

A current carrying conductor maintains a magnetic field around the conductor which is proportional to the current flowing. When this magnetic field interacts with another magnetic field, forces are exerted which describe the basic principles of electric motors.

Definition

Michael Faraday demonstrated on 29 August 1831 that electricity could be produced by magnetism. He stated that *'When a conductor cuts or is cut by a magnetic field an emf is induced in that conductor. The amount of induced emf is proportional to the rate or speed at which the magnetic field cuts the conductor'.*

Michael Faraday demonstrated on 29 August 1831 that electricity could be produced by magnetism. He stated that **'When a conductor cuts or is cut by a magnetic field an emf is induced in that conductor. The amount of induced emf is proportional to the rate or speed at which the magnetic field cuts the conductor'**. This basic principle laid down the laws of present-day electricity generation where a strong magnetic field is rotated inside a coil of wire to generate electricity.

This law can be translated into a formula as follows:

$$\text{Induced emf} = Blv \text{ (V)}$$

where B is the magnetic flux density, measured in tesla, to commemorate Nikola Tesla (1856–1943) a famous Yugoslav who invented the two-phase and three-phase alternator and motor; l is the length of conductor in the magnetic field, measured in metres; and v is the velocity or speed at which the conductor cuts the magnetic flux (measured in metres per second).

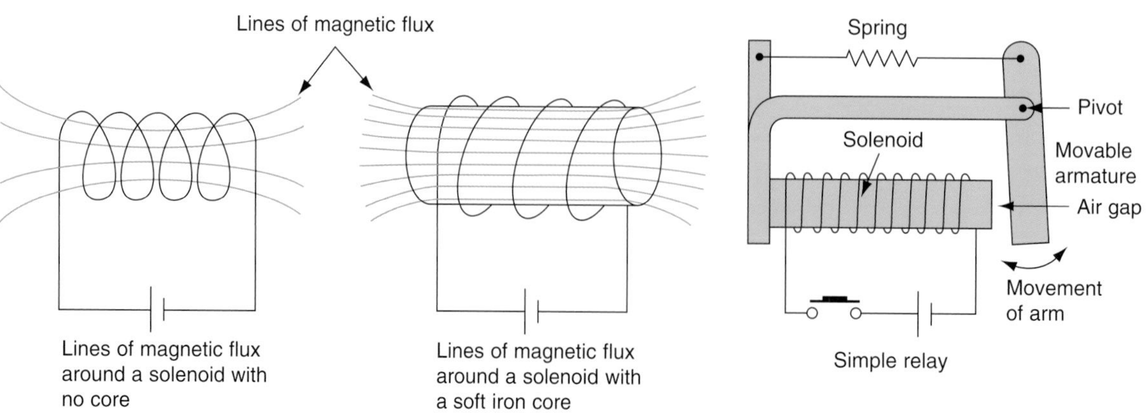

FIGURE 4.18
The solenoid and one practical application: the relay.

Example

A 15 cm length of conductor is moved at 20 m/s through a magnetic field of flux density 2T. Calculate the induced emf.

$$\text{emf} = Blv\ (\text{V})$$
$$\therefore \text{emf} = 2\ \text{T} \times 0.15 \times 20\ \text{m/s}$$
$$\text{emf} = 6\ \text{V}$$

SELF AND MUTUAL INDUCTANCE

If a coil of wire is wound on to an iron core as shown in Fig. 4.19, a magnetic field will become established in the core when a current flows in the coil due to the switch being closed.

When the switch is opened the current stops flowing and, therefore, the magnetic flux collapses. The collapsing magnetic flux induces an emf into the coil and this voltage appears across the switch contacts. The effect is known as *self-inductance*, or just **inductance**, and is one property of any coil. The unit of inductance is the henry (symbol H), to commemorate the work of the American physicist Joseph Henry (1797–1878), and a circuit is said to possess an inductance of 1 H when an emf of 1 V is induced in the circuit by a current changing at the rate of 1 A/s.

Definition

Inductance is one property of any coil in which a current establishes a magnetic field and the storage of magnetic enegy.

Fluorescent light fittings contain a choke or inductive coil in series with the tube and starter lamp. The starter lamp switches on and off very quickly, causing rapid current changes which induce a large voltage across the tube electrodes sufficient to strike an arc in the tube.

When two separate coils are placed close together, as they are in a transformer, a current in one coil produces a magnetic flux which links with the second coil. This induces a voltage in the second coil and is the basic principle of the transformer action which is described later in this chapter. The two coils in this case are said to possess **mutual inductance**, as shown in Fig. 4.20. A mutual inductance of 1 H exists between two coils when a uniformly varying current of 1 A/s in one coil produces an emf of 1 V in the other coil.

Definition

A *mutual inductance* of 1 H exists between two coils when a uniformly varying current of 1 A/s in one coil produces an emf of 1 V in the other coil.

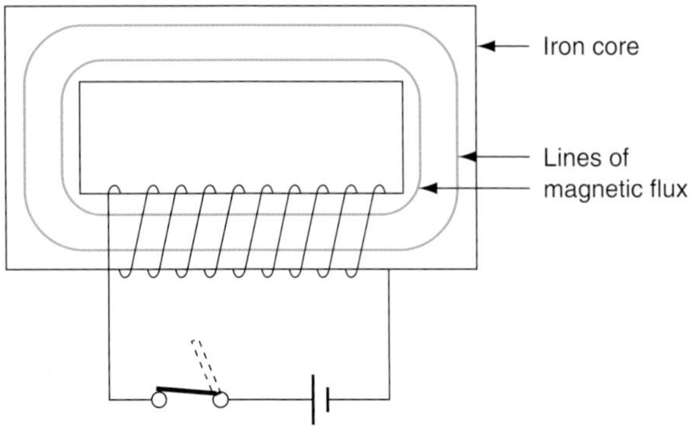

FIGURE 4.19
An inductive coil or choke.

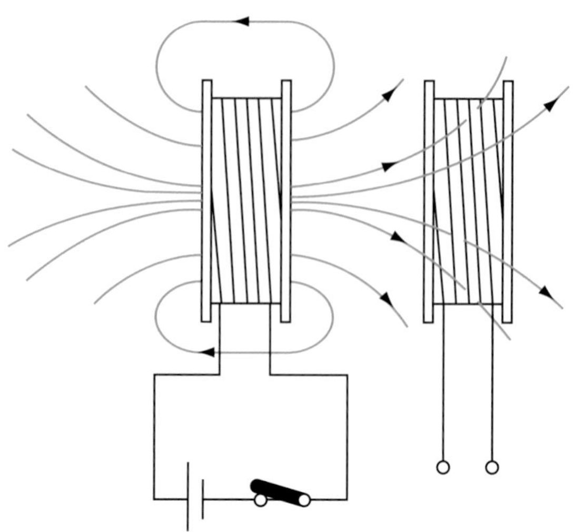

FIGURE 4.20
Mutual inductance between two coils.

The emf induced in the right-hand coil of Fig. 4.20 is dependent upon the rate of change of magnetic flux and the number of turns on the coil. The average induced emf is, therefore, given by:

$$\text{emf} = \frac{-(\Phi_2 - \Phi_1)}{t} \times N \ (V)$$

where Φ is the magnetic flux measured in webers, to commemorate the work of the German physicist, Wilhelm Weber (1804–1891), t is the time in seconds and N the number of turns. The minus sign indicates that the emf is a back emf opposing the rate of change of current as described later by Lenz's law.

Example

The magnetic flux linking 2000 turns of electromagnetic relay changes from 0.6 to 0.4 mWb in 50 ms. Calculate the average value of the induced emf.

$$\text{emf} = -\frac{(\Phi_2 - \Phi_1)}{t} \times N \ (V)$$

$$\therefore \text{emf} = -\frac{(0.6 - 0.4) \times 10^{-3}}{50 \times 10^{-3}} \times 2000$$

$$\text{emf} = -8 \text{ V}$$

ENERGY STORED IN A MAGNETIC FIELD

When we open the switch of an inductive circuit such as an electric motor or fluorescent light circuit the magnetic flux collapses and produces an arc across the switch contacts. The arc is produced by the stored magnetic energy being discharged across the switch contacts. The stored magnetic energy (symbol W) is expressed in joules and given by the following formula:

$$\text{Energy} = W = \tfrac{1}{2}\, LI^2 \text{ (J)}$$

where L is the inductance of the coil in henrys and I is the current flowing in amperes.

Example

The field windings of a motor have an inductance of 3 H and carry a current of 2 A. Calculate the magnetic energy stored in the coils.

$$W = \tfrac{1}{2}\, LI^2 \text{ (J)}$$

$$W = \tfrac{1}{2} \times 3\text{H} \times (2\text{A})^2$$

$$W = 6\text{J}$$

Definition

Some materials magnetize easily, and some are difficult to magnetize. Some materials retain their magnetism, while others lose it. The result will look like the graphs shown in Fig. 4.21 and are called *hysteresis loops*.

MAGNETIC HYSTERESIS

There are many different types of magnetic material and they all respond differently to being magnetized. Some materials magnetize easily, and some are difficult to magnetize. Some materials retain their magnetism, while others lose it. The result will look like the graphs shown in Fig. 4.21 and are called **hysteresis loops**.

Materials from which permanent magnets are made should display a wide hysteresis loop, as shown by loop (b) in Fig. 4.21.

The core of an electromagnet is required to magnetize easily, and to lose its magnetism equally easily when switched off. Suitable materials will, therefore, display a narrow hysteresis loop, as shown in loop (a) in Fig. 4.21.

The hysteresis effect causes an energy loss whenever the magnetic flux changes. This energy loss is converted to heat in the iron. The energy lost during a complete cycle of flux change is proportional to the area enclosed by the hysteresis loop.

When an iron core is subjected to alternating magnetization, as in a transformer, the energy loss occurs at every cycle and so constitutes a continuous power loss, and, therefore, for applications such as transformers, a material with a narrow hysteresis loop is required.

B
H
(a)

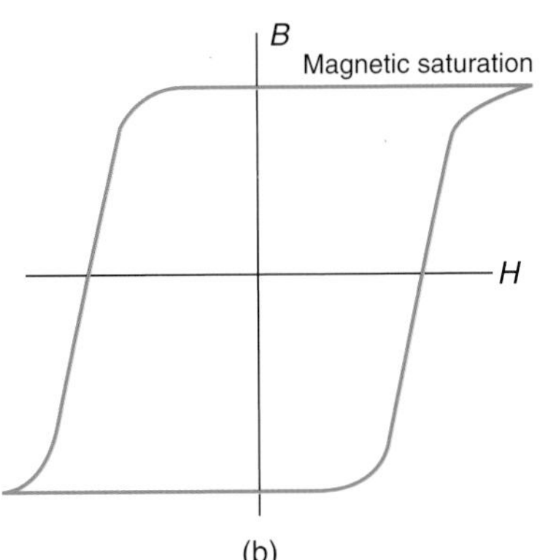

FIGURE 4.21
Magnetic hysteresis loops: (a) electromagnetic material; (b) permanent magnetic material.

Electrostatics

If a battery is connected between two insulated plates, the emf of the battery forces electrons from one plate to another until the p.d. between the plates is equal to the battery emf.

The electrons flowing through the battery constitute a current, I (in amperes), which flows for a time, t (in seconds). The plates are then said to be charged.

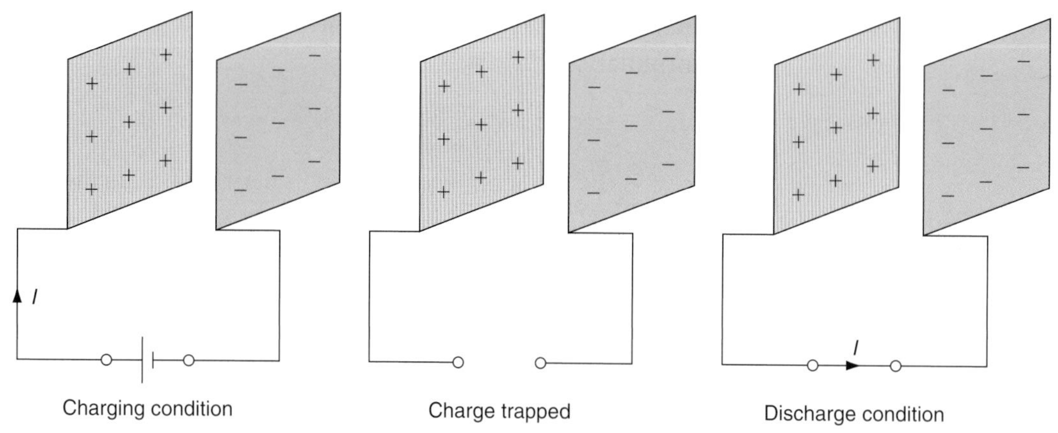

FIGURE 4.22
The charge on a capacitor's plates.

Definition

The property of a pair of plates to store an electric charge is called its *capacitance*.

Definition

By definition, a *capacitor* has a capacitance (C) of one farad (symbol F) when a p.d. of one volt maintains a charge of one coulomb on that capacitor.

The amount of charge transferred is given by:

$$Q = It \text{ (coulomb [Symbol C])}$$

Figure 4.22 shows the charges on a capacitor's plates.

When the voltage is removed the charge Q is trapped on the plates, but if the plates are joined together, the same quantity of electricity, $Q = It$, will flow back from one plate to the other, so discharging them. The property of a pair of plates to store an electric charge is called its **capacitance**.

By definition, a **capacitor** has a capacitance (C) of one farad (symbol F) when a p.d. of one volt maintains a charge of one coulomb on that capacitor, or

$$C = \frac{Q}{V} \text{ (F)}$$

Collecting these important formulae together, we have:

$$Q = It = CV$$

CAPACITORS

A capacitor consists of two metal plates, separated by an insulating layer called the dielectric. It has the ability of storing a quantity of electricity as an excess of electrons on one plate and a deficiency on the other.

Example

A 100 μF capacitor is charged by a steady current of 2 mA flowing for 5 seconds. Calculate the total charge stored by the capacitor and the p.d. between the plates.

$$Q = It \text{ (C)}$$

$$\therefore Q = 2 \times 10^{-3}\,\text{A} \times 5\,\text{s} = 10\,\text{mC}$$

$$Q = CV$$

$$\therefore V = \frac{Q}{C} \text{ (V)}$$

$$V = \frac{10 \times 10^{-3}\,\text{C}}{100 \times 10^{-6}\,\text{F}} = 100\,\text{V}$$

The p.d. or voltage which may be maintained across the plates of a capacitor is determined by the type and thickness of the dielectric medium. Capacitor manufacturers usually indicate the maximum safe working voltage for their products.

Capacitors are classified by the type of dielectric material used in their construction. Fig. 4.23 shows the general construction and appearance of some capacitor types to be found in installation work.

AIR-DIELECTRIC CAPACITORS

Air-dielectric capacitors are usually constructed of multiple aluminium vanes of which one section moves to make the capacitance variable. They are often used for radio-tuning circuits.

MICA-DIELECTRIC CAPACITORS

Mica-dielectric capacitors are constructed of thin aluminium foils separated by a layer of mica. They are expensive, but this dielectric is very

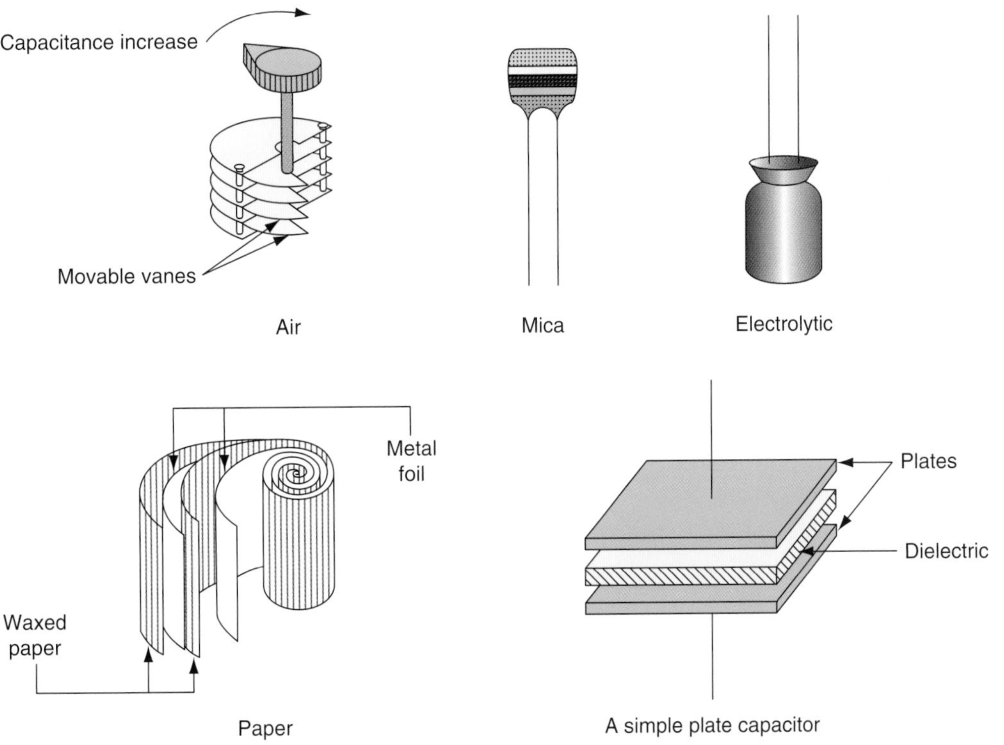

FIGURE 4.23
Construction and appearance of capacitors.

stable and has low dielectric loss. They are often used in high-frequency electronic circuits.

PAPER-DIELECTRIC CAPACITORS

Paper-dielectric capacitors usually consist of thin aluminium foils separated by a layer of waxed paper. This paper–foil sandwich is rolled into a cylinder and usually contained in a metal cylinder. These capacitors are used in fluorescent lighting fittings and motor circuits.

Key Fact

Capacitors

A capacitor is a device which stores an electric charge.

ELECTROLYTIC CAPACITORS

The construction of these is similar to that of the paper-dielectric capacitors, but the dielectric material in this case is an oxide skin formed electrolytically by the manufacturers. Since the oxide skin is very thin, a large capacitance is achieved for a small physical size, but if a voltage of the wrong polarity is applied, the oxide skin is damaged and the gas inside the sealed container explodes. For this reason electrolytic capacitors must be connected to the correct voltage polarity. They are used where a large capacitance is required from a small physical size and where the terminal voltage never reverses polarity.

The practical considerations of capacitors and the use of colour codes to determine capacitor values are dealt with later in this chapter.

Energy stored in a capacitor

Following a period of charge, the capacitor will store a small amount of energy as an electrostatic charge which, we will see later, can be made to do work. The energy stored (symbol W) in a capacitor is expressed in joules and given by the formula:

$$\text{Energy} = W = \tfrac{1}{2}CV^2 \text{ (J)}$$

where C is the capacitance of the capacitor and V is the applied voltage.

Example 1

A 60 μF capacitor is used for power-factor correction in a fluorescent luminaire. Calculate the energy stored in the capacitor when it is connected to the 230 V mains supply.

$$\text{Energy} = W = \tfrac{1}{2}CV^2 \text{ (J)}$$

$$\therefore\ W = \tfrac{1}{2} \times 60 \times 10^{-6}\ \text{F} \times (230\ \text{V})^2$$

$$W = 3.17\ \text{J}$$

Example 2

The energy stored in a certain capacitor when connected across a 400 V supply is 0.3 J. Calculate (a) the capacitance and (b) the charge on the capacitor.

For (a),

$$W = \tfrac{1}{2}CV^2 \text{ (J)}$$

Transposing,

$$C = \frac{2W}{V^2} \text{ (F)}$$

$$\therefore C = \frac{2 \times 0.3 \text{ J}}{(400 \text{ V})^2}$$

$$C = 3.75 \ \mu\text{F}$$

For (b), the charge is given by:

$$Q = CV \text{ (C)}$$

$$\therefore Q = 3.75 \times 10^{-6} \text{ F} \times 400 \text{ V}$$

$$Q = 1500 \ \mu\text{C}$$

CR CIRCUITS

As we have discussed earlier in this chapter, connecting a voltage to the plates of a capacitor causes it to charge up to the potential of the supply. This involves electrons moving around the circuit to create the necessary charge conditions and, therefore, this action does not occur instantly, but takes some time, depending upon the size of the capacitor and the resistance of the circuit. Such circuits are called capacitor–resistor (CR) circuits, and have many applications in electronics as timers and triggers and for controlling the time base sweeps of a cathode ray oscilloscope.

Figure 4.24 shows the circuit diagram for a simple CR circuit and the graphs drawn from the meter readings. It can be seen that:

(a) initially the current has a maximum value and decreases slowly to zero as the capacitor charges and

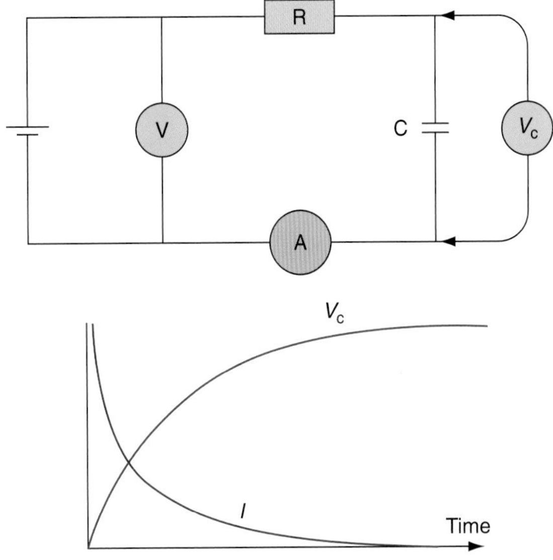

FIGURE 4.24
A CR circuit.

(b) initially the capacitor voltage rises rapidly but then slows down, increasing gradually until the capacitor voltage is equal to the supply voltage when fully charged.

The mathematical name for the shape of these curves is an *exponential* curve and, therefore, we say that the capacitor voltage is growing exponentially while the current is decaying exponentially during the charging period. The *rate* at which the capacitor charges is dependent upon the *size* of the capacitor and resistor. The bigger the values of C and R, the longer it will take to charge the capacitor. The time taken to charge a capacitor by a *constant* current is given by the time constant of the circuit which is expressed mathematically as $T = CR$, where T is the time in seconds.

Example 1

A 60 μF capacitor is connected in series with a 20 kΩ resistor across a 12 V supply. Determine the time constant of this circuit.

$$T = CR \text{ (s)}$$

$$\therefore\ T = 60 \times 10^{-6}\ \text{F} \times 20 \times 10^{3}\ \Omega$$

$$T = 1.2\,\text{s}$$

We have already seen that in practice the capacitor is not charged by a *constant* current but, in fact, charges exponentially. However, it can be shown by experiment that in one time constant the capacitor will have reached about 63% of its final steady value, taking about five times the time constant to become fully charged. Therefore, in 1.2 seconds the 60 μF capacitor of Example 1 will have reached about 63% of 12V and after 5 T, that is 6 seconds, will be fully charged at 12V.

GRAPHICAL DERIVATION OF CR CIRCUIT

The exponential charging and discharging curves of the CR circuit described in Example 1 may also be drawn to scale by following the procedure described below and is shown in Fig. 4.25.

1. We have calculated the time constant for the circuit (T) and found it to be 1.2 seconds.
2. We know that the maximum voltage of the fully charged capacitor will be 12V because the supply voltage is 12V.
3. To draw the graph we must first select suitable scales: 0–12 on the voltage axis would be appropriate for this example and 0–6 seconds on the time axis because we know that the capacitor must be fully charged in five time constants.
4. Next draw a horizontal dotted line along the point of maximum voltage, 12V in this example.
5. Along the time axis measure off one time constant (T), distance OA in Fig. 4.25. This corresponds to 1.2 seconds because in this example T is equal to 1.2 seconds.

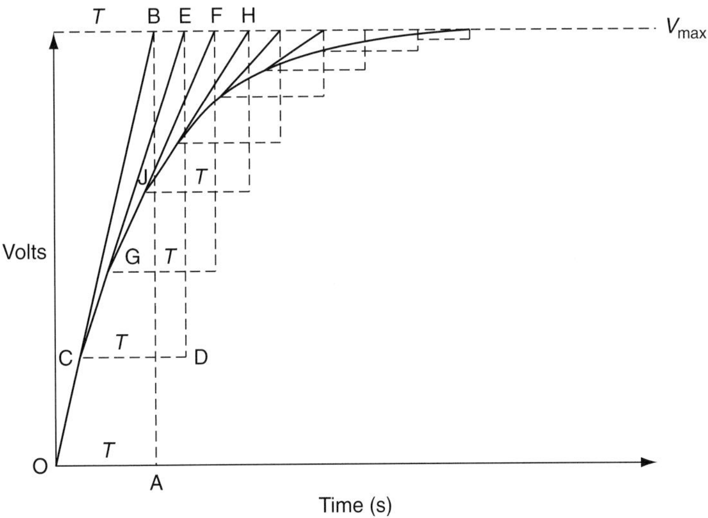

FIGURE 4.25
Graphical derivation of CR growth curve.

6. Draw the vertical dotted line AB.
7. Next, draw a full line OB; this is the start of the charging curve.
8. Select a point C, somewhere convenient and close to O along line OB.
9. Draw a horizontal line CD equal to the length of the time constant (*T*).
10. Draw the dotted vertical line DE.
11. Draw the line CE, the second line of our charging curve.
12. Select another point G close to C along line CE and repeat the procedures 9 to 12 to draw lines GF, JH and so on as shown in Fig. 4.25.
13. Finally, join together with a smooth curving line the points OCGJ, etc., and we have the exponential growth curve of the voltage across the capacitor.

Switching off the supply and discharging the capacitor through the 20 kΩ resistor will produce the exponential decay of the voltage across the capacitor which will be a mirror image of the growth curve. The decay curve can be derived graphically in the same way as the growth curve and is shown in Fig. 4.26.

SELECTING A CAPACITOR

There are two broad categories of capacitor, the non-polarized and the polarized type.

The non-polarized type is often found in electrical installation work for power-factor correction. A paper-dielectric capacitor is non-polarized and can be connected either way round.

The polarized type must be connected to the polarity indicated otherwise the capacitor will explode. Electrolytic capacitors are polarized and are used where a large value of capacitance is required in a relatively small

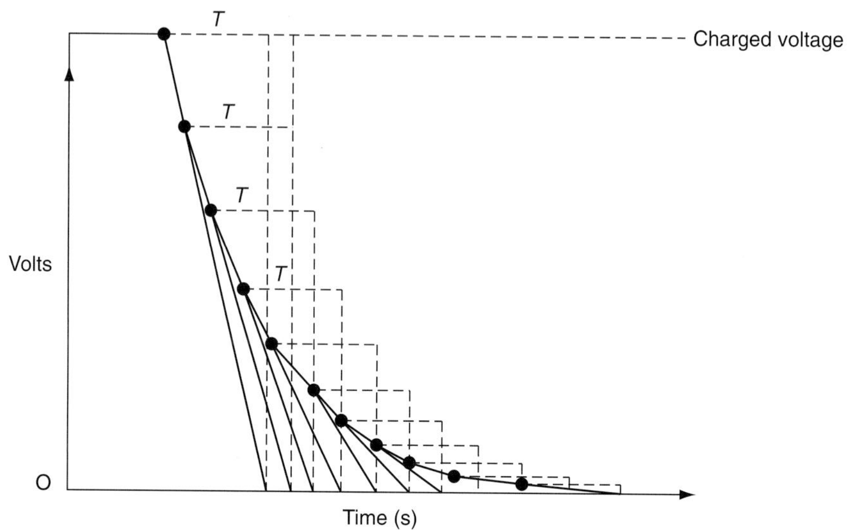

FIGURE 4.26
Graphical derivation of CR decay curve.

package. We therefore find polarized capacitors in electronic equipment such as smoothing or stabilized supplies, emergency lighting and alarm systems, so be careful when working on these systems.

When choosing a capacitor for a particular application, three factors must be considered: value, working voltage and leakage current.

The unit of capacitance is the *farad* (symbol F), to commemorate the name of the English scientist Michael Faraday. However, for practical purposes the farad is much too large and in electrical installation work and electronics we use fractions of a farad as follows:

$$\begin{aligned} 1 \text{ microfarad} &= 1\ \mu\text{F} = 1 \times 10^{-6}\ \text{F} \\ 1 \text{ nanofarad} &= 1\ \text{nF} = 1 \times 10^{-9}\ \text{F} \\ 1 \text{ picofarad} &= 1\ \text{pF} = 1 \times 10^{-12}\ \text{F} \end{aligned}$$

The p.f. correction capacitor used in a domestic fluorescent luminaire would typically have a value of 8 μF at a working voltage of 400 V. In an electronic filter circuit a typical capacitor value might be 100 pF at 63 V.

One microfarad is 1 million times greater than one picofarad. It may be useful to remember that:

$$1000\ \text{pF} = 1\ \text{nF}, \quad \text{and} \quad 1000\ \text{nF} = 1\ \mu\text{F}$$

The working voltage of a capacitor is the *maximum* voltage that can be applied between the plates of the capacitor without breaking down the dielectric insulating material. This is a d.c. rating and, therefore, a capacitor with a 200 V rating must only be connected across a maximum of 200 V d.c. Since a.c. voltages are usually given as rms values, a 200 V a.c. supply would have a maximum value of about 283 V which would damage the 200 V capacitor. When connecting a capacitor to the 230 V mains supply we must choose a working voltage of about 400 V because 230 V rms is approximately 325 V maximum. The 'factor of safety' is small and, therefore, the working voltage of the capacitor must not be exceeded.

An ideal capacitor which is isolated will remain charged forever, but in practice no dielectric insulating material is perfect, and the charge will slowly *leak* between the plates, gradually discharging the capacitor. The loss of charge by leakage through it should be very small for a practical capacitor. However, the capacitors used in electrical installation work for power-factor correction are often fitted with a high-value discharge resistor to encourage the charge to leak away safely when not in use. This is to prevent anyone getting a shock from touching the terminals of a charged capacitor.

Alternating current theory

Earlier in this chapter in Figs. 4.13 and 4.14 we looked at the generation of an a.c. waveform and the calculation of average and rms values. In this section we will first of all consider the theoretical circuits of pure resistance, inductance and capacitance acting alone in an a.c. circuit before going on to consider the practical circuits of resistance, inductance and capacitance acting together. Let us first define some of our terms of reference.

Definition

In any circuit, *resistance* is defined as opposition to current flow.

RESISTANCE

In any circuit, **resistance** is defined as opposition to current flow. From Ohm's law,

$$R = \frac{V_R}{I_R}\ (\Omega)$$

However, in an a.c. circuit, resistance is only part of the opposition to current flow. The inductance and capacitance of an a.c. circuit also cause an opposition to current flow, which we call *reactance.*

Definition

Inductive reactance (X_L) is the opposition to an a.c. current in an inductive circuit. It causes the current in the circuit to lag behind the applied voltage.

Inductive reactance (X_L) is the opposition to an a.c. current in an inductive circuit. It causes the current in the circuit to lag behind the applied voltage, as shown in Fig. 4.27. It is given by the formula

$$X_L = 2\pi fL\ (\Omega)$$

where

$\pi = 3.142$ a constant

$f =$ the frequency of the supply

$L =$ the inductance of the circuit

or by:

$$X_L = \frac{V_L}{I_L}$$

Definition

Capacitive reactance (X_C) is the opposition to an a.c. current in a capacitive circuit. It causes the current in the circuit to lead ahead of the voltage.

Capacitive reactance (X_C) is the opposition to an a.c. current in a capacitive circuit. It causes the current in the circuit to lead ahead of the voltage, as

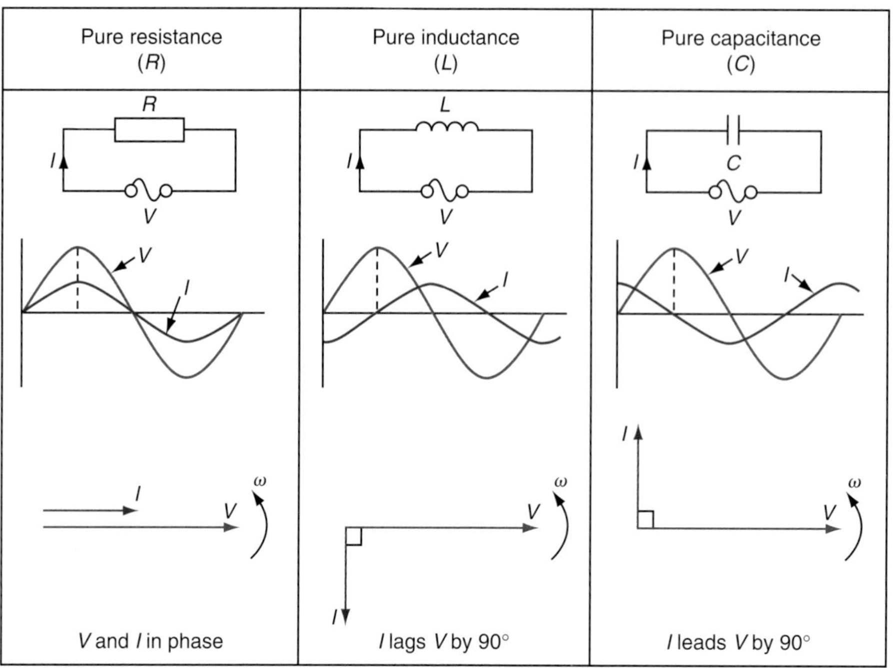

FIGURE 4.27
Voltage and current relationships in resistive, capacitive and inductive circuits.

shown in Fig. 4.27. It is given by the formula

$$X_C = \frac{1}{2\pi fC} \quad (\Omega)$$

where π and f are defined as before and C is the capacitance of the circuit. It can also be expressed as:

$$X_C = \frac{V_C}{I_C}$$

Example

Calculate the reactance of a 150 μF capacitor and a 0.05 H inductor if they were separately connected to the 50 Hz mains supply.

For capacitive reactance:

$$X_C = \frac{1}{2\pi fC}$$

where $f = 50\,\text{Hz}$ and $C = 150\,\mu\text{F} = 150 \times 10^{-6}\,\text{F}$.

$$\therefore X_C = \frac{1}{2 \times 3.142 \times 50\ \text{Hz} \times 150 \times 10^{-6}} = 21.2\ \Omega$$

For inductive reactance:

$$X_L = 2\pi fL$$

where $f = 50$ Hz and $L = 0.05$ H.

$$\therefore X_L = 2 \times 3.142 \times 50 \text{ Hz} \times 0.05 \text{ H} = 15.7\ \Omega$$

Definition

The total opposition to current flow in an a.c. circuit is called *impedance* and given the symbol Z.

IMPEDANCE

The total opposition to current flow in an a.c. circuit is called **impedance** and given the symbol Z. Thus impedance is the combined opposition to current flow of the resistance, inductive reactance and capacitive reactance of the circuit and can be calculated from the formula

$$Z = \sqrt{R^2 + X^2}\ (\Omega)$$

or

$$Z = \frac{V_T}{I_T}$$

Example 1

Calculate the impedance when a 5 Ω resistor is connected in series with a 12 Ω inductive reactance.

$$Z = \sqrt{R^2 + X_L^2}\ (\Omega)$$
$$\therefore Z = \sqrt{5^2 + 12^2}$$
$$Z = \sqrt{25 + 144}$$
$$Z = \sqrt{169}$$
$$Z = 13\ \Omega$$

Example 2

Calculate the impedance when a 48 Ω resistor is connected in series with a 55 Ω capacitive reactance.

$$Z = \sqrt{R^2 + X_C^2}\ (\Omega)$$
$$\therefore Z = \sqrt{48^2 + 55^2}$$
$$Z = \sqrt{2304 + 3025}$$
$$Z = \sqrt{5329}$$
$$Z = 73\ \Omega$$

RESISTANCE, INDUCTANCE AND CAPACITANCE IN AN A.C. CIRCUIT

When a resistor only is connected to an a.c. circuit the current and voltage waveforms remain together, starting and finishing at the same time. We say that the waveforms are *in phase*.

When a pure inductor is connected to an a.c. circuit the current lags behind the voltage waveform by an angle of 90°. We say that the current *lags* the voltage by 90°. When a pure capacitor is connected to an a.c. circuit the current *leads* the voltage by an angle of 90°. These various effects can be observed on an oscilloscope, but the circuit diagram, waveform diagram and phasor diagram for each circuit are shown in Fig. 4.27.

Phasor diagrams

Phasor diagrams and a.c. circuits are an inseparable combination. Phasor diagrams allow us to produce a model or picture of the circuit under consideration which helps us to understand the circuit. A phasor is a straight line, having definite length and direction, which represents to scale the magnitude and direction of a quantity such as a current, voltage or impedance.

Try This

Definitions

To help you to remember definitions, try writing them down. Write down a definition of resistance, inductive reactance and capacitive reactance.

To find the combined effect of two quantities we combine their phasors by adding the beginning of the second phasor to the end of the first. The combined effect of the two quantities is shown by the resultant phasor, which is measured from the original zero position to the end of the last phasor.

Example

Find by phasor addition the combined effect of currents *A* and *B* acting in a circuit. Current *A* has a value of 4 A, and current *B* a value of 3 A, leading *A* by 90°. We usually assume phasors to rotate anticlockwise and so the complete diagram will be as shown in Fig. 4.28. Choose a scale of, for example, 1 A = 1 cm and draw the phasors to scale, that is $A = 4$ cm and $B = 3$ cm, leading *A* by 90°.

The magnitude of the resultant phasor can be measured from the phasor diagram and is found to be 5 A acting at a phase angle ϕ of about 37° leading *A*. We therefore say that the combined effect of currents *A* and *B* is a current of 5 A at an angle of 37° leading *A*.

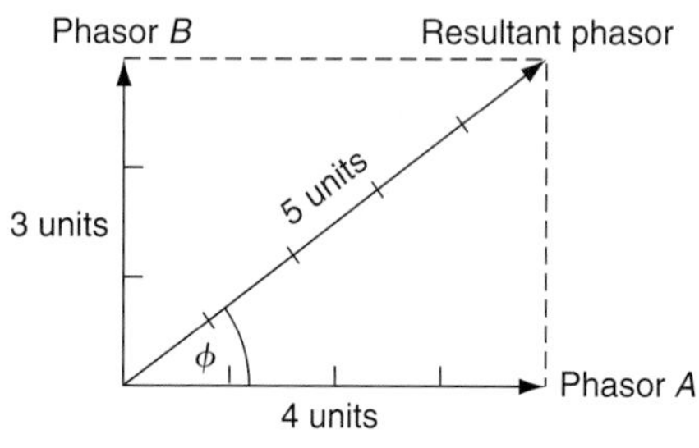

FIGURE 4.28
The phasor addition of currents A and B.

Phase angle ϕ

In an a.c. circuit containing resistance only, such as a heating circuit, the voltage and current are in phase, which means that they reach their peak and zero values together, as shown in Fig. 4.29(a).

In an a.c. circuit containing inductance, such as a motor or discharge lighting circuit, the current often reaches its maximum value after the voltage, which means that the current and voltage are out of phase with each other, as shown in Fig. 4.29(b). The phase difference, measured in degrees between the current and voltage, is called the phase angle of the circuit, and is denoted by the symbol ϕ, the Greek letter phi.

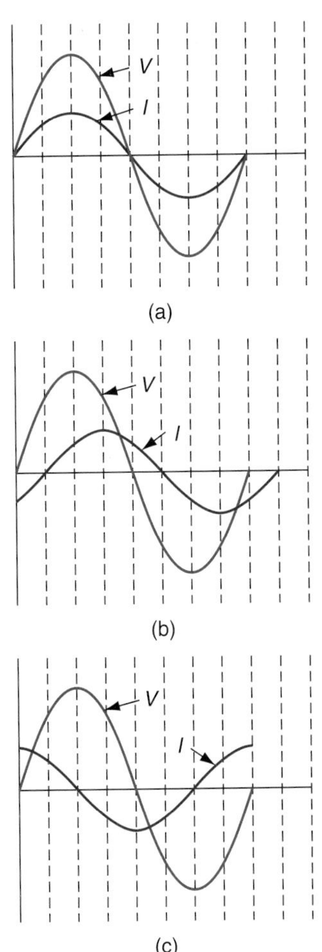

FIGURE 4.29
Phase relationship of a.c. waveform: (a) V and I in phase, phase angle $\phi = 0°$ and power factor (p.f.) = cos ϕ = 1; (b) V and I displaced by 45°, f = 45° and p.f. = 0.707; (c) V and I displaced by 90°, ϕ = 90° and p.f. = 0.

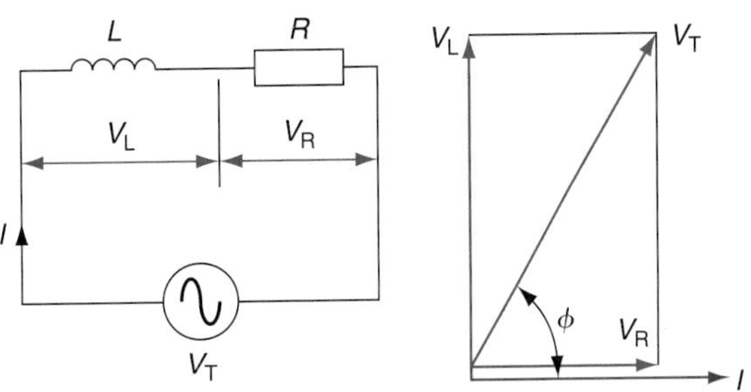

FIGURE 4.30
A series RL circuit and phasor diagram.

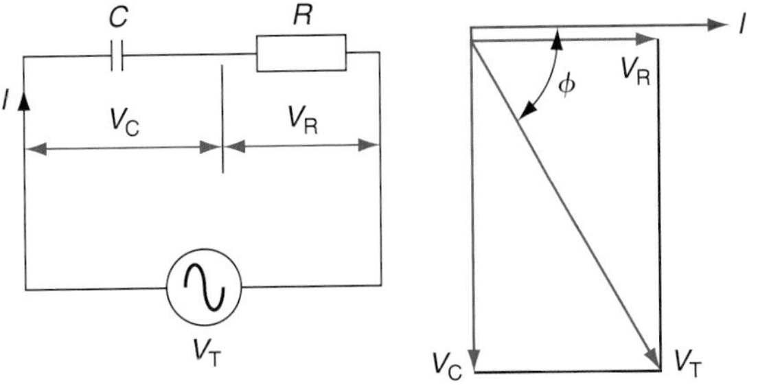

FIGURE 4.31
A series RC circuit and phasor diagram.

When circuits contain two or more separate elements, such as RL, RC or RLC, the phase angle between the total voltage and total current will be neither 0° nor 90° but will be determined by the relative values of resistance and reactance in the circuit. In Fig. 4.30 the phase angle between applied voltage and current is angle ϕ.

ALTERNATING CURRENT SERIES CIRCUIT

In a circuit containing a resistor and inductor connected in series as shown in Fig. 4.30, the current I will flow through the resistor and the inductor causing the voltage V_R to be dropped across the resistor and V_L to be dropped across the inductor. The sum of these voltages will be equal to the total voltage V_T but because this is an a.c. circuit the voltages must be added by phasor addition. The result is shown in Fig. 4.30, where V_R is drawn to scale and in phase with the current and V_L is drawn to scale and leading the current by 90°. The phasor addition of these two voltages gives us the magnitude and direction of V_T, which leads the current by angle ϕ.

In a circuit containing a resistor and capacitor connected in series as shown in Fig. 4.31, the current I will flow through the resistor and capacitor causing voltage drops V_R and V_C. The voltage V_R will be in phase with the current and V_C will lag the current by 90°. The phasor addition of these voltages is equal to the total voltage V_T which, as can be seen in Fig. 4.31, is lagging the current by some angle ϕ.

THE IMPEDANCE TRIANGLE

We have now established the general shape of the phasor diagram for a series a.c. circuit. Figures 4.30 and 4.31 show the voltage phasors, but we know that $V_R = IR$, $V_L = IX_L$, $V_C = IX_C$ and $V_T = IZ$, and therefore the phasor diagrams (a) and (b) of Fig. 4.32 must be equal. From Fig. 4.32(b), by the theorem of Pythagoras, we have:

$$(IZ)^2 = (IR)^2 + (IX)^2$$
$$I^2Z^2 = I^2R^2 + I^2X^2$$

If we now divide throughout by I^2 we have:

$$Z^2 = R^2 + X^2$$
$$\text{or } Z = \sqrt{R^2 + X^2}\ \Omega$$

The phasor diagram can be simplified to the impedance triangle given in Fig. 4.32(c).

Example 1

A coil of 0.15 H is connected in series with a 50 Ω resistor across a 100 V 50 Hz supply. Calculate (a) the reactance of the coil, (b) the impedance of the circuit and (c) the current.

For (a),

$$X_L = 2\pi fL\ (\Omega)$$
$$\therefore X_L = 2 \times 3.142 \times 50\text{ Hz} \times 0.15\text{ H} = 47.1\ \Omega$$

For (b),

$$Z = \sqrt{R^2 + X^2}\ (\Omega)$$
$$\therefore Z = \sqrt{(50\ \Omega)^2 + (47.1\ \Omega)^2} = 68.69\ \Omega$$

For (c),

$$I = V/Z\,(\text{A})$$
$$\therefore I = \frac{100\text{ V}}{68.69\ \Omega} = 1.46\text{ A}$$

Example 2

A 60 μF capacitor is connected in series with a 100 Ω resistor across a 230 V 50 Hz supply. Calculate (a) the reactance of the capacitor, (b) the impedance of the circuit and (c) the current.

For (a),

$$X_C = \frac{1}{2\pi fC}\ (\Omega)$$
$$\therefore X_C = \frac{1}{2\pi \times 50\text{Hz} \times 60 \times 10^{-6}\text{ F}} = 53.05\ \Omega$$

For (b),

$$Z = \sqrt{R^2 + X^2}\ (\Omega)$$

$$\therefore\ Z = \sqrt{(100\,\Omega)^2 + (53.05\,\Omega)^2} = 113.2\,\Omega$$

For (c),

$$I = V/\,Z(\text{A})$$

$$\therefore\ I = \frac{230\,\text{V}}{113.2\,\Omega} = 2.03\,\text{A}$$

POWER AND POWER FACTOR

Power factor (p.f.) is defined as the cosine of the phase angle between the current and voltage:

$$\text{p.f.} = \cos\phi$$

If the current lags the voltage as shown in Fig. 4.30, we say that the p.f. is lagging, and if the current leads the voltage as shown in Fig. 4.31, the p.f. is said to be leading. From the trigonometry of the impedance triangle shown in Fig. 4.32, p.f. is also equal to:

$$\text{p.f.} = \cos\phi = \frac{R}{Z} = \frac{V_R}{V_T}$$

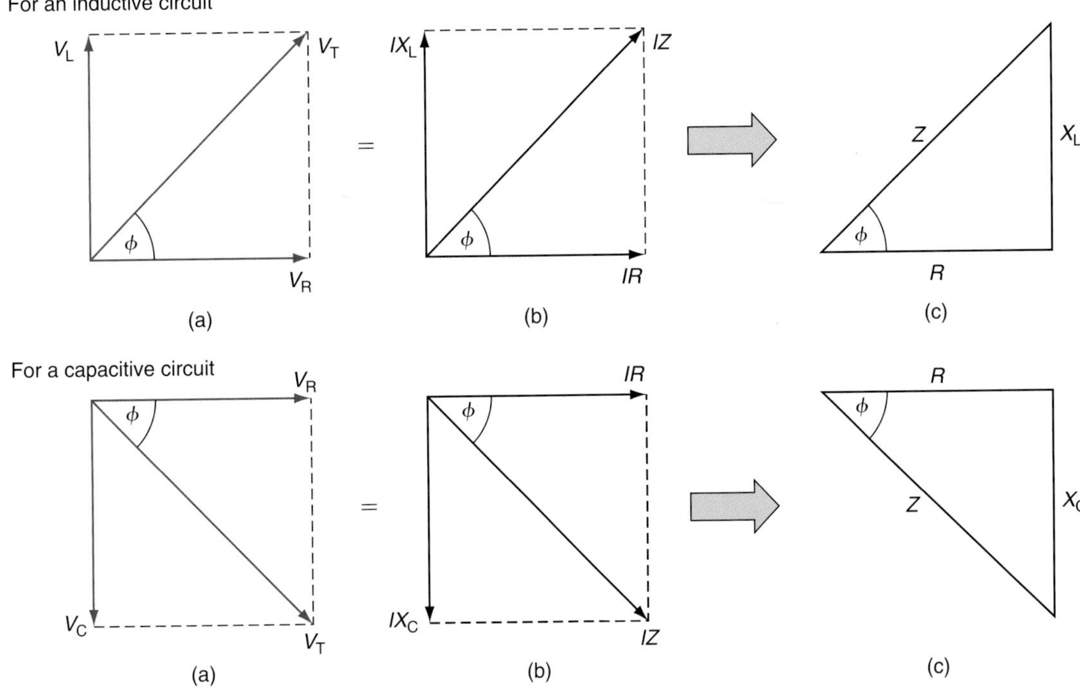

FIGURE 4.32
Phasor diagram and impedance triangle.

Pure inductor

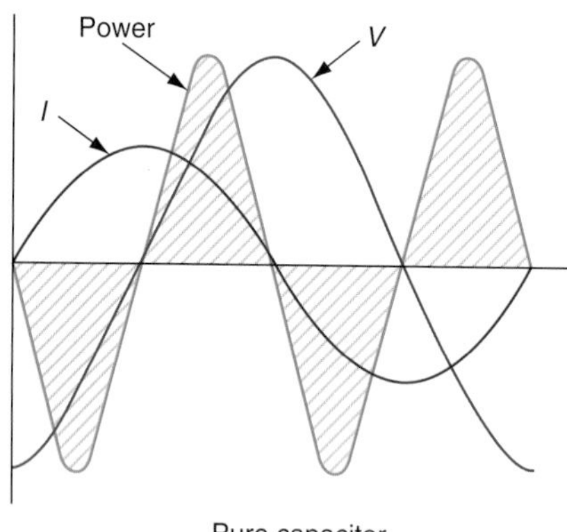

FIGURE 4.33
Waveform for the a.c. power in purely inductive and purely capacitive circuits.

The electrical power in a circuit is the product of the instantaneous values of the voltage and current. Fig. 4.33 shows the voltage and current waveform for a pure inductor and pure capacitor. The power waveform is obtained from the product of V and I at every instant in the cycle. It can be seen that the power waveform reverses every quarter cycle, indicating that energy is alternately being fed into and taken out of the inductor and capacitor. When considered over one complete cycle, the positive and negative portions are equal, showing that the average power consumed by a pure inductor or capacitor is zero. This shows that inductors and capacitors store energy during one part of the voltage cycle and feed it back into the supply later in the cycle. Inductors store energy as a magnetic field and capacitors as an electric field.

In an electric circuit more power is taken from the supply than is fed back into it, since some power is dissipated by the resistance of the circuit, and therefore:

$$P = I^2R \text{ (W)}$$

In any d.c. circuit the power consumed is given by the product of the voltage and current, because in a d.c. circuit voltage and current are in phase. In an a.c. circuit the power consumed is given by the product of the current and that part of the voltage which is in phase with the current. The in-phase component of the voltage is given by $V \cos \phi$, and so power can also be given by the equation:

$$P = VI \cos \phi \text{ (W)}$$

Example 1

A coil has a resistance of 30 Ω and a reactance of 40 Ω when connected to a 250 V supply. Calculate (a) the impedance, (b) the current, (c) the p.f. and (d) the power.

For (a),

$$Z = \sqrt{R^2 + X^2} \ (\Omega)$$

$$\therefore Z = \sqrt{(30\ \Omega)^2 + (40\ \Omega)^2} = 50\ \Omega$$

For (b),

$$I = V/Z \text{(A)}$$

$$\therefore I = \frac{250\text{ V}}{50\ \Omega} = 5\text{ A}$$

For (c),

$$\text{p.f.} = \cos \phi = \frac{R}{Z}$$

$$\therefore \text{p.f.} = \frac{30\ \Omega}{50\ \Omega} = 0.6 \text{ lagging}$$

For (d),

$$P = VI\cos\phi\ (\text{W})$$

$$\therefore\ P = 250\ \text{V} \times 5\ \text{A} \times 0.6 = 750\ \text{W}$$

Example 2

A capacitor of reactance 12 Ω is connected in series with a 9 Ω resistor across a 150 V supply. Calculate (a) the impedance of the circuit, (b) the current, (c) the p.f. and (d) the power.

For (a),

$$Z = \sqrt{R^2 + X^2}\ (\Omega)$$

$$\therefore\ Z = \sqrt{(9\ \Omega)^2 + (12\ \Omega)^2} = 15\ \Omega$$

For (b),

$$I = V/Z(\text{A})$$

$$\therefore\ I = \frac{150\ \text{V}}{15\ \Omega} = 10\ \text{A}$$

For (c),

$$\text{p.f.} = \cos\phi = \frac{R}{Z}$$

$$\therefore\ \text{p.f.} = \frac{9\ \Omega}{15\ \Omega} = 0.6\ \text{leading}$$

For (d),

$$P = VI\cos\phi\ (\text{W})$$

$$\therefore\ P = 150\ \text{V} \times 10\ \text{A} \times 0.6 = 900\ \text{W}$$

The power factor of most industrial loads is lagging because the machines and discharge lighting used in industry are mostly inductive. This causes an additional magnetizing current to be drawn from the supply, which does not produce power, but does need to be supplied, making supply cables larger.

Example 3

A 230 V supply feeds three 1.84 kW loads with power factors of 1, 0.8 and 0.4. Calculate the current at each power factor.

The current is given by:

$$I = \frac{P}{V\cos\phi}$$

where $\boldsymbol{P}$ = 1.84 kW = 1840 W and $\boldsymbol{V}$ = 230 V. If the p.f. is 1, then:

$$I = \frac{1840\ \text{W}}{230\ \text{V} \times 1} = 8\ \text{A}$$

For a p.f. of 0.8,

$$I = \frac{1840\ \text{W}}{230\ \text{V} \times 0.8} = 10\ \text{A}$$

For a p.f. of 0.4,

$$I = \frac{1840\ \text{W}}{230\ \text{V} \times 0.4} = 20\ \text{A}$$

It can be seen from these calculations that a 1.84 kW load supplied at a power factor of 0.4 would require a 20 A cable, while the same load at unity power factor could be supplied with an 8 A cable. There may also be the problem of higher voltage drops in the supply cables. As a result, the supply companies encourage installation engineers to improve their power factor to a value close to 1 and sometimes charge penalties if the power factor falls below 0.8.

Power-factor improvement

Most installations have a low or bad power factor because of the inductive nature of the load. A capacitor has the opposite effect of an inductor, and so it seems reasonable to add a capacitor to a load which is known to have a lower power factor.

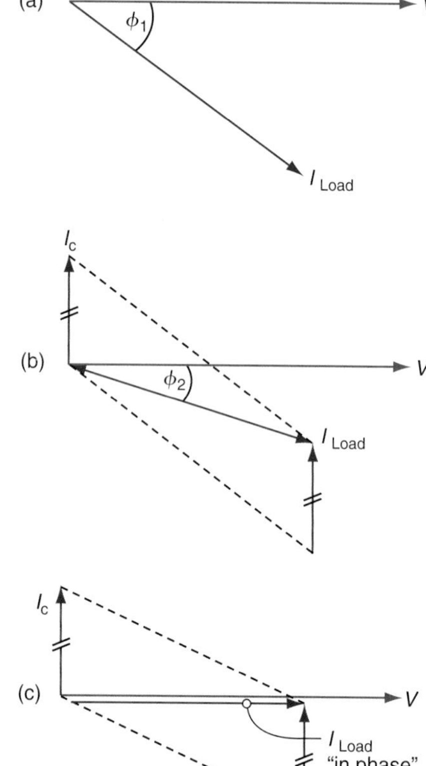

FIGURE 4.34
Power-factor improvement using capacitors.

Figure 4.34(a) shows an industrial load with a low power factor. If a capacitor is connected in parallel with the load, the capacitor current I_C leads the applied voltage by 90°. When this capacitor current is added to the load current as shown in 4.34(b) the resultant current has a much improved power factor. However, using a slightly bigger capacitor, the load current can be pushed up until it is 'in phase' with the voltage as can be seen in Fig. 4.34(c).

Capacitors may be connected across the main busbars of industrial loads in order to provide power-factor improvement, but smaller capacitors may also be connected across an individual piece of equipment, as is the case for fluorescent light fittings or a small electric motor.

Basic electronics

There are numerous types of electronic component – diodes, transistors, thyristors and integrated circuits (ICs) – each with its own limitations, characteristics and designed application. When repairing electronic circuits it is important to replace a damaged component with an identical or equivalent component. Manufacturers issue comprehensive catalogues with details of working voltage, current, power dissipation, etc., and the reference numbers of equivalent components, and some of this information is included in the Appendices. These catalogues of information, together with a high-impedance multimeter should form a part of the extended tool-kit for anyone in the electrotechnical industries proposing to repair electronic circuits.

Electronic circuit symbols

The British Standard BS EN 60617 recommends that particular graphical symbols should be used to represent a range of electronic components on circuit diagrams. The same British Standard recommends a range of symbols suitable for electrical installation circuits with which electricians will already be familiar. Figure 4.35 shows a selection of electronic symbols.

Definition

All materials have some resistance to the flow of an electric current but, in general, the term *resistor* describes a conductor specially chosen for its resistive properties.

Resistors

All materials have some resistance to the flow of an electric current but, in general, the term **resistor** describes a conductor specially chosen for its resistive properties.

Resistors are the most commonly used electronic component and they are made in a variety of ways to suit the particular type of application. They

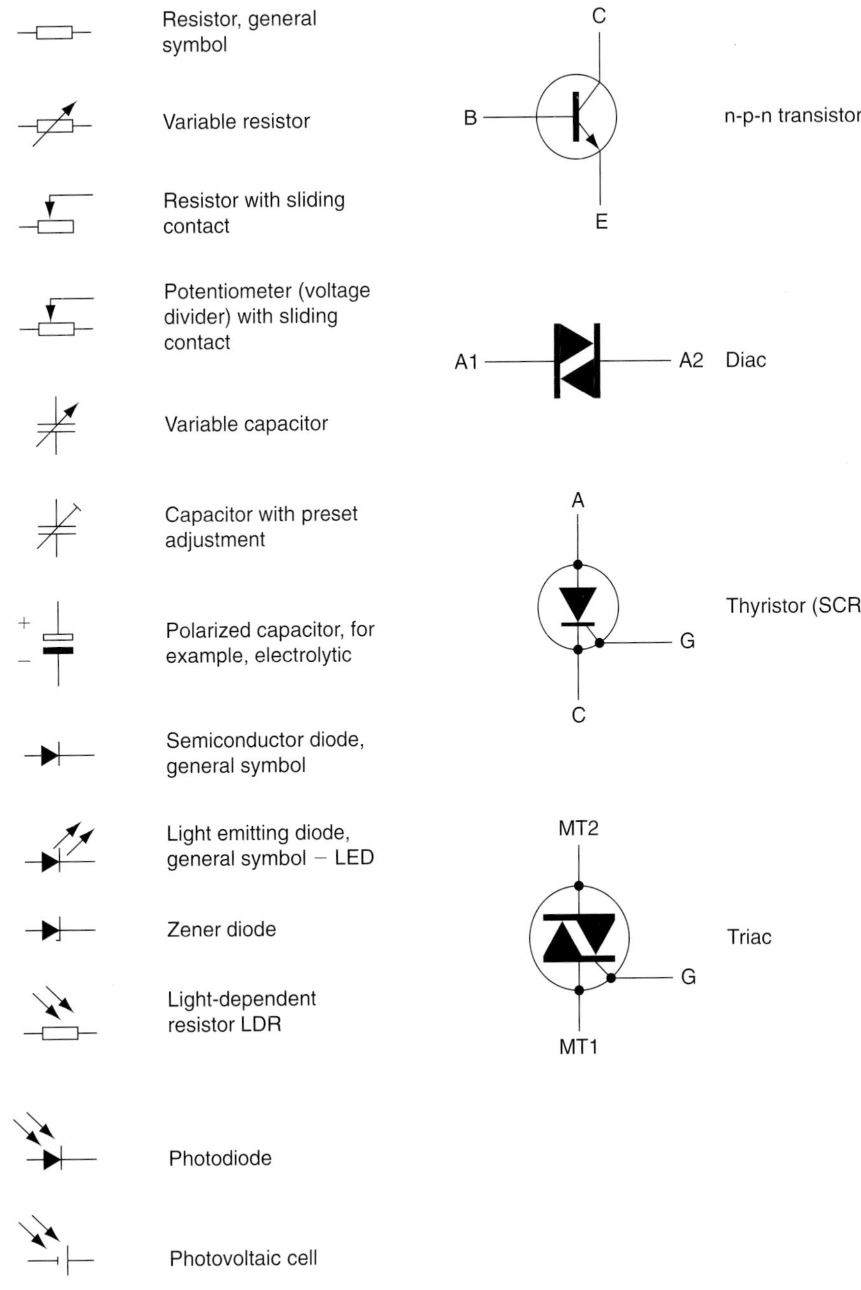

FIGURE 4.35
Some BS EN 60617 graphical symbols used in electronics.

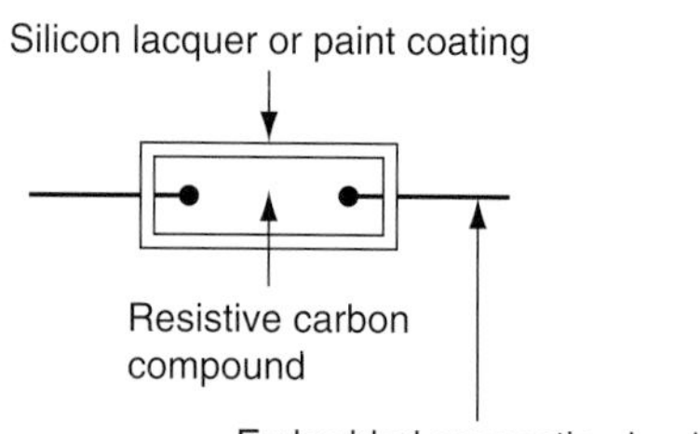

(a) Carbon-composition resistor

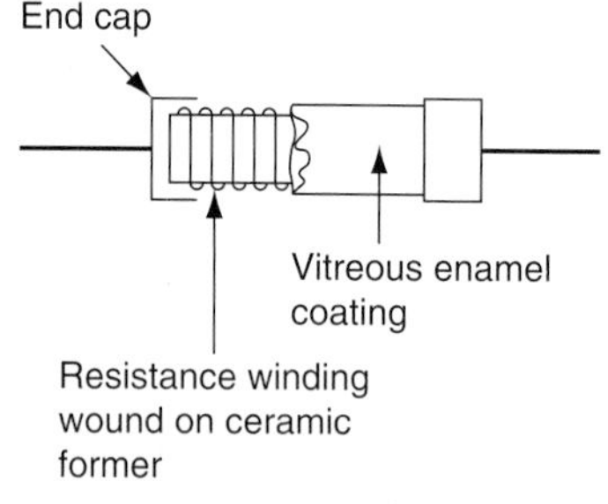

(b) Wire-wound resistor

FIGURE 4.36
Construction of resistors.

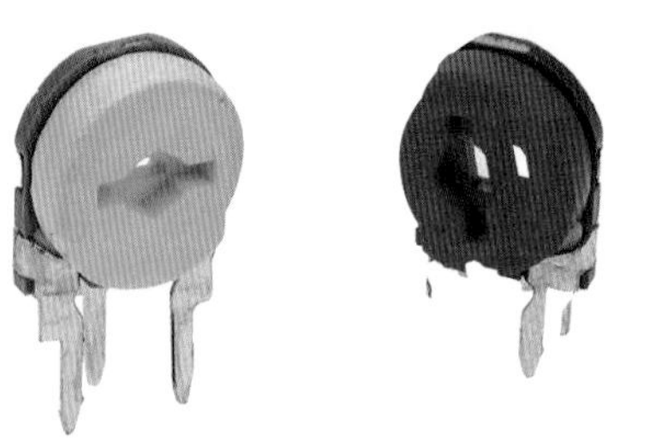

FIGURE 4.37
Types of variable resistor.

are usually manufactured as either carbon composition or carbon film. In both cases the base resistive material is carbon and the general appearance is of a small cylinder with leads protruding from each end, as shown in Fig. 4.36(a).

If subjected to overload, carbon resistors usually decrease in resistance since carbon has a negative temperature coefficient. This causes more current to flow through the resistor, so that the temperature rises and failure occurs, usually by fracturing. Carbon resistors have a power rating between 0.1 and 2W, which should not be exceeded.

When a resistor of a larger power rating is required a wire-wound resistor should be chosen. This consists of a resistance wire of known value wound on a small ceramic cylinder which is encapsulated in a vitreous enamel coating, as shown in Fig. 4.36(b). Wire-wound resistors are designed to run hot and have a power rating up to 20W. Care should be taken when mounting wire-wound resistors to prevent the high operating temperature affecting any surrounding components.

A variable resistor is one which can be varied continuously from a very low value to the full rated resistance. This characteristic is required in tuning circuits to adjust the signal or voltage level for brightness, volume or tone. The most common type used in electronic work has a circular carbon track contacted by a metal wiper arm. The wiper arm can be adjusted by means of an adjusting shaft (rotary type) or by placing a screwdriver in a slot (preset type), as shown in Fig. 4.37. Variable resistors are also known as potentiometers because they can be used to adjust the potential difference (voltage) in a circuit. The variation in resistance can be either a logarithmic or a linear scale.

The value of the resistor and the tolerance may be marked on the body of the component either by direct numerical indication or by using a standard colour code. The method used will depend upon the type, physical size and manufacturer's preference, but in general the larger components have values marked directly on the body and the smaller components use the standard resistor colour code.

ABBREVIATIONS USED IN ELECTRONICS

Where the numerical value of a component includes a decimal point, it is standard practice to include the prefix for the multiplication factor in place of the decimal point, to avoid accidental marks being mistaken for decimal points. Multiplication factors and prefixes are dealt with in Chapter 8.

The abbreviation R means $\times$ 1
k means $\times$ 1000
M means $\times$ 1,000,000

Therefore, a 4.7 kΩ resistor would be abbreviated to 4k7, a 5.6 Ω resistor to 5R6 and a 6.8 MΩ resistor to 6 M8.

Tolerances may be indicated by adding a letter at the end of the printed code.

The abbreviation F means ±1%, G means ±2%, J means ±5%, K means ±10% and M means ±20%. Therefore a 4.7 kΩ resistor with a tolerance of 2% would be abbreviated to 4k7G. A 5.6 Ω resistor with a tolerance of 5% would be abbreviated to 5R6J. A 6.8 MΩ resistor with a 10% tolerance would be abbreviated to 6 M8 K.

This is the British Standard BS 1852 code which is recommended for indicating the values of resistors on circuit diagrams and components when their physical size permits.

THE STANDARD COLOUR CODE

Small resistors are marked with a series of coloured bands, as shown in Table 4.3. These are read according to the standard colour code to determine the resistance. The bands are located on the component towards one end. If the resistor is turned so that this end is towards the left, the bands are then read from left to right. Band (a) gives the first number of the component value, band (b) the second number, band (c) the number of zeros to be added after the first two numbers and band (d) the resistor tolerance. If the bands are not clearly oriented towards one end, first identify the tolerance band and turn the resistor so that this is towards the right before commencing to read the colour code as described.

139

Table 4.3 The Resistor Colour Code

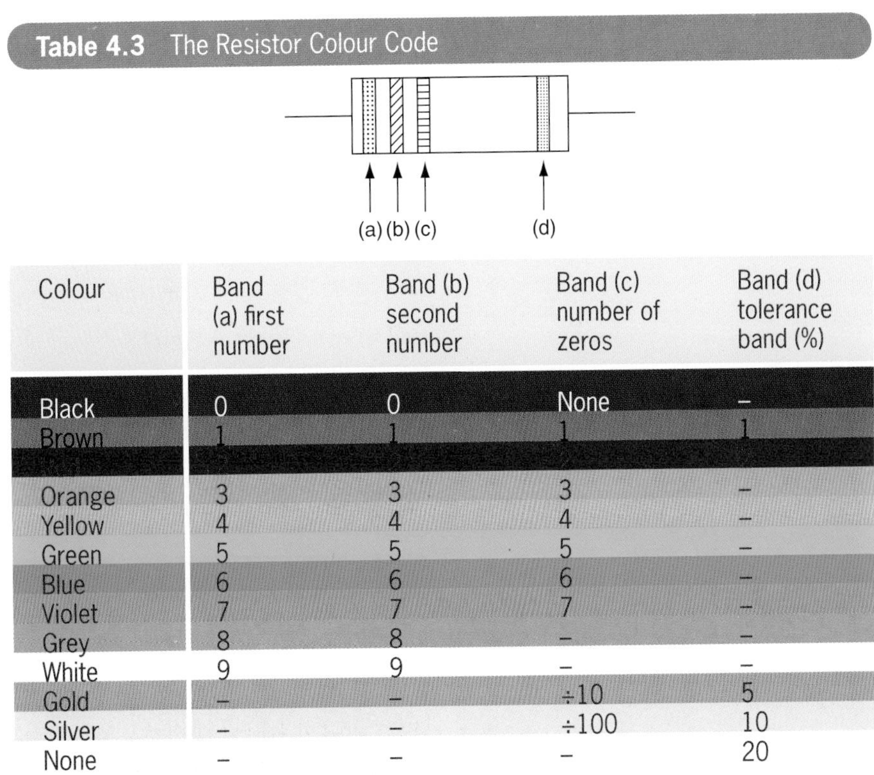

Colour	Band (a) first number	Band (b) second number	Band (c) number of zeros	Band (d) tolerance band (%)
Black	0	0	None	–
Brown	1	1	1	1
Red	2	2	2	2
Orange	3	3	3	–
Yellow	4	4	4	–
Green	5	5	5	–
Blue	6	6	6	–
Violet	7	7	7	–
Grey	8	8	–	–
White	9	9	–	–
Gold	–	–	÷10	5
Silver	–	–	÷100	10
None	–	–	–	20

The tolerance band indicates the maximum tolerance variation in the declared value of resistance. Thus a 100 Ω resistor with a 5% tolerance will have a value somewhere between 95 and 105 Ω, since 5% of 100 Ω is 5 Ω.

Example 1

A resistor is colour coded yellow, violet, red, gold. Determine the value of the resistor.

Band (a) – yellow has a value of 4.
Band (b) – violet has a value of 7.
Band (c) – red has a value of 2.
Band (d) – gold indicates a tolerance of 5%.
The value is therefore 4700 ± 5%.
This could be written as 4.7 kΩ ± 5% or 4 k7J.

Example 2

A resistor is colour coded green, blue, brown, silver. Determine the value of the resistor.

Band (a) – green has a value of 5.
Band (b) – blue has a value of 6.
Band (c) – brown has a value of 1.
Band (d) – silver indicates a tolerance of 10%.

The value is therefore 560 ± 10% and could be written as 560 Ω± 10% or 560 RK.

Example 3

A resistor is colour coded blue, grey, green, gold. Determine the value of the resistor.

Band (a) – blue has a value of 6.
Band (b) – grey has a value of 8.
Band (c) – green has a value of 5.
Band (d) – gold indicates a tolerance of 5%.

The value is therefore 6,800,000 ± 5% and could be written as 6.8 MΩ ± 5% or 6 M8J.

Example 4

A resistor is colour coded orange, white, silver, silver. Determine the value of the resistor.

Band (a) – orange has a value of 3.
Band (b) – white has a value of 9.
Band (c) – silver indicates divide by 100 in this band.
Band (d) – silver indicates a tolerance of 10%.

The value is therefore 0.39 ± 10% and could be written as 0.39 Ω ± 10% or R39 K.

Try This

Electronics

Electricians are increasingly coming across electronic components and equipment. Make a list in the margin of some of the electronic components that you have come across at work.

PREFERRED VALUES

It is difficult to manufacture small electronic resistors to exact values by mass production methods. This is not a disadvantage as in most electronic circuits the value of the resistors is not critical. Manufacturers produce a limited range of *preferred* resistance values rather than an overwhelming number of individual resistance values. Therefore, in electronics, we use the preferred value closest to the actual value required.

A resistor with a preferred value of 100 Ω and a 10% tolerance could have any value between 90 and 110 Ω. The next larger preferred value which would give the maximum possible range of resistance values without too much overlap would be 120 Ω. This could have any value between 108 and 132 Ω. Therefore, these two preferred value resistors cover all possible resistance values between 90 and 132 Ω. The next preferred value would be 150 Ω, then 180, 220 Ω and so on.

There is a series of preferred values for each tolerance level, as shown in Table 4.4, so that every possible numerical value is covered. Table 4.4 indicates the values between 10 and 100, but larger values can be obtained by multiplying these preferred values by some multiplication factor. Resistance values of 47 Ω, 470 Ω, 4.7 kΩ, 470 kΩ, 4.7 MΩ, etc., are available in this way.

Table 4.4 Preferred Values

E6 series 20% tolerance	E12 series 10% tolerance	E24 series 5% tolerance
10	10	10
		11
	12	12
		13
15	15	15
		16
	18	18
		20
22	22	22
		24
	27	27
		30
33	33	33
		36
	39	39
		43
47	47	47
		51
	56	56
		62
68	68	68
		75
	82	82
		91

TESTING RESISTORS

The resistor being tested should have a value close to the preferred value and within the tolerance stated by the manufacturer. To measure the resistance of a resistor which is not connected into a circuit, the leads of a suitable ohmmeter should be connected to each resistor connection lead and a reading obtained.

If the resistor to be tested is connected into an electronic circuit it is *always necessary* to disconnect one lead from the circuit before the test leads are connected, otherwise the components in the circuit will provide parallel paths, and an incorrect reading will result.

Capacitors

The fundamental principles of capacitors were discussed earlier in this chapter under the subheading 'Electrostatics'. In this section we shall consider the practical aspects associated with capacitors in electronic circuits.

A capacitor stores a small amount of electric charge; it can be thought of as a small rechargeable battery which can be quickly recharged. In electronics we are not only concerned with the amount of charge stored by the capacitor but in the way the value of the capacitor determines the performance of timers and oscillators by varying the time constant of a simple capacitor–resistor circuit.

CAPACITORS IN ACTION

If a test circuit is assembled as shown in Fig. 4.38 and the changeover switch connected to d.c. the signal lamp will only illuminate for a very short pulse as the capacitor charges. The charged capacitor then blocks any further d.c. current flow. If the changeover switch is then connected to a.c. the lamp will illuminate at full brilliance because the capacitor will charge and discharge continuously at the supply frequency. Current is *apparently* flowing through the capacitor because electrons are moving to and fro in the wires joining the capacitor plates to the a.c. supply.

COUPLING AND DECOUPLING CAPACITORS

Capacitors can be used to separate a.c. and d.c. in an electronic circuit. If the output from circuit A, shown in Fig. 4.39(a), contains both a.c. and d.c.

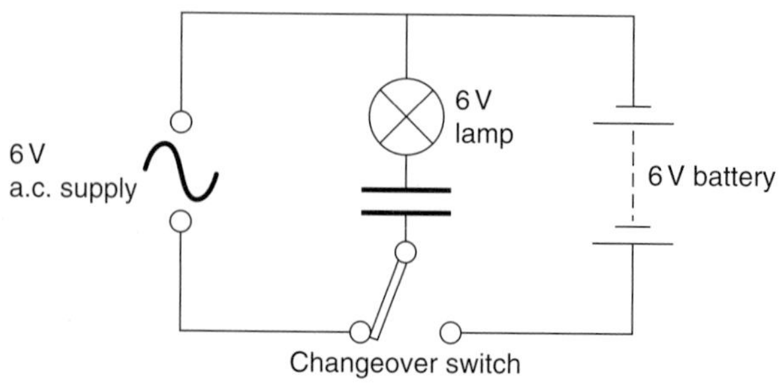

FIGURE 4.38
Test circuit showing capacitors in action.

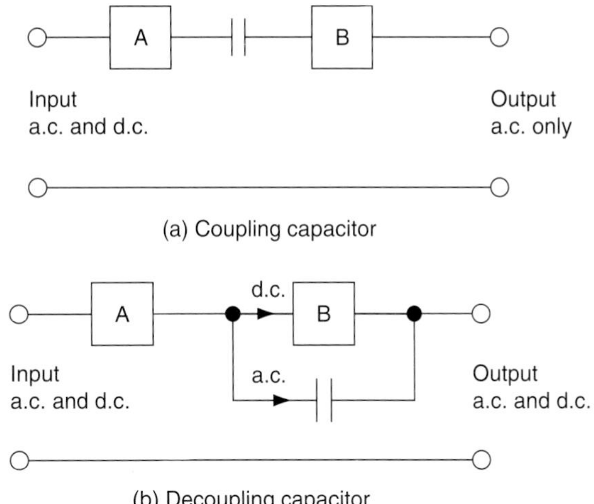

FIGURE 4.39
(a) Coupling and (b) decoupling capacitors.

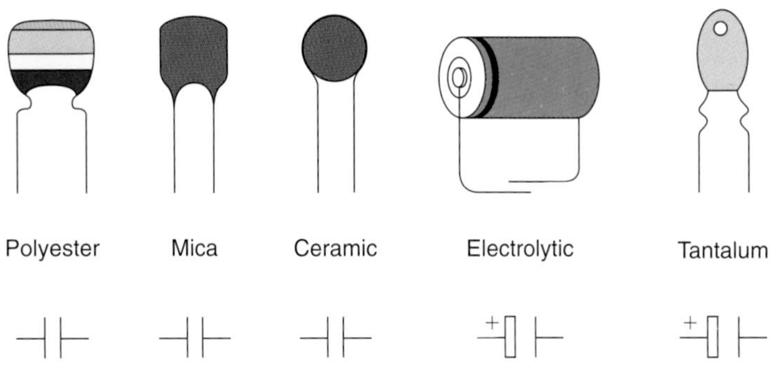

FIGURE 4.40
Capacitors and their symbols used in electronic circuits.

but only an a.c. input is required for circuit B then a *coupling* capacitor is connected between them. This blocks the d.c. while offering a low reactance to the a.c. component. Alternatively, if it is required that only d.c. be connected to circuit B, shown in Fig. 4.39(b), a *decoupling* capacitor can be connected in parallel with circuit B. This will provide a low reactance path for the a.c. component of the supply and only d.c. will be presented to the input of B. This technique is used to *filter out* unwanted a.c. in, for example, d.c. power supplies.

TYPES OF CAPACITOR

There are two broad categories of capacitor, the non-polarized and polarized type. The non-polarized type can be connected either way round, but polarized capacitors *must* be connected to the polarity indicated otherwise a short circuit and consequent destruction of the capacitor will result. There are many different types of capacitor, each one being distinguished by the type of dielectric used in its construction. Fig. 4.40 shows some of the capacitors used in electronics.

Table 4.5 Colour Code for Plastic Film Capacitors (Values in Picofarads)

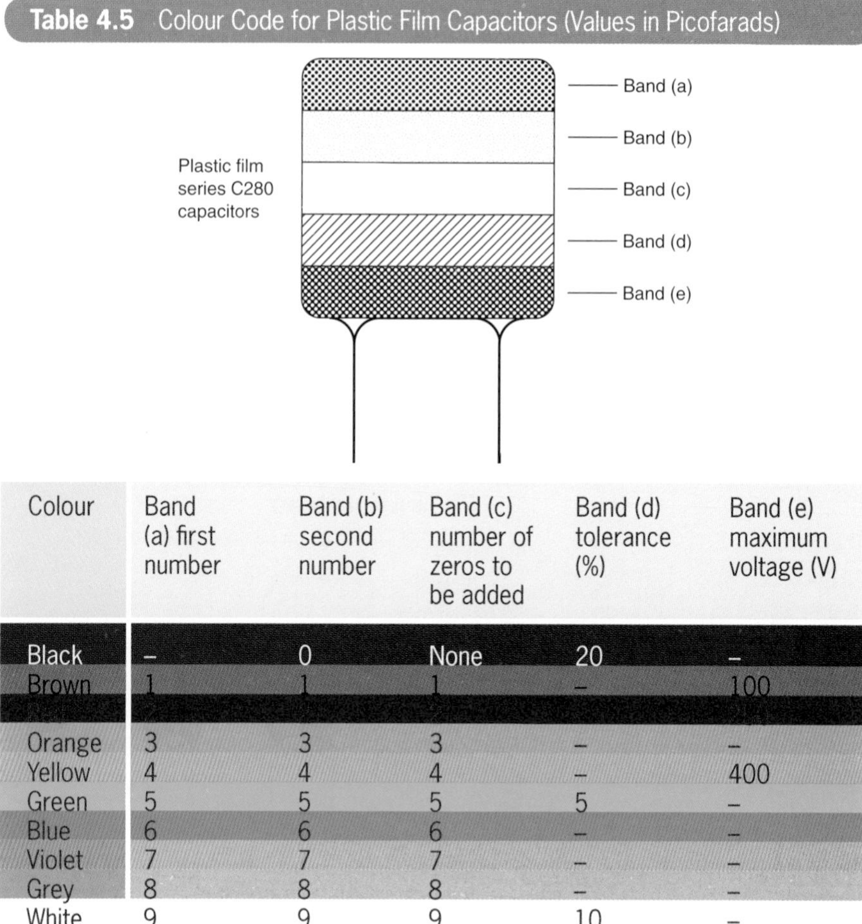

Colour	Band (a) first number	Band (b) second number	Band (c) number of zeros to be added	Band (d) tolerance (%)	Band (e) maximum voltage (V)
Black	–	0	None	20	–
Brown	1	1	1	–	100
[illegible]	[illegible]	[illegible]	[illegible]	[illegible]	[illegible]
Orange	3	3	3	–	–
Yellow	4	4	4	–	400
Green	5	5	5	5	–
Blue	6	6	6	–	–
Violet	7	7	7	–	–
Grey	8	8	8	–	–
White	9	9	9	10	–

Polyester capacitors

Polyester capacitors are an example of the plastic film capacitor. Polypropylene, polycarbonate and polystyrene capacitors are other types of plastic film capacitor. The capacitor value may be marked on the plastic film, or the capacitor colour code given in Table 4.5 may be used. This dielectric material gives a compact capacitor with good electrical and temperature characteristics. They are used in many electronic circuits, but are not suitable for high-frequency use.

Mica capacitors

Mica capacitors have excellent stability and are accurate to $\pm 1\%$ of the marked value. Since costs usually increase with increased accuracy, they tend to be more expensive than plastic film capacitors. They are used where high stability is required, for example in tuned circuits and filters.

Ceramic capacitors

Ceramic capacitors are mainly used in high-frequency circuits subjected to wide temperature variations. They have high stability and low loss.

Electrolytic capacitors

Electrolytic capacitors are used where a large value of capacitance coupled with a small physical size is required. They are constructed on the 'Swiss roll' principle as are the paper dielectric capacitors used for power-factor correction in electrical installation circuits. The electrolytic capacitors' high capacitance for very small volume is derived from the extreme thinness of the dielectric coupled with a high dielectric strength. Electrolytic capacitors have a size gain of approximately 100 times over the equivalent non-electrolytic type. Their main disadvantage is that they are polarized and must be connected to the correct polarity in a circuit. Their large capacity makes them ideal as smoothing capacitors in power supplies.

Tantalum capacitors

Tantalum capacitors are a new type of electrolytic capacitor using tantalum and tantalum oxide to give a further capacitance/size advantage. They look like a 'raindrop' or 'blob' with two leads protruding from the bottom. The polarity and values may be marked on the capacitor, or a colour code may be used. The voltage ratings available tend to be low, as with all electrolytic capacitors. They are also extremely vulnerable to reverse voltages in excess of 0.3 V. This means that even when testing with an ohmmeter, extreme care must be taken to ensure correct polarity.

Variable capacitors

Variable capacitors are constructed so that one set of metal plates moves relative to another set of fixed metal plates as shown in Fig. 4.41. The plates are separated by air or sheet mica, which acts as a dielectric. Air dielectric variable capacitors are used to tune radio receivers to a chosen station, and small variable capacitors called *trimmers* or *presets* are used to make fine, infrequent adjustments to the capacitance of a circuit.

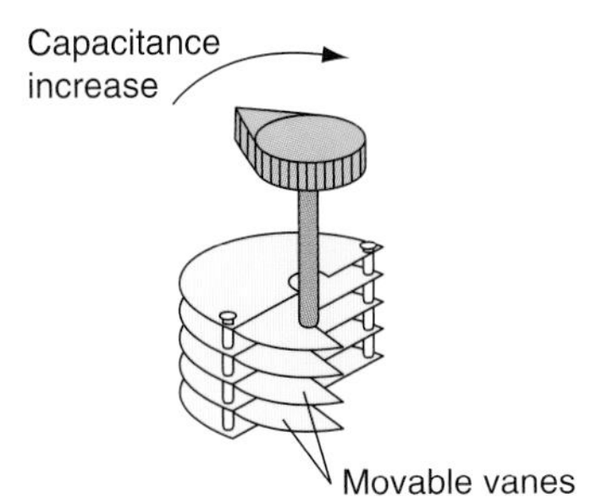

(a) Variable type

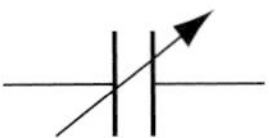

(b) Trimmer or preset type

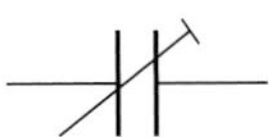

FIGURE 4.41
Variable capacitors and their symbols: (a) variable type; (b) trimmer or preset type.

SELECTING A CAPACITOR

When choosing a capacitor for a particular application, three factors must be considered: value, working voltage and leakage current.

The unit of capacitance is the *farad* (symbol F), to commemorate the name of the English scientist Michael Faraday. However, for practical purposes the farad is much too large and in electrical installation work and electronics we use fractions of a farad as follows:

$$1\text{ microfarad} = 1\mu\text{F} = 1\times10^{-6}\text{ F}$$
$$1\text{ nanofarad} = 1\text{nF} = 1\times10^{-9}\text{ F}$$
$$1\text{ picofarad} = 1\text{pF} = 1\times10^{-12}\text{ F}$$

The power-factor correction capacitor used in a domestic fluorescent luminaire would typically have a value of 8 μF at a working voltage of 400 V. In an electronic filter circuit a typical capacitor value might be 100 pF at 63 V.

One microfarad is one million times greater than one picofarad. It may be useful to remember that:

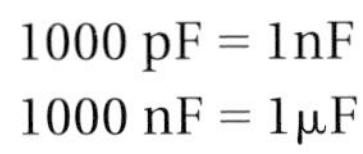

$$1000\text{ pF} = 1\text{nF}$$
$$1000\text{ nF} = 1\mu\text{F}$$

The working voltage of a capacitor is the *maximum* voltage that can be applied between the plates of the capacitor without breaking down the dielectric insulating material. This is a d.c. rating and, therefore, a capacitor with a 200 V rating must only be connected across a maximum of 200V d.c. Since a.c. voltages are usually given as rms values, a 200 V a.c. supply would have a maximum value of about 283 V, which would damage the 200 V capacitor. When connecting a capacitor to the 230 V mains supply we must choose a working voltage of about 400 V because 230 V rms is approximately 325 V maximum. The 'factor of safety' is small and, therefore, the working voltage of the capacitor must not be exceeded.

An ideal capacitor which is isolated will remain charged forever, but in practice no dielectric insulating material is perfect, and the charge will slowly *leak* between the plates, gradually discharging the capacitor. The loss of charge by leakage through it should be very small for a practical capacitor.

Capacitor colour code

The actual value of a capacitor can be identified by using the colour codes given in Table 4.5 in the same way that the resistor colour code was applied to resistors.

Example 1

A plastic film capacitor is colour coded, from top to bottom, brown, black, yellow, black, red. Determine the value of the capacitor, its tolerance and working voltage.

From Table 4.5 we obtain the following:

Band (a) – brown has a value 1.
Band (b) – black has a value 0.
Band (c) – yellow indicates multiply by 10,000.
Band (d) – black indicates 20%.
Band (e) – red indicates 250 V.

The capacitor has a value of 1,00,000 pF or 0.1 μF with a tolerance of 20% and a maximum working voltage of 250 V.

Example 2

Determine the value, tolerance and working voltage of a polyester capacitor colour-coded, from top to bottom, yellow, violet, yellow, white, yellow.

From Table 4.5 we obtain the following:

Band (a) – yellow has a value 4.
Band (b) – violet has a value 7.
Band (c) – yellow indicates multiply by 10,000.
Band (d) – white indicates 10%.
Band (e) – yellow indicates 400 V.

The capacitor has a value of 4,70,000 pF or 0.47 μF with a tolerance of 10% and a maximum working voltage of 400V.

Example 3

A plastic film capacitor has the following coloured bands from its top down to the connecting leads: blue, grey, orange, black, brown. Determine the value, tolerance and voltage of this capacitor.

From Table 4.5 we obtain the following:

Band (a) – blue has a value 6.
Band (b) – grey has a value 8.
Band (c) – orange indicates multiply by 1000.
Band (d) – black indicates 20%.
Band (e) – brown indicates 100 V.

The capacitor has a value of 68,000 pF or 68 nF with a tolerance of 20% and a maximum working voltage of 100 V.

CAPACITANCE VALUE CODES

Where the numerical value of the capacitor includes a decimal point, it is standard practice to use the prefix for the multiplication factor in place of the decimal point. This is the same practice as we used earlier for resistors.

The abbreviation μ means microfarad, n means nanofarad and p means picofarad. Therefore, a 1.8 pF capacitor would be abbreviated to 1p8, a 10 pF capacitor to 10 p, a 150 pF capacitor to 150 p or n15, a 2200 pF capacitor to 2n2 and a 10,000 pF capacitor to 10 n.

$$1000\ \text{pF} = 1\text{nF} = 0.001\mu\text{F}$$

Packaging electronic components

When we talk about packaging electronic components we are not referring to the parcel or box which contains the components for storage and delivery, but to the type of encapsulation in which the tiny semiconductor material is contained. Figure 4.42 shows three different package outlines for just one type of discrete component, the transistor. Identification of the pin connections for different packages is given within the text as each separate or discrete component is considered, particularly later in this chapter when we discuss semiconductor devices. However, the Appendices aim

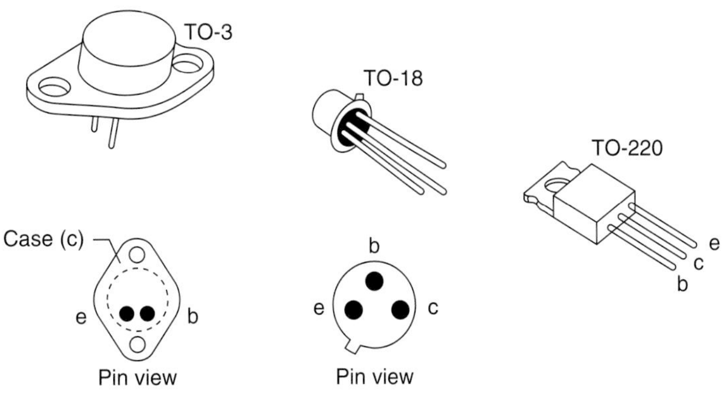

FIGURE 4.42
Three different package outlines for transistors.

to draw together all the information on pin connections and packages for easy reference.

Obtaining information and components

Electricians use electrical wholesalers and suppliers to purchase electrical cable, equipment and accessories. Similar facilities are available in most towns and cities for the purchase of electronic components and equipment. There are also a number of national suppliers who employ representatives who will call at your workshop to offer technical advice and take your order. Some of these national companies also offer a 24-hour telephone order and mail order service. Their full-colour, fully illustrated catalogues also contain an enormous amount of technical information. The names and addresses of these national companies are given in Appendix A. For local suppliers you must consult your local phone book and *Yellow Pages*. The Appendices of this book also contain some technical reference information.

SEMICONDUCTOR MATERIALS

Modern electronic devices use the semiconductor properties of materials such as silicon or germanium. The atoms of pure silicon or germanium are arranged in a lattice structure, as shown in Fig. 4.43. The outer electron orbits contain four electrons known as *valence* electrons. These electrons are all linked to other valence electrons from adjacent atoms, forming a covalent bond. There are no free electrons in pure silicon or germanium and, therefore, no conduction can take place unless the bonds are broken and the lattice framework is destroyed.

To make conduction possible without destroying the crystal it is necessary to replace a four-valent atom with a three- or five-valent atom. This process is known as *doping*.

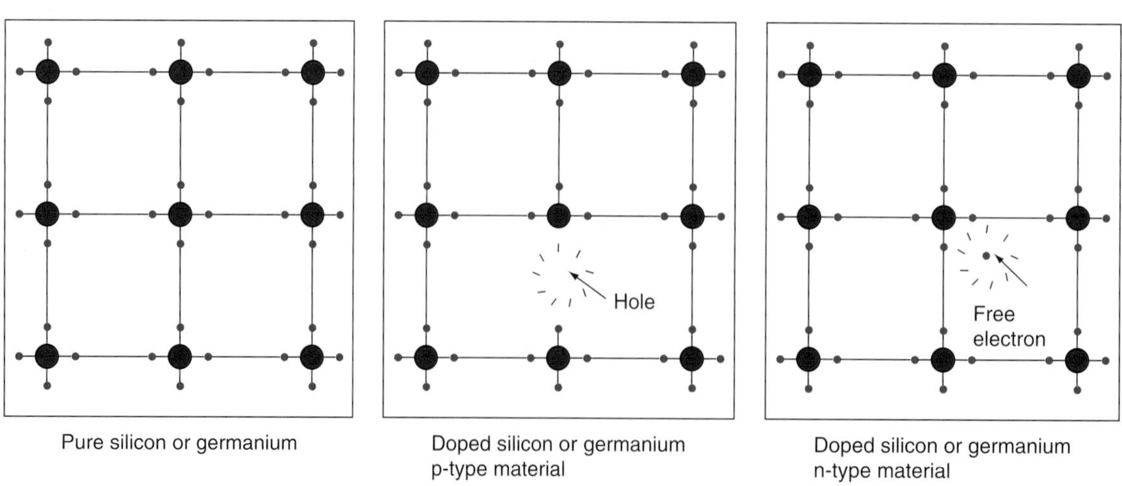

FIGURE 4.43
Semiconductor material.

If a three-valent atom is added to silicon or germanium a hole is left in the lattice framework. Since the material has lost a negative charge, the material becomes positive and is known as a p-type material (p for positive).

If a five-valent atom is added to silicon or germanium, only four of the valence electrons can form a bond and one electron becomes mobile or free to carry charge. Since the material has gained a negative charge it is known as an n-type material (n for negative).

Bringing together a p-type and n-type material allows current to flow in one direction only through the p–n junction. Such a junction is called a diode, since it is the semiconductor equivalent of the vacuum diode valve used by Fleming to rectify radio signals in 1904.

SEMICONDUCTOR DIODE

A semiconductor or junction diode consists of a p-type and n-type material formed in the same piece of silicon or germanium. The p-type material forms the anode and the n-type the cathode, as shown in Fig. 4.44. If the anode is made positive with respect to the cathode, the junction will have very little resistance and current will flow. This is referred to as forward bias. However, if reverse bias is applied, that is, the anode is made negative with respect to the cathode, the junction resistance is high and no current can flow, as shown in Fig. 4.45. The characteristics for a forward and reverse bias p–n junction are given in Fig. 4.46.

It can be seen that a small voltage is required to forward bias the junction before a current can flow. This is approximately 0.6V for silicon and 0.2V for germanium. The reverse bias potential of silicon is about 1200V and for germanium about 300V. If the reverse bias voltage is exceeded the diode

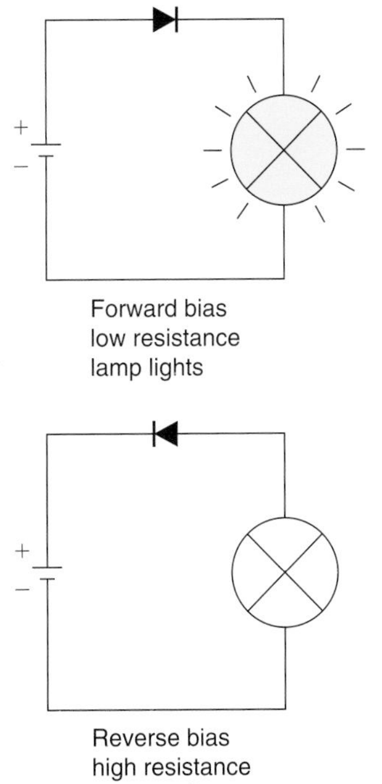

FIGURE 4.45
Forward and reverse bias of a diode.

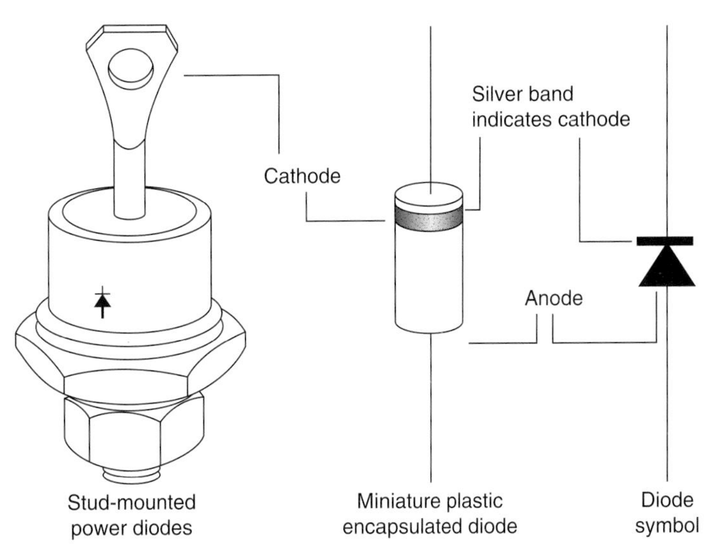

FIGURE 4.44
Symbol for and appearance of semiconductor diodes.

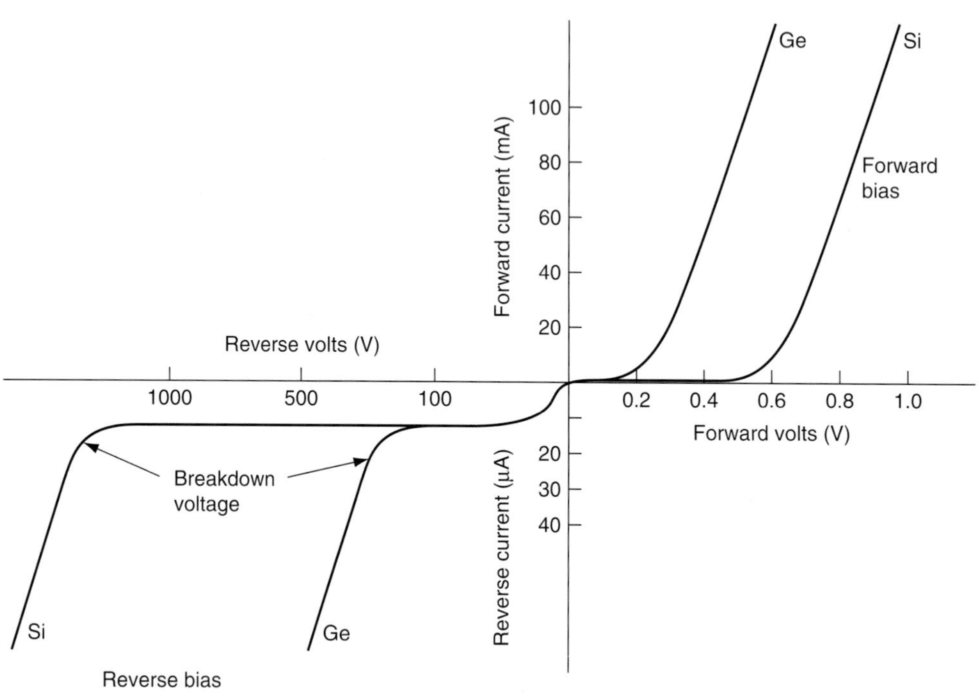

FIGURE 4.46
Forward and reverse bias characteristic of silicon and germanium.

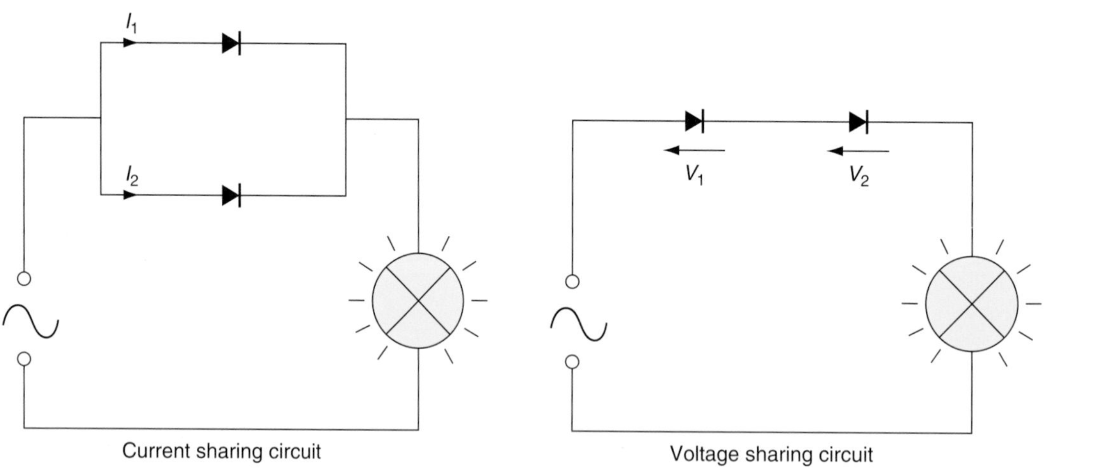

FIGURE 4.47
Using two diodes to reduce the current or voltage applied to a diode.

will break down and current will flow in both directions. Similarly, the diode will break down if the current rating is exceeded, because excessive heat will be generated. Manufacturer's information therefore gives maximum voltage and current ratings for individual diodes which must not be exceeded. However, it is possible to connect a number of standard diodes in series or parallel, thereby sharing current or voltage, as shown in Fig. 4.47, so that the manufacturers' maximum values are not exceeded by the circuit.

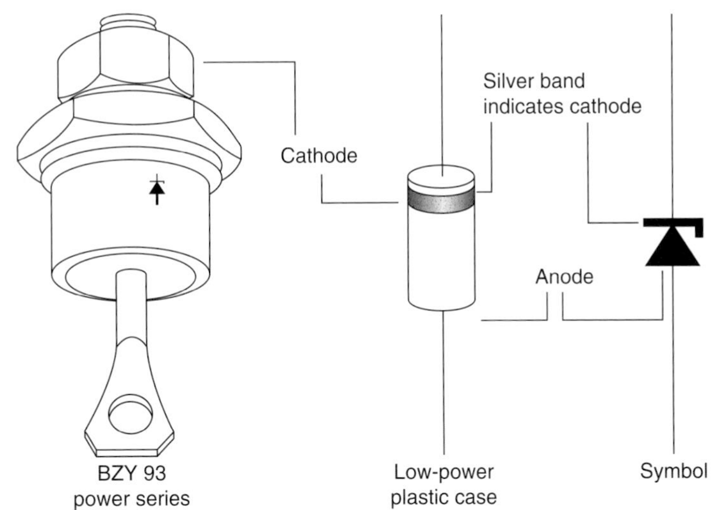

FIGURE 4.48
Symbol for and appearance of Zener diodes.

DIODE TESTING

The p–n junction of the diode has a low resistance in one direction and a very high resistance in the reverse direction.

Connecting an ohmmeter, with the red positive lead to the anode of the junction diode and the black negative lead to the cathode, would give a very low reading. Reversing the lead connections would give a high resistance reading in a 'good' component.

ZENER DIODE

A Zener diode is a silicon junction diode but with a different characteristic than the semiconductor diode considered previously. It is a special diode with a predetermined reverse breakdown voltage, the mechanism for which was discovered by Carl Zener in 1934. Its symbol and general appearance are shown in Fig. 4.48. In its forward bias mode, that is, when the anode is positive and the cathode negative, the Zener diode will conduct at about 0.6 V, just like an ordinary diode, but it is in the reverse mode that the Zener diode is normally used. When connected with the anode made negative and the cathode positive, the reverse current is zero until the reverse voltage reaches a predetermined value, when the diode switches on, as shown by the characteristics given in Fig. 4.49. This is called the Zener voltage or reference voltage. Zener diodes are manufactured in a range of preferred values, for example, 2.7, 4.7, 5.1, 6.2, 6.8, 9.1, 10, 11, 12V, etc., up to 200V at various ratings. The diode may be damaged by overheating if the current is not limited by a series resistor, but when this is connected, the voltage across the diode remains constant. It is this property of the Zener diode which makes it useful for stabilizing power supplies and these circuits are considered at the end of this chapter.

If a test circuit is constructed as shown in Fig. 4.50, the Zener action can be observed. When the supply is less than the Zener voltage (5.1V in this case) no current will flow and the output voltage will be equal to the input

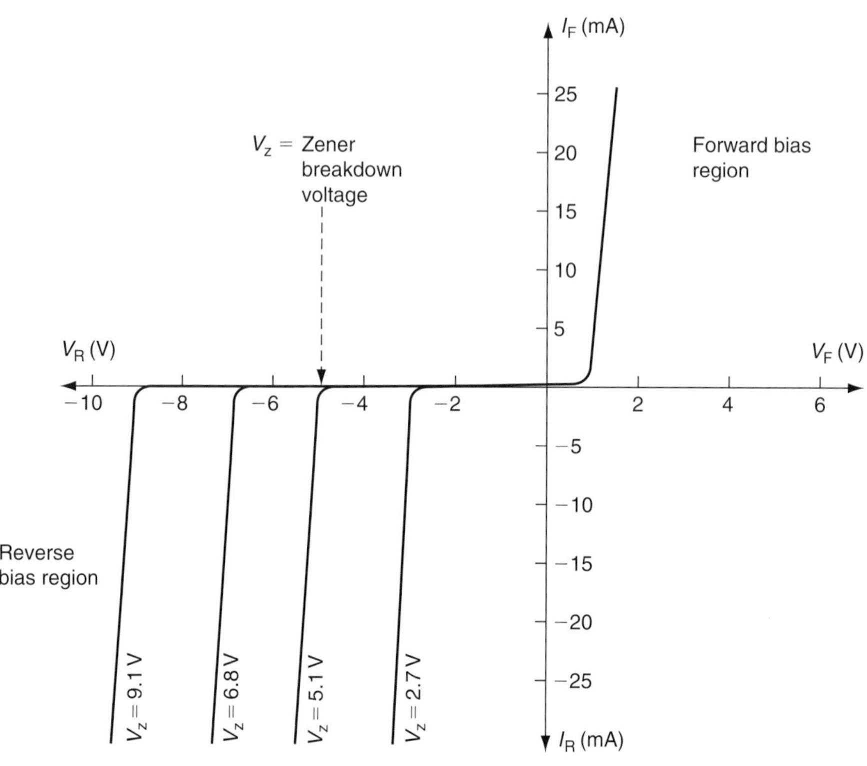

FIGURE 4.49
Zener diode characteristics.

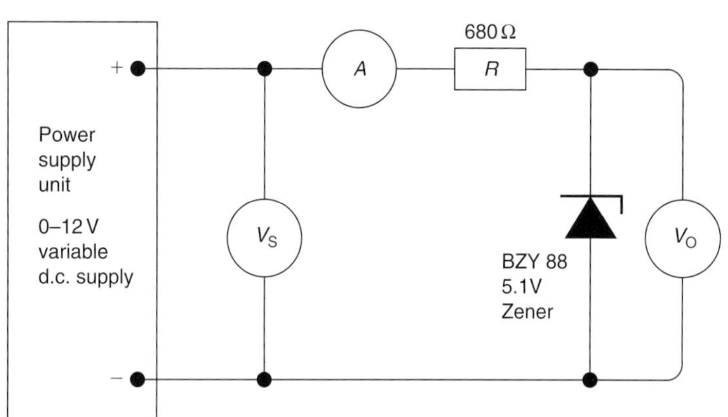

P.S.U. Supply Volts (V_S)	Current (A)	Output Volts (V_O)
1		
2		
3		
4		
5		
6		
7		
8		
9		
10		
11		
12		

FIGURE 4.50
Experiment to demonstrate the operation of a Zener diode.

voltage. When the supply is equal to or greater than the Zener voltage, the diode will conduct and any excess voltage will appear across the 680 Ω resistor, resulting in a very stable voltage at the output. When connecting this and other electronic circuits you must take care to connect the polarity of the Zener diode as shown in the diagram. Note that current must flow through the diode to enable it to stabilize.

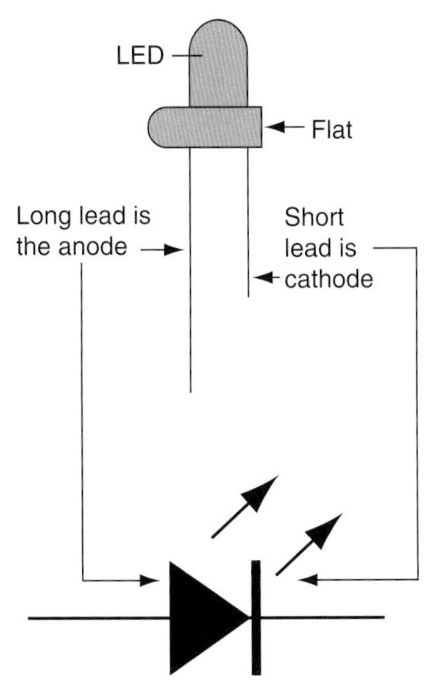

FIGURE 4.51
Symbol for and general appearance of an LED.

LIGHT-EMITTING DIODE

The light-emitting diode (LED) is a p–n junction especially manufactured from a semiconducting material which emits light when a current of about 10 mA flows through the junction.

No light is emitted when the junction is reverse biased and if this exceeds about 5V the LED may be damaged.

The general appearance and circuit symbol are shown in Fig. 4.51.

The LED will emit light if the voltage across it is about 2V. If a voltage greater than 2V is to be used then a resistor must be connected in series with the LED.

To calculate the value of the series resistor we must ask ourselves what we know about LEDs. We know that the diode requires a forward voltage of about 2V and a current of about 10 mA must flow through the junction to give sufficient light. The value of the series resistor *R* will, therefore, be given by:

$$R = \frac{\text{Supply voltage} - 2\text{ V}}{10\text{ mA}}\ \Omega$$

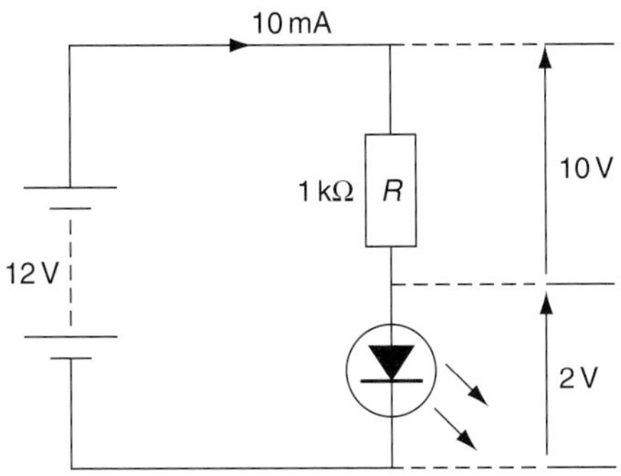

FIGURE 4.52
Circuit diagram for LED example.

Example

Calculate the value of the series resistor required when an LED is to be used to show the presence of a 12V supply.

$$R = \frac{12\text{ V} - 2\text{ V}}{10\text{ mA}}\ \Omega$$

$$R = \frac{10\text{ V}}{10\text{ mA}} = 1\text{ k}\Omega$$

The circuit is, therefore, as shown in Fig. 4.52.

LEDs are available in red, yellow and green and, when used with a series resistor, may replace a filament lamp. They use less current than a filament lamp, are smaller, do not become hot and last indefinitely. A filament lamp, however, is brighter and emits white light. LEDs are often used as indicator lamps, to indicate the presence of a voltage. They do not, however, indicate the *precise* amount of voltage present at that point.

Try This

LEDs

Make a list in the margin of examples where you have seen LEDs being used.

Another application of the LED is the seven-segment display used as a numerical indicator in calculators, digital watches and measuring instruments. Seven LEDs are arranged as a figure 8 so that when various segments are illuminated, the numbers 0–9 are displayed as shown in Fig. 4.53.

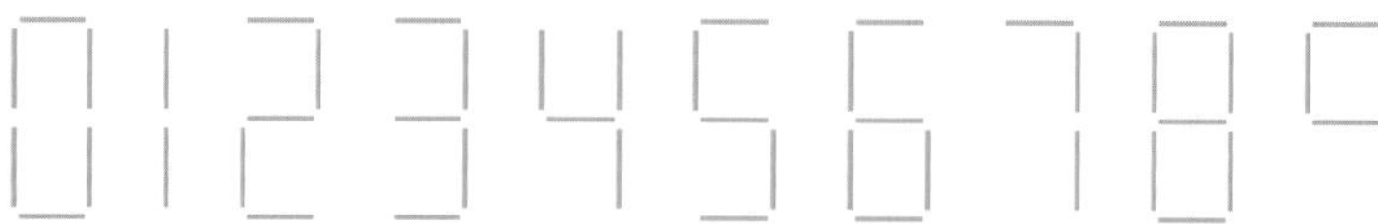

FIGURE 4.53
LED used in seven-segment display.

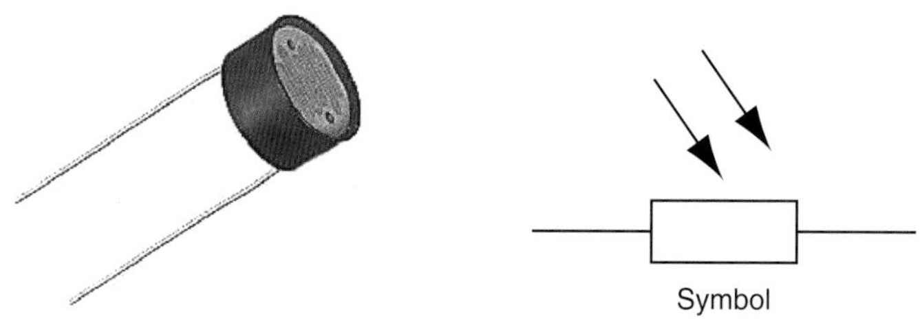

FIGURE 4.54
Symbol and appearance of an LDR.

LIGHT-DEPENDENT RESISTOR

Almost all materials change their resistance with a change in temperature. Light energy falling on a suitable semiconductor material also causes a change in resistance. The semiconductor material of a light-dependent resistor (LDR) is encapsulated as shown in Fig. 4.54 together with the circuit symbol. The resistance of an LDR in total darkness is about 10 MΩ, in normal room lighting about 5 kΩ and in bright sunlight about 100 Ω. They can carry tens of milliamperes, an amount which is sufficient to operate a relay. The LDR uses this characteristic to switch on automatically street lighting and security alarms.

PHOTODIODE

The photodiode is a normal junction diode with a transparent window through which light can enter. The circuit symbol and general appearance are shown in Fig. 4.55. It is operated in reverse bias mode and the leakage current increases in proportion to the amount of light falling on the junction. This is due to the light energy breaking bonds in the crystal lattice of the semiconductor material to produce holes and electrons.

Photodiodes will only carry microamperes of current but can operate much more quickly than LDRs and are used as 'fast' counters when the light intensity is changing rapidly.

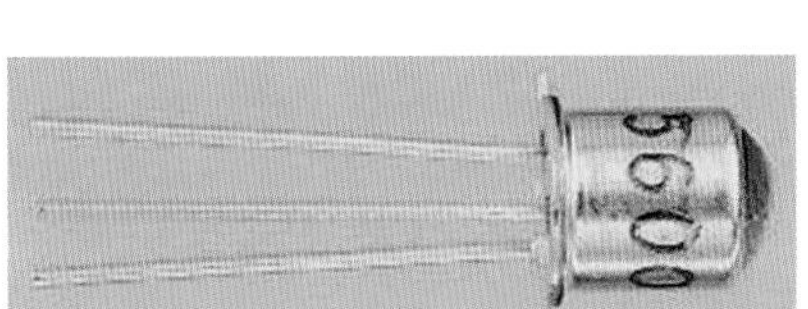

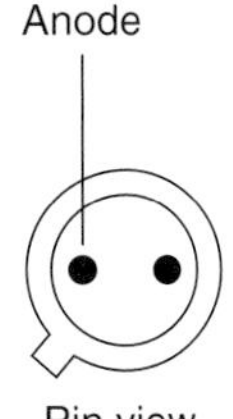

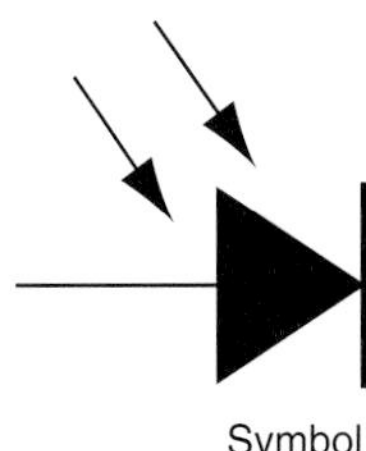

FIGURE 4.55
Symbol for, pin connections of and appearance of a photodiode.

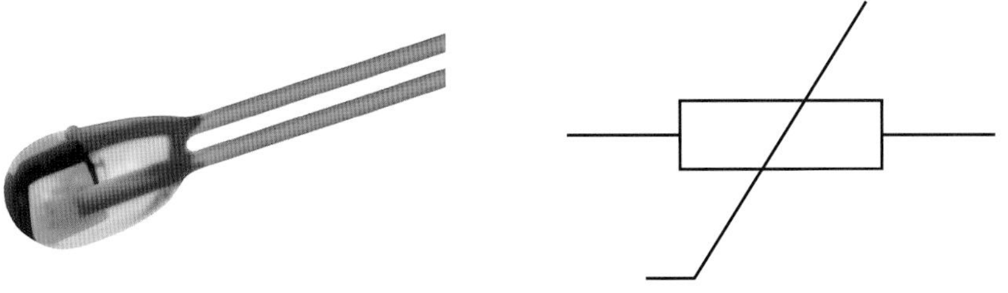

FIGURE 4.56
Symbol for and appearance of a thermistor.

THERMISTOR

The thermistor is a thermal resistor, a semiconductor device whose resistance varies with temperature. Its circuit symbol and general appearance are shown in Fig. 4.56. They can be supplied in many shapes and are used for the measurement and control of temperature up to their maximum useful temperature limit of about 300°C. They are very sensitive and because the bead of semiconductor material can be made very small, they can measure temperature in the most inaccessible places with very fast response times. Thermistors are embedded in high-voltage underground transmission cables in order to monitor the temperature of the cable. Information about the temperature of a cable allows engineers to load the cables more efficiently. A particular cable can carry a larger load in winter for example, when heat from the cable is being dissipated more efficiently. A thermistor is also used to monitor the water temperature of a motor car.

TRANSISTORS

The transistor has become the most important building block in electronics. It is the modern, miniature, semiconductor equivalent of the thermionic valve and was invented in 1947 by Bardeen, Shockley and Brattain at the Bell Telephone Laboratories in the United States. Transistors are packaged as separate or *discrete* components, as shown in Fig. 4.57.

There are two basic types of transistor, the *bipolar* or junction transistor and the *field-effect transistor* (FET).

The FET has some characteristics which make it a better choice in electronic switches and amplifiers. It uses less power and has a higher resistance and frequency response. It takes up less space than a bipolar transistor and, therefore, more of them can be packed together on a given area of

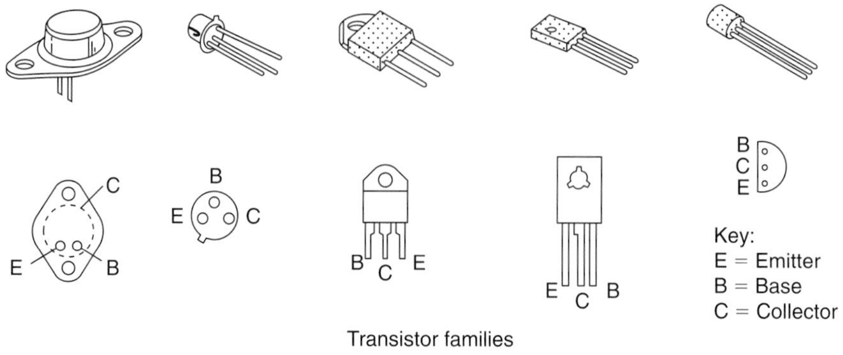

FIGURE 4.57
The appearance and pin connections of the transistor family.

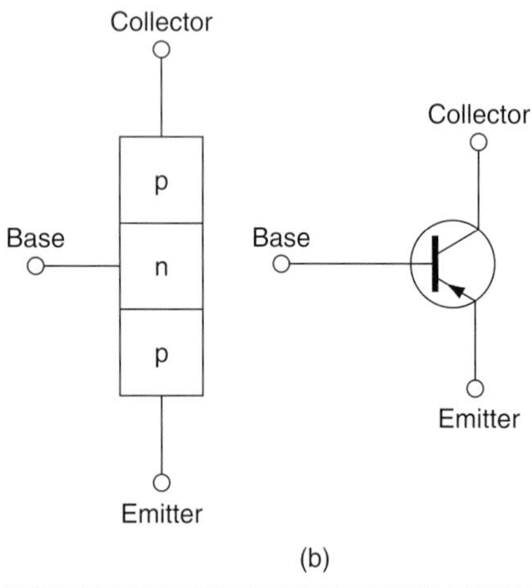

FIGURE 4.58
Structure of and symbol for (a) n-p-n and (b) p-n-p transistors.

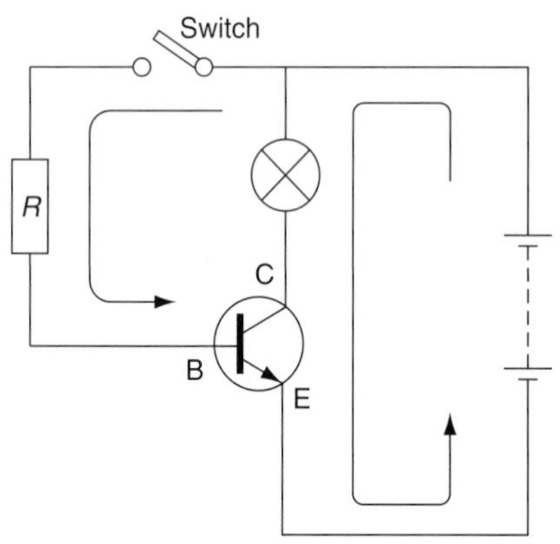

FIGURE 4.59
Operation of the transistor.

silicon chip. It is, therefore, the FET which is used when many transistors are integrated on to a small area of silicon chip as in the IC that will be discussed later.

When packaged as a discrete component the FET looks much the same as the bipolar transistor. Its circuit symbol and connections are given in the Appendix. However, it is the bipolar transistor which is much more widely used in electronic circuits as a discrete component.

The bipolar transistor

The bipolar transistor consists of three pieces of semiconductor material sandwiched together as shown in Fig. 4.58. The structure of this transistor makes it a three-terminal device having a base, collector and emitter terminal. By varying the current flowing into the base connection a much larger current flowing between collector and emitter can be controlled. Apart from the supply connections, the n-p-n and p-n-p types are essentially the same but the n-p-n type is more common.

A transistor is generally considered a current-operated device. There are two possible current paths through the transistor circuit, shown in Fig. 4.59: the base–emitter path when the switch is closed; and the collector–emitter path. Initially, the positive battery supply is connected to the n-type material of the collector, the junction is reverse biased and, therefore, no current will flow. Closing the switch will forward bias the base–emitter junction and current flowing through this junction causes current to flow across the collector–emitter junction and the signal lamp will light.

A small base current can cause a much larger collector current to flow. This is called the *current gain* of the transistor, and is typically about 100. When I say a much larger collector current, I mean a large current in electronic terms, up to about half an ampere.

We can, therefore, regard the transistor as operating in two ways: as a switch because the base current turns on and controls the collector current; and as a current amplifier because the collector current is greater than the base current.

We could also consider the transistor to be operating in a similar way to a relay. However, transistors have many advantages over electrically operated switches such as relays. They are very small, reliable, have no moving parts and, in particular, they can switch millions of times a second without arcing occurring at the contacts.

Table 4.6 Transistor Testing Using an Ohmmeter

A 'good' n-p-n transistor will give the following readings:
Red to base and black to collector = low resistance Red to base and black to emitter = low resistance Reversed connections on the above terminals will result in a high resistance reading, as will connections of either polarity between the collector and emitter terminals.
A 'good' p-n-p transistor will give the following readings:
Black to base and red to collector = low resistance Black to base and red to emitter = low resistance Reversed connections on the above terminals will result in a high resistance reading, as will connections of either polarity between the collector and emitter terminals.

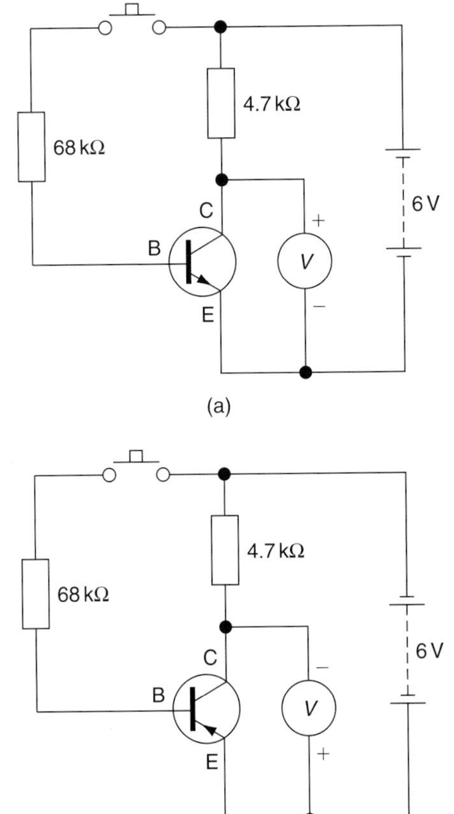

FIGURE 4.60
Transistor test circuits (a) n-p-n transistor test; (b) p-n-p transistor test.

Transistor testing

A transistor can be thought of as two diodes connected together and, therefore, a transistor can be tested using an ohmmeter in the same way as was described for the diode.

Assuming that the red lead of the ohmmeter is positive, the transistor can be tested in accordance with Table 4.6.

When many transistors are to be tested, a simple test circuit can be assembled as shown in Fig. 4.60.

With the circuit connected, as shown in Fig. 4.60 a 'good' transistor will give readings on the voltmeter of 6V with the switch open and about 0.5 V when the switch is made. The voltmeter used for the test should have a high internal resistance, about ten times greater than the value of the resistor being tested – in this case 4.7 kΩ – and this is usually indicated on the back of a multi-range meter or in the manufacturers' information supplied with a new meter.

INTEGRATED CIRCUITS

ICs were first developed in the 1960s. They are densely populated miniature electronic circuits made up of hundreds and sometimes thousands of microscopically small transistors, resistors, diodes and capacitors, all connected together on a single chip of silicon no bigger than a baby's fingernail. When assembled in a single package, as shown in Fig. 4.61, we call the device an IC.

There are two broad groups of IC: digital ICs and linear ICs. Digital ICs contain simple switching-type circuits used for logic control and calculators,

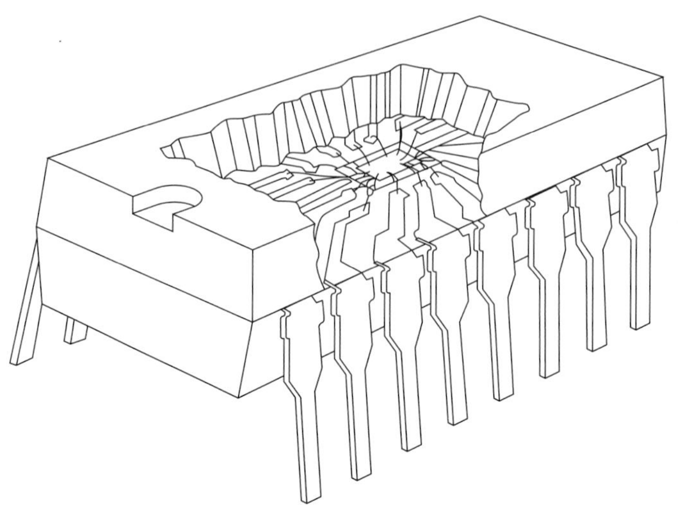

FIGURE 4.61
Exploded view of an IC.

linear ICs incorporate amplifier-type circuits which can respond to audio and radio frequency signals. The most versatile linear IC is the operational amplifier which has applications in electronics, instrumentation and control.

The IC is an electronic revolution. ICs are more reliable, cheaper and smaller than the same circuit made from discrete or separate transistors, and electronically superior. One IC behaves differently than another because of the arrangement of the transistors within the IC.

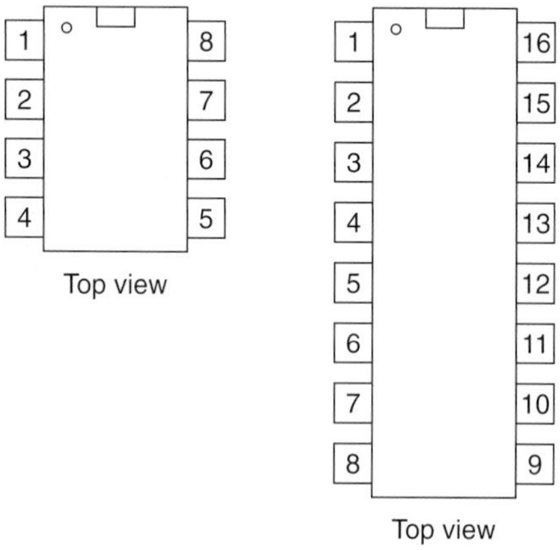

Manufacturers' data sheets describe the characteristics of the different ICs, which have a reference number stamped on the top.

When building circuits, it is necessary to be able to identify the IC pin connection by number. The number 1 pin of any IC is indicated by a dot pressed into the encapsulation; it is also the pin to the left of the cutout (Fig. 4.62). Since the packaging of ICs has two rows of pins they are called DIL (dual in line) packaged ICs and their appearance is shown in Fig. 4.63.

FIGURE 4.62
IC pin identification.

ICs are sometimes connected into DIL sockets and at other times are soldered directly into the circuit. The testing of ICs is beyond the scope of a practising electrician, and when they are suspected of being faulty an identical or equivalent replacement should be connected into the circuit, ensuring that it is inserted the correct way round, which is indicated by the position of pin number 1 as described earlier.

THE THYRISTOR

The *thyristor* was previously known as a 'silicon controlled rectifier' since it is a rectifier which controls the power to a load. It consists of four pieces of semiconductor material sandwiched together and connected to three terminals, as shown in Fig. 4.64.

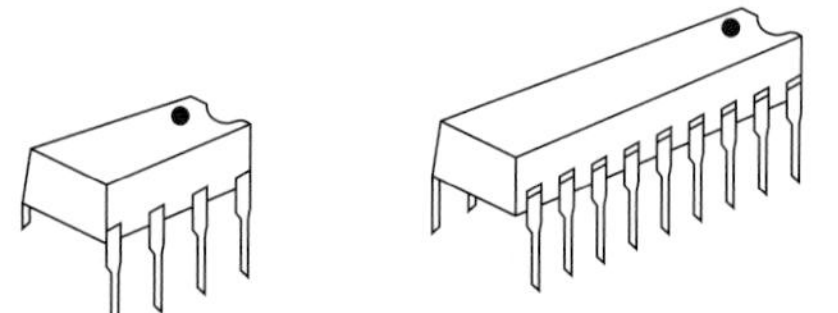

FIGURE 4.63
DIL packaged ICs.

The word thyristor is derived from the Greek word *thyra* meaning door, because the thyristor behaves like a door. It can be open or shut, allowing

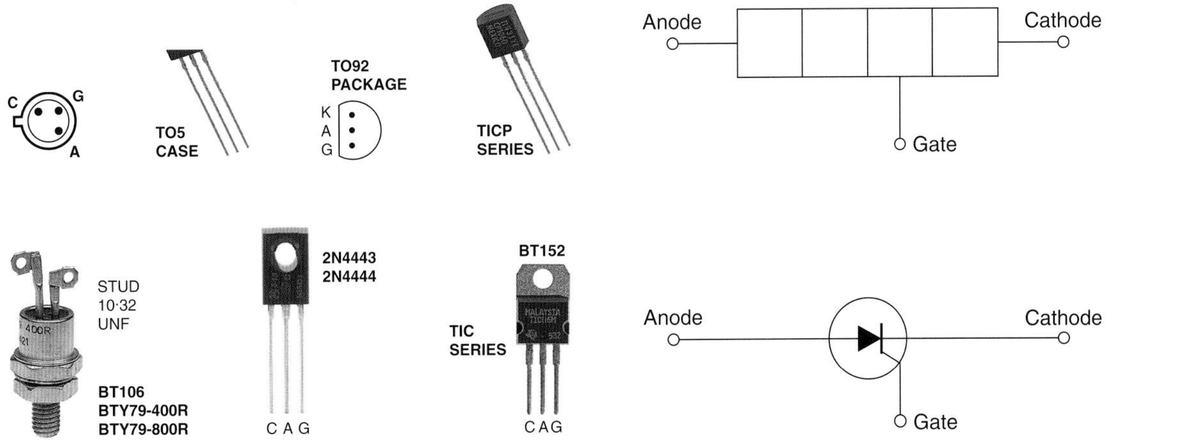

FIGURE 4.64
Symbol for and structure and appearance of a thyristor.

Table 4.7 Thyristor Testing Using an Ohmmeter
A 'good' thyristor will give the following readings:
Black to cathode and red on gate = low resistance Red to cathode and black on gate = a higher resistance value The value of the second reading will depend upon the thyristor, and may vary from only slightly greater to very much greater. Connecting the test instrument leads from cathode to anode will result in a very high resistance reading, whatever polarity is used.

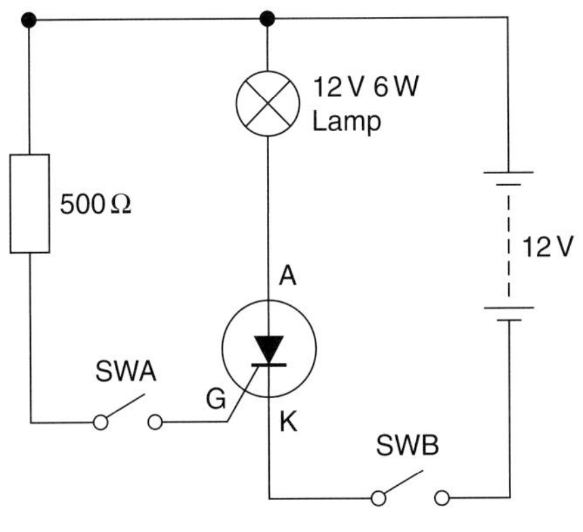

FIGURE 4.65
Thyristor test circuit.

or preventing current flow through the device. The door is opened – we say the thyristor is triggered – to a conducting state by applying a pulse voltage to the gate connection. Once the thyristor is in the conducting state, the gate loses all control over the devices. The only way to bring the thyristor back to a non-conducting state is to reduce the voltage across the anode and cathode to zero or apply reverse voltage across the anode and cathode.

We can understand the operation of a thyristor by considering the circuit shown in Fig. 4.65. This circuit can also be used to test suspected faulty components.

When SWB only is closed the lamp will not light, but when SWA is also closed, the lamp lights to full brilliance. The lamp will remain illuminated even when SWA is opened. This shows that the thyristor is operating correctly. Once a voltage has been applied to the gate the thyristor becomes forward conducting, like a diode, and the gate loses control.

A thyristor may also be tested using an ohmmeter as described in Table 4.7, assuming that the red lead of the ohmmeter is positive.

The thyristor has no moving parts and operates without arcing. It can operate at extremely high speeds, and the currents used to operate the gate are very small. The most common application for the thyristor is to control the power supply to a load, for example, lighting dimmers and motor speed control.

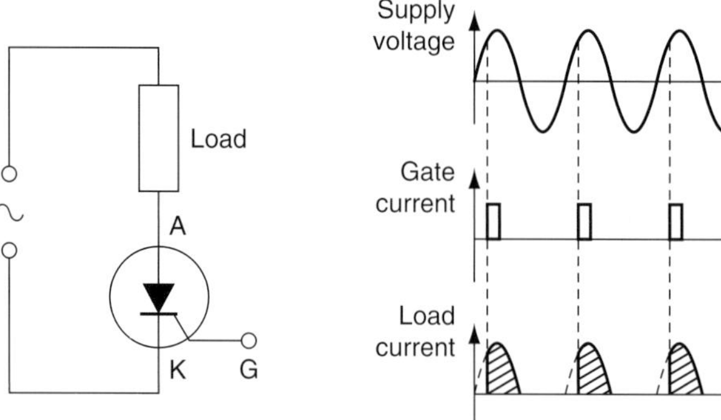

FIGURE 4.66
Waveforms to show the control effect of a thyristor.

The power available to an a.c. load can be controlled by allowing current to be supplied to the load during only a part of each cycle. This can be achieved by supplying a gate pulse automatically at a chosen point in each cycle, as shown by Fig. 4.66. Power is reduced by triggering the gate later in the cycle.

The thyristor is only a half-wave device (like a diode) allowing control of only half the available power in an a.c. circuit. This is very uneconomical, and a further development of this device has been the triac which is considered next.

THE TRIAC

The triac was developed following the practical problems experienced in connecting two thyristors in parallel, to obtain full-wave control, and in providing two separate gate pulses to trigger the two devices.

The triac is a single device containing a back-to-back, two-directional thyristor which is triggered on both halves of each cycle of the a.c. supply by the same gate signal. The power available to the load can, therefore, be varied between zero and full load.

Its symbol and general appearance are shown in Fig. 4.67. Power to the load is reduced by triggering the gate later in the cycle, as shown by the waveforms of Fig. 4.68.

The triac is a three-terminal device, just like the thyristor, but the terms anode and cathode have no meaning for a triac. Instead, they are called main terminal one (MT_1) and main terminal two (MT_2). The device is triggered by applying a small pulse to the gate (G). A gate current of 50 mA is sufficient to trigger a triac switching up to 100A. They are used for many commercial applications where control of a.c. power is required, for example, motor speed control and lamp dimming.

THE DIAC

The diac is a two-terminal device containing a two-directional Zener diode. It is used mainly as a trigger device for the thyristor and triac. The symbol is shown in Fig. 4.69.

FIGURE 4.67
Appearance of a triac.

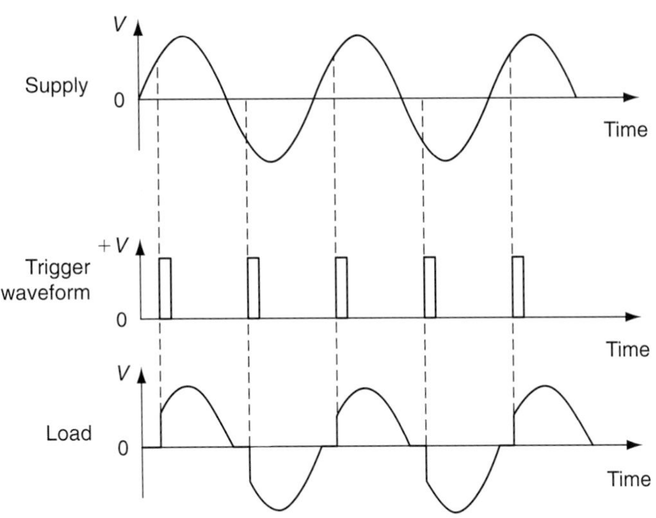

FIGURE 4.68
Waveforms to show the control effect of a triac.

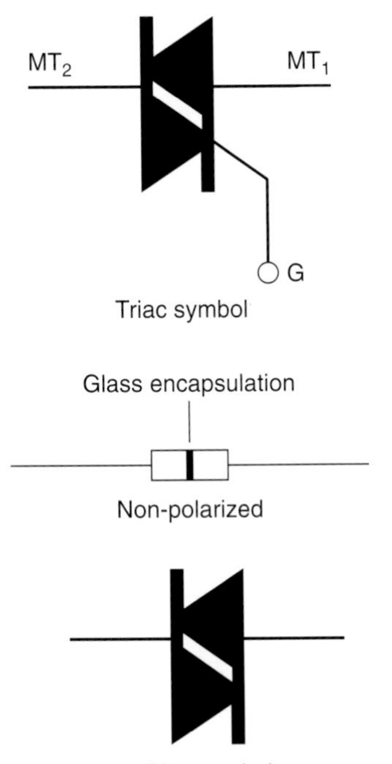

FIGURE 4.69
Symbol for and appearance of a diac used in triac firing circuits.

The device turns on when some predetermined voltage level is reached, say 30V, and, therefore, it can be used to trigger the gate of a triac or thyristor each time the input waveform reaches this predetermined value. Since the device contains back-to-back Zener diodes it triggers on both the positive and negative half-cycles.

Voltage divider

At the beginning of this chapter we considered the distribution of voltage across resistors connected in series. We found that the supply voltage was divided between the series resistors in proportion to the size of the resistor. If two identical resistors were connected in series across a 12 V supply, as shown in Fig. 4.70(a), both common sense and a simple calculation would confirm that 6V would be measured across the output. In the circuit shown in Fig. 4.70(b), the 1 and 2 kΩ resistors divide the input voltage into three equal parts. One part, 4 V, will appear across the 1 kΩ resistor and two parts, 8 V, will appear across the 2 kΩ resistor. In Fig. 4.70(c) the situation is reversed and, therefore, the voltmeter will read 4 V. The division of the voltage is proportional to the ratio of the two resistors and, therefore, we call this simple circuit a *voltage divider* or *potential divider*. The values of the resistors R_1 and R_2 determine the output voltage as follows:

$$V_{OUT} = V_{IN} \times \frac{R_2}{R_1 + R_2} \text{ (V)}$$

1 kΩ R_1

V_{IN} = 12 V

1 kΩ R_2

V_{OUT}

(a)

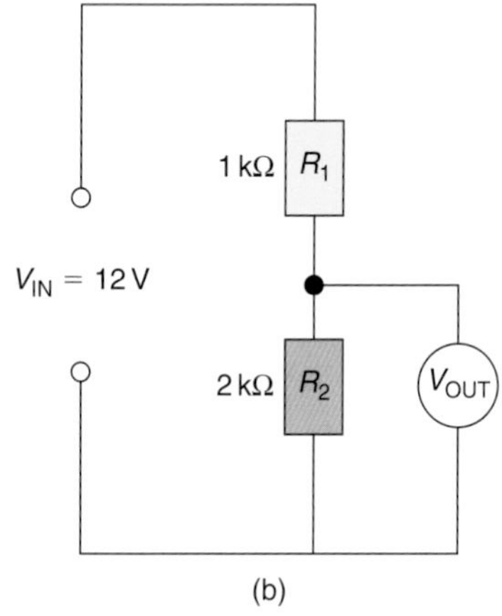

(b)

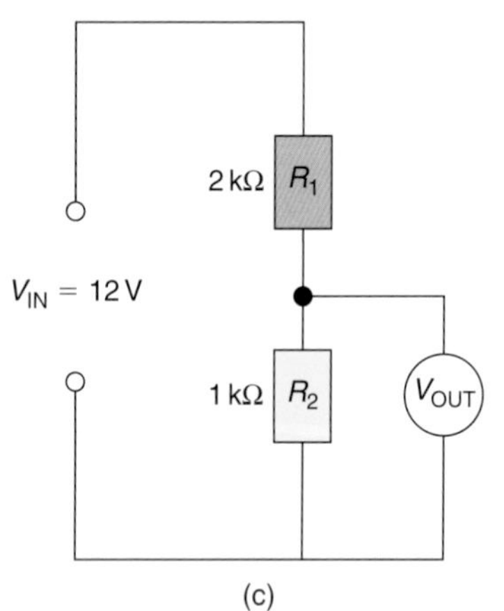

(c)

FIGURE 4.70
Voltage divider circuit.

For the circuit shown in Fig. 4.70(b),

$$V_{OUT} = 12\ \text{V} \times \frac{2\ \text{k}\Omega}{1\ \text{k}\Omega + 2\ \text{k}\Omega} = 8\ \text{V}$$

For the circuit shown in Fig. 4.70(c),

$$V_{OUT} = 12\text{V} \times \frac{1\ \text{k}\Omega}{2\ \text{k}\Omega + 1\ \text{k}\Omega} = 4\text{V}$$

Example 1

For the circuit shown in Fig. 4.71, calculate the output voltage.

$$V_{OUT} = 6\ \text{V} \times \frac{2.2\ \text{k}\Omega}{10\ \text{k}\Omega + 2.2\ \text{k}\Omega} = 1.08\ \text{V}$$

Example 2

For the circuit shown in Fig. 4.72(a), calculate the output voltage.

We must first calculate the equivalent resistance of the parallel branch:

$$\frac{1}{R_T} = \frac{1}{R_1} + \frac{1}{R_2}$$

$$\frac{1}{R_T} = \frac{1}{10\ \text{k}\Omega} + \frac{1}{10\ \text{k}\Omega} = \frac{1+1}{10\ \text{k}\Omega} = \frac{2}{10\ \text{k}\Omega}$$

$$R_T = \frac{10\ \text{k}\Omega}{2} = 5\ \text{k}\Omega$$

The circuit may now be considered as shown in Fig. 4.72(b):

$$V_{OUT} = 6\ \text{V} \times \frac{10\ \text{k}\Omega}{5\ \text{k}\Omega + 10\ \text{k}\Omega} = 4\ \text{V}$$

Voltage dividers are used in electronic circuits to produce a reference voltage which is suitable for operating transistors and ICs. The volume control in a radio or the brightness control of a cathode-ray oscilloscope requires a continuously variable voltage divider and this can be achieved by connecting a variable resistor or potentiometer, as shown in Fig. 4.73. With the wiper arm making a connection at the bottom of the resistor, the output would be zero. When connection is made at the centre, the voltage would be 6 V, and at the top of the resistor the voltage would be 12 V. The voltage is continuously variable between 0 and 12 V simply by moving the wiper arm of a suitable variable resistor such as those shown in Fig. 4.37.

When a load is connected to a voltage divider it 'loads' the circuit, causing the output voltage to fall below the calculated value. To avoid this, the resistance of the load should be at least ten times as great as the value of the resistor across which it is connected. For example, the load connected across the voltage divider shown in Fig. 4.70(b) must be greater than 20 kΩ

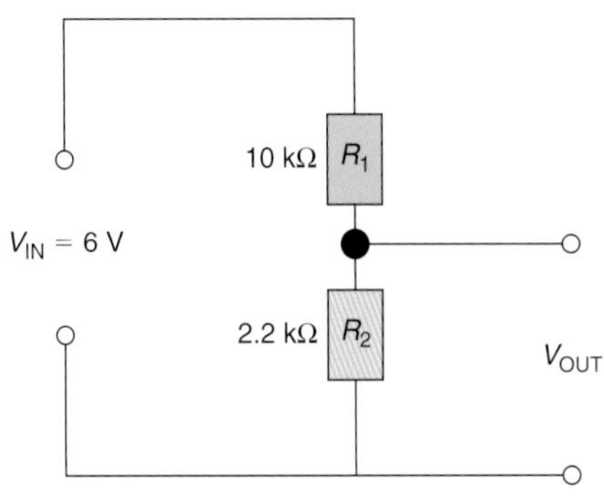

FIGURE 4.71
Voltage divider circuit for Example 1.

and across 4.70(c) greater than 10 kΩ. This problem of loading the circuit also occurs when taking voltage readings, as discussed later in this chapter under the subheading Instrument Errors.

Rectification of a.c.

When a d.c. supply is required, batteries or a rectified a.c. supply can be provided. Batteries have the advantage of portability, but a battery supply is more expensive than using the a.c. mains supply suitably rectified. **Rectification** is the conversion of an a.c. supply into a unidirectional or d.c. supply. This is one of the many applications for a diode which will conduct in one direction only, that is when the anode is positive with respect to the cathode.

Definition

Rectification is the conversion of an a.c. supply into a unidirectional or d.c. supply.

HALF-WAVE RECTIFICATION

The circuit is connected as shown in Fig. 4.74. During the first half-cycle the anode is positive with respect to the cathode and, therefore, the diode will conduct. When the supply goes negative during the second half-cycle, the anode is negative with respect to the cathode and, therefore, the diode will not allow current to flow. Only the positive half of the waveform will be available at the load and the lamp will light at reduced brightness.

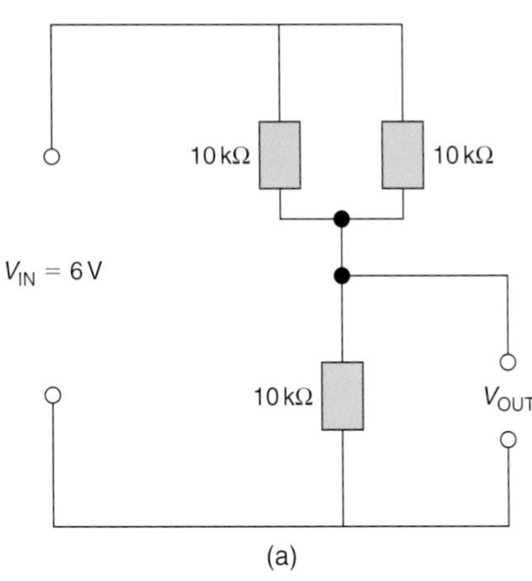

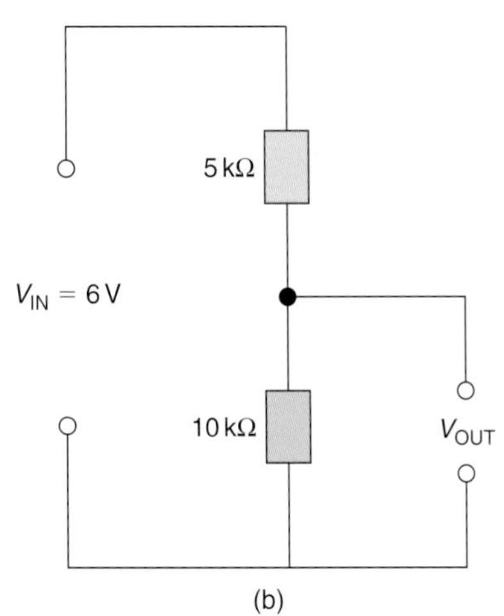

FIGURE 4.72
(a) Voltage divider circuit for Example 2; (b) Equivalent circuit for Example 2.

FULL-WAVE RECTIFICATION

Figure 4.75 shows an improved rectifier circuit which makes use of the whole a.c. waveform and is, therefore, known as a full-wave rectifier. When the four diodes are assembled in this diamond-shaped configuration, the circuit is also known as a *bridge rectifier*. During the first half-cycle diodes D_1 and D_3 conduct, and diodes D_2 and D_4 conduct during the second half-cycle. The lamp will light to full brightness.

Full-wave and half-wave rectification can be displayed on the screen of a CRO and will appear as shown in Figs. 4.74 and 4.75.

Smoothing

The circuits of Figs. 4.74 and 4.75 convert an alternating waveform into a waveform which never goes negative, but they cannot be called continuous d.c. because they contain a large alternating component. Such a waveform is too bumpy to be used to supply electronic equipment but may be used for battery charging. To be useful in electronic circuits the output must be smoothed. The simplest way to smooth an output is to connect a large-value capacitor across the output terminals as shown in Fig. 4.76.

When the output from the rectifier is increasing, as shown by the dotted lines of Fig. 4.77, the capacitor charges up. During the second quarter of the cycle, when the output from the rectifier is falling to zero, the capacitor discharges into the load. The output voltage falls until the output from the rectifier once again charges the capacitor. The capacitor connected to the full-wave rectifier circuit is charged up twice as often as the

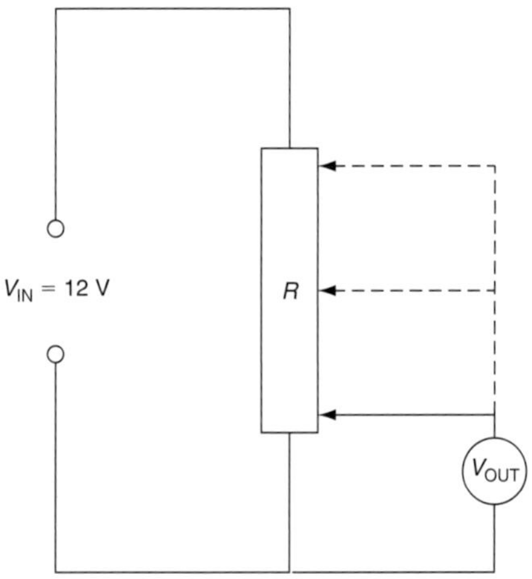

FIGURE 4.73
Constantly variable voltage divider circuit.

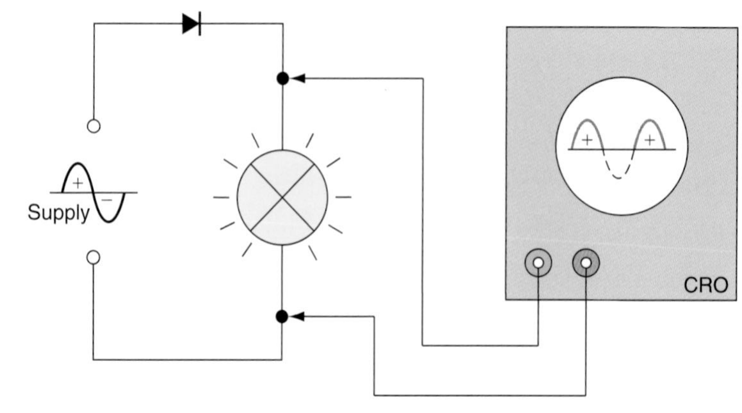

FIGURE 4.74
Half-wave rectification.

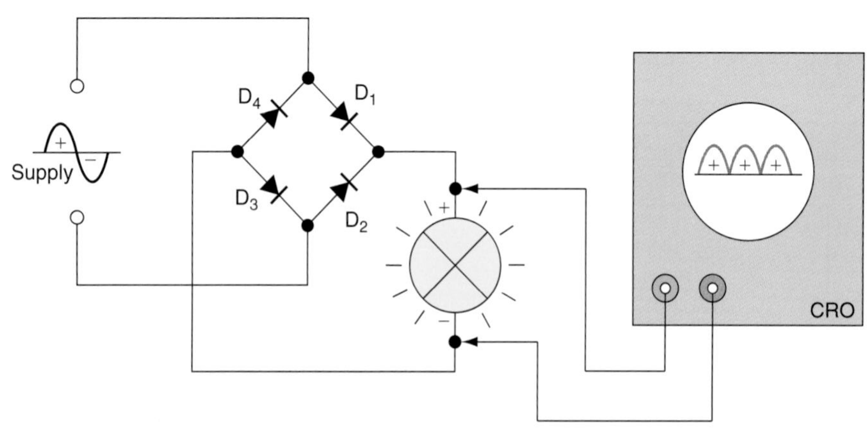

FIGURE 4.75
Full-wave rectification using a bridge circuit.

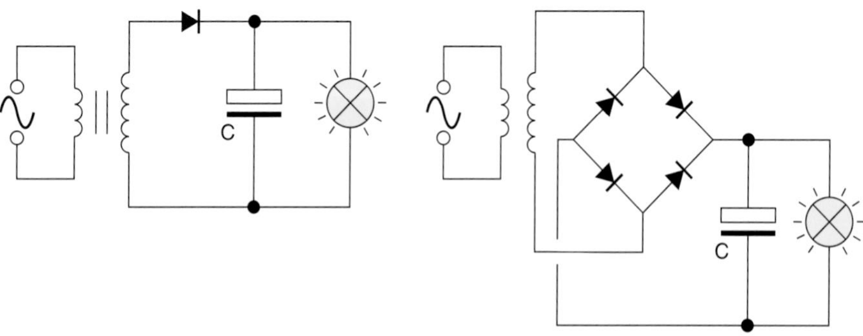

FIGURE 4.76
Rectified a.c. with smoothing capacitor connected.

capacitor connected to the half-wave circuit and, therefore, the output ripple on the full-wave circuit is smaller, giving better smoothing. Increasing the current drawn from the supply increases the size of the ripple. Increasing the size of the capacitor reduces the amount of ripple.

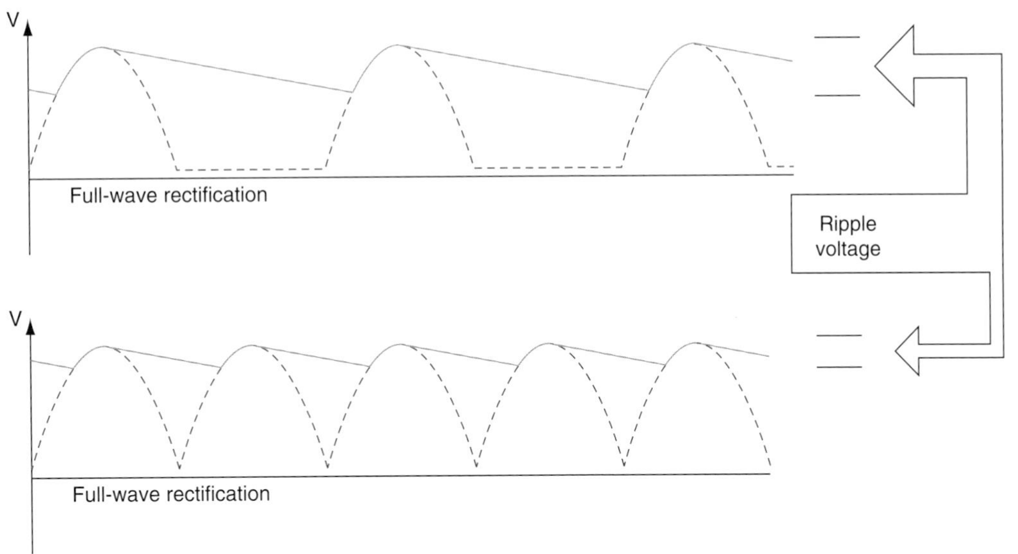

FIGURE 4.77
Output waveforms with smoothing showing reduced ripple with full wave.

Try This

Battery Charger

- Do you have a battery charger for your car battery?
- What type of circuit do you think is inside?
- Carefully look inside and identify the components.

LOW-PASS FILTER

The ripple voltage of the rectified and smoothed circuit shown in Fig. 4.76 can be further reduced by adding a low-pass filter, as shown in Fig. 4.78. A low-pass filter allows low frequencies to pass while blocking higher frequencies. Direct current has a frequency of zero hertz, while the ripple voltage of a full-wave rectifier has a frequency of 100 Hz. Connecting the low-pass filter will allow the d.c. to pass while blocking the ripple voltage, resulting in a smoother output voltage.

The low-pass filter shown in Fig. 4.78 does, however, increase the output resistance, which encourages the voltage to fall as the load current increases. This can be reduced if the resistor is replaced by a choke, which has a high-impedance to the ripple voltage but a low resistance, which reduces the output ripple without increasing the output resistance.

Stabilized power supplies

The power supplies required for electronic circuits must be ripple-free, stabilized and have good regulation, that is the voltage must not change in value over the whole load range. A number of stabilizing circuits are available which, when connected across the output of the circuit shown in Fig. 4.76, give a constant or stabilized voltage output. These circuits use the

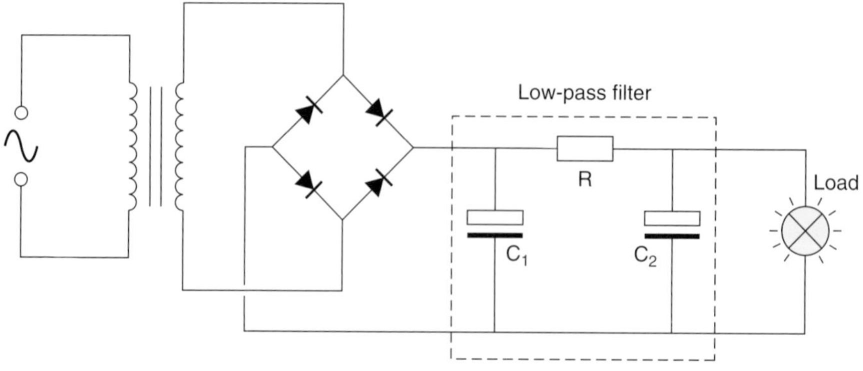

FIGURE 4.78
Rectified a.c. with low-pass filter connected.

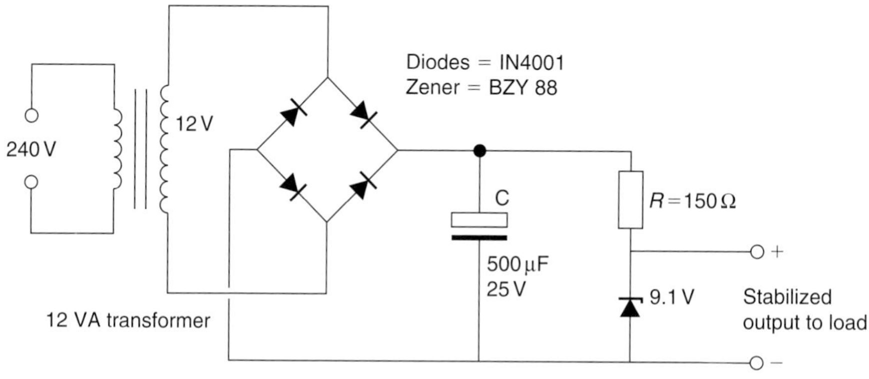

FIGURE 4.79
Stabilized d.c. supply.

characteristics of the Zener diode which was described by the experiment in Fig. 4.50.

Figure 4.79 shows an a.c. supply which has been rectified, smoothed and stabilized. You could build and test this circuit at college if your lecturers agree.

Lighting and luminares

In ancient times, much of the indoor work done by humans depended upon daylight being available to light the interior. Today almost all buildings have electric lighting installed and we automatically assume that we can work indoors or out of doors at any time of the day or night, and that light will always be available.

Good lighting is important in all building interiors, helping work to be done efficiently and safely and also playing an important part in creating pleasant and comfortable surroundings.

Lighting schemes are designed using many different types of light fitting or luminaire. 'Luminaire' is the modern term given to the equipment which supports and surrounds the lamp and may control the distribution of the light. Modern lamps use the very latest technology to provide illumination cheaply and efficiently. To begin to understand the lamps and lighting technology used today, we must first define some of the terms we will be using.

LUMINOUS INTENSITY – SYMBOL I

This is the illuminating power of the light source to radiate luminous flux in a particular direction. The earliest term used for the unit of luminous intensity was the candle power because the early standard was the wax candle. The SI unit is the candela (pronounced candeela and abbreviated as cd).

LUMINOUS FLUX – SYMBOL F

This is the flow of light which is radiated from a source. The SI unit is the lumen, one lumen being the light flux which is emitted within a unit solid angle (volume of a cone) from a point source of 1 candela.

ILLUMINANCE – SYMBOL E

This is a measure of the light falling on a surface, which is also called the incident radiation. The SI unit is the lux (lx) and is the illumination produced by 1 lumen over an area of 1 m^2.

LUMINANCE – SYMBOL L

Since this is a measure of the brightness of a surface it is also a measure of the light which is reflected from a surface. The objects we see vary in appearance according to the light which they emit or reflect towards the eye.

The SI units of luminance vary with the type of surface being considered. For a diffusing surface such as blotting paper or a matt white painted surface the unit of luminance is the lumen per square metre. With polished surfaces such as a silvered glass reflector, the brightness is specified in terms of the light intensity and the unit is the candela per square metre.

Illumination laws

Rays of light falling upon a surface from some distance d will illuminate that surface with an illuminance of say 1 lx. If the distance d is doubled as shown in Fig. 4.80, the illuminance of 1 lx will fall over four square units of area. Thus the illumination of a surface follows the **inverse square law**, where

$$E = \frac{I}{d^2} \text{(lx)}$$

Definition

Thus the illumination of a surface follows the *inverse square law*, where

$$E = \frac{I}{d^2} \text{(lx)}$$

Example 1

A lamp of luminous intensity 1000 cd is suspended 2 m above a laboratory bench. Calculate the illuminance directly below the lamp:

$$E = \frac{I}{d^2} \text{(lx)}$$

$$\therefore E = \frac{1000 \text{ cd}}{(2 \text{ m})^2} = 250 \text{ lx}$$

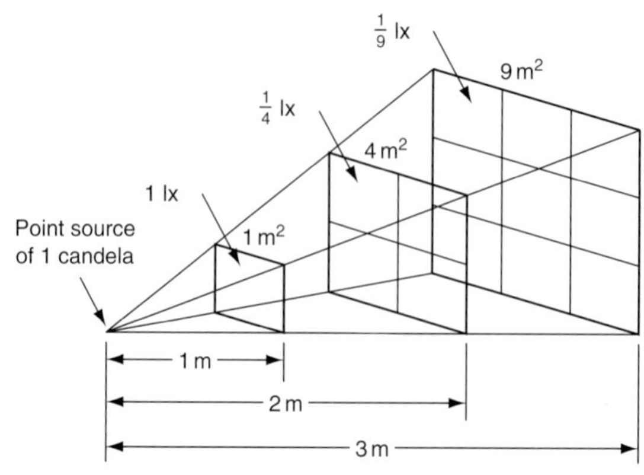

FIGURE 4.80
The inverse square law.

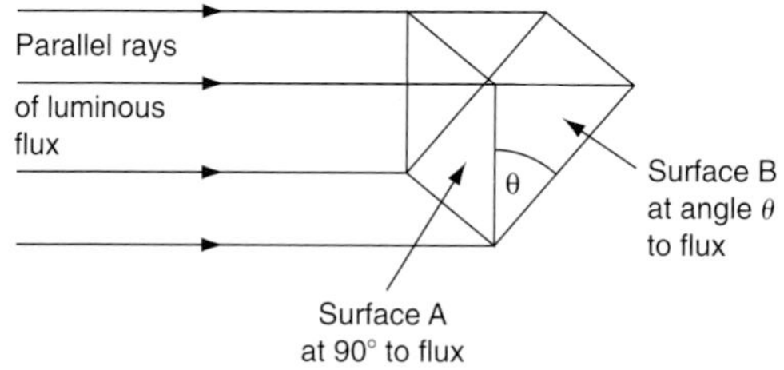

FIGURE 4.81
The cosine law.

The illumination of surface A in Fig. 4.81 will follow the inverse square law described above. If this surface were removed, the same luminous flux would then fall on surface B. Since the parallel rays of light falling on the inclined surface B are spread over a larger surface area, the illuminance will be reduced by a factor θ, and therefore:

$$E = \frac{I \cos \theta}{d^2} \text{(lx)}$$

Since the two surfaces are joined together by the trigonometry of the cosine rules this equation is known as the **cosine law**.

Definition

$$E = \frac{I \cos \theta}{d^2} \text{(lx)}$$

Since the two surfaces are joined together by the trigonometry of the cosine rules this equation is known as the *cosine law*.

Example 2

A street lantern suspends a 2000 cd light source 4 m above the ground. Determine the illuminance directly below the lamp and 3 m to one side of the lamp base.

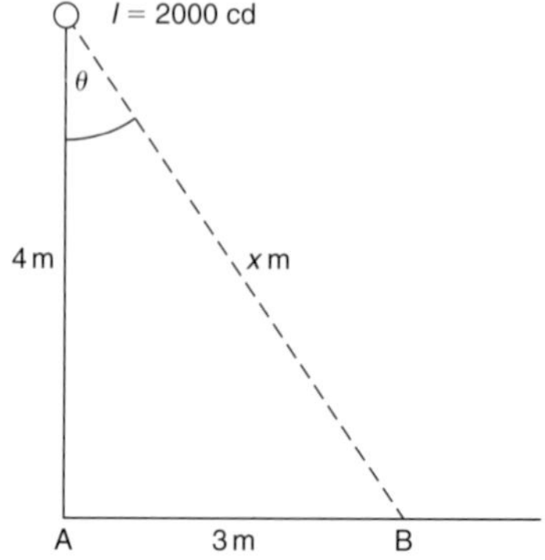

The illuminance below the lamp, E_A, is:

$$E_A = \frac{I}{d^2}(\text{lx})$$

$$\therefore E_A = \frac{2000\text{ cd}}{(4\text{ m})^2} = 125\text{ lx}$$

To work out the illuminance at 3 m to one side of the lantern, E_B, we need the distance between the light source and the position on the ground at B; this can be found by Pythagoras' theorem:

$$x\text{ (m)} = \sqrt{(4\text{ m})^2 + (3\text{ m})^2} = \sqrt{25\text{ m}}$$

$$x = 5\text{ m}$$

$$\therefore E_B = \frac{I\cos\theta}{d^2}(\text{lx}) \text{ and } \cos\theta = \frac{4}{5}$$

$$\therefore E_B = \frac{2000\text{ cd} \times 4}{(5\text{ m})^2 \times 5} = 64\text{ lx}$$

Example 3

A discharge lamp is suspended from a ceiling 4 m above a bench. The illuminance on the bench below the lamp was 300 lx. Find:

(a) the luminous intensity of the lamp
(b) the distance along the bench where the illuminance falls to 153.6 lx.

For (a),

$$E_A = \frac{I}{d^2}(\text{lx})$$

$$\therefore I = E_A\, d^2(\text{cd})$$

$$I = 300\text{ lx} \times 16\text{ m} = 4800\text{ cd}$$

For (b),

$$E_B = \frac{I}{d^2}\cos\theta\,(\text{lx})$$

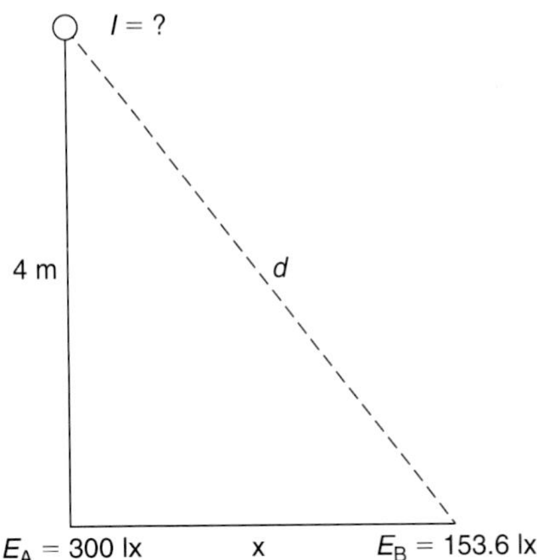

t0080

$$\therefore d^2 = \frac{I \cos\theta}{E_B}(\text{m}^2)$$

$$d^2 = \frac{4800 \text{ cd}}{153.6 \text{ lx}} \times \frac{4 \text{ m}}{d \text{ m}}$$

$$d^3 = 125$$

$$\therefore \text{d} = \sqrt[3]{125} = 5 \text{ m}$$

By Pythagoras,

$$x = \sqrt{5^2 - 4^2} = 3\text{m}$$

The recommended levels of illuminance for various types of installation are given by the IES (Illumination Engineers Society). Some examples are given in Table 4.8.

Table 4.8 Illuminance Values

Task	Working situation	Illuminance (lx)
Casual vision	Storage rooms, stairs and washrooms	100
Rough assembly	Workshops and garages	300
Reading, writing and drawing	Classrooms and offices	500
Fine assembly	Electronic component assembly	1000
Minute assembly	Watchmaking	3000

The activities being carried out in a room will determine the levels of illuminance required since different levels of illumination are required for the successful operation or completion of different tasks. The assembly of electronic components in a factory will require a higher level of illumination than, say, the assembly of engine components in a garage because the electronic components are much smaller and finer detail is required for their successful assembly.

The inverse square law calculations considered earlier are only suitable for designing lighting schemes where there are no reflecting surfaces producing secondary additional illumination. This method could be used to design an outdoor lighting scheme for a cathedral, bridge or public building.

Interior luminaires produce light directly on to the working surface but additionally there is a secondary source of illumination from light reflected from the walls and ceilings. When designing interior lighting schemes the method most frequently used depends upon a determination of the total flux required to provide a given value of illuminance at the working place. This method is generally known as the **lumen method**.

Definition

When designing interior lighting schemes the method most frequently used depends upon a determination of the total flux required to provide a given value of illuminance at the working place. This method is generally known as the *lumen method*.

THE LUMEN METHOD

To determine the total number of luminaires required to produce a given illuminance by the lumen method we apply the following formula:

$$\text{Total number of luminaires required to provide a chosen level of illumination at a surface} = \frac{\text{Illuminance level (lx)} \times \text{Area (m}^2\text{)}}{\text{Lumen output of each luminaire (lm)} \times \text{UF} \times \text{LLF}}$$

where

- the illuminance level is chosen after consideration of the IES code,
- the area is the working area to be illuminated,
- the lumen output of each luminaire is that given in the manufacturer's specification and may be found by reference to tables such as Table 4.9,
- UF is the utilization factor,
- LLF is the light loss factor.

Utilization factor

The light flux reaching the working plane is always less than the lumen output of the lamp since some of the light is absorbed by the various surface textures. The method of calculating the utilization factor (UF) is detailed in Chartered Institution of Building Services Engineers (CIBSE) Technical Memorandum No 5, although lighting manufacturers' catalogues give factors for standard conditions. The UF is expressed as a number which is always less than unity; a typical value might be 0.9 for a modern office building.

Light Loss Factor

The light output of a luminaire is reduced during its life because of an accumulation of dust and dirt on the lamp and fitting. Decorations also

Table 4.9 Characteristics of a Thorn Lighting 1500 mm 65 W Bi-Pin Tube

Tube colour	Initial lamp lumens*	Lighting design lumens†	Colour rendering quality	Colour appearance
Artifical daylight	2600	2100	Excellent	Cool
De Luxe Natural	2900	2500	Very Good	Intermediate
De Luxe Warm white	3500	3200	Good	Warm
Natural	3700	3400	Good	Intermediate
Daylight	4800	4450	Fair	Cool
Warm white	4950	4600	Fair	Warm
White	5100	4750	Fair	Warm
Red	250*	250	Poor	Deep red

Coloured tubes are intended for decorative purposes only.
*The initial lumens are the measured lumens after 100 hours of life.
†The lighting design lumens are the output lumens after 2000 hours.

Burning position: Lamp may be operated in any position
Rated life: 7500 hours
Efficacy: 30–70 lm/W depending upon the tube colour

deteriorate with time, and this results in more light flux being absorbed by the walls and ceiling.

You can see from Table 4.9 that the output lumens of the lamp decrease with time – for example, a warm white tube gives out 4950 lumens after the first 100 hours of its life but this falls to 4600 lumens after 2000 hours.

The total light loss can be considered under four headings:

1. light loss due to luminaire dirt depreciation (LDD),
2. light loss due to room dirt depreciation (RDD),
3. light loss due to lamp failure factor (LFF),
4. light loss due to lamp lumen depreciation (LLD).

The LLF is the total loss due to these four separate factors and typically has a value between 0.8 and 0.9.

When using the LLF in lumen method calculations we always use the manufacturer's initial lamp lumens for the particular lamp because the LLF takes account of the depreciation in lumen output with time. Let us now consider a calculation using the lumen method.

Example

It is proposed to illuminate an electronic workshop of dimensions 9 × 8 × 3 m to an illuminance of 550 lx at the bench level. The specification calls for luminaires having one 1500 mm 65 W natural tube with an initial output of 3700 lumens (see Table 4.9). Determine the number of luminaires required for this installation when the UF and LLF are 0.9 and 0.8, respectively.

$$\text{The number of luminaires required} = \frac{E\ (\text{lx}) \times \text{area}\ (\text{m}^2)}{\text{lumens from each luminaire} \times \text{UF} \times \text{LLF}}$$

$$\text{The number of luminaires} = \frac{550\ \text{lx} \times 9\ \text{m} \times 8\ \text{m}}{3700 \times 0.9 \times 0.8} = 14.86$$

Therefore 15 luminaires will be required to illuminate this workshop to a level of 550 lx.

Comparison of light sources

When comparing one light source with another we are interested in the colour reproducing qualities of the lamp and the efficiency with which the lamp converts electricity into illumination. These qualities are expressed by the lamp's efficacy and colour rendering qualities.

Definition

The performance of a lamp is quoted as a ratio of the number of lumens of light flux which it emits to the electrical energy input which it consumes. Thus *efficacy* is measured in lumens per watt; the greater the efficacy the better is the lamp's performance in converting electrical energy into light energy.

Lamp efficacy

The performance of a lamp is quoted as a ratio of the number of lumens of light flux which it emits to the electrical energy input which it consumes. Thus **efficacy** is measured in lumens per watt; the greater the efficacy the better is the lamp's performance in converting electrical energy into light energy.

A general lighting service (GLS) lamp, for example, has an efficacy of 14 lumens per watt, while a fluorescent tube, which is much more efficient at converting electricity into light, has an efficacy of about 50 lumens per watt.

Colour rendering

We recognize various materials and surfaces as having a particular colour because luminous flux of a frequency corresponding to that colour is reflected from the surface to our eye which is then processed by our brain. White light is made up of the combined frequencies of the colours red, orange, yellow, green, blue, indigo and violet. Colours can only be seen if the lamp supplying the illuminance is emitting light of that particular frequency. The ability to show colours faithfully as they would appear in daylight is a measure of the colour rendering property of the light source.

Definition

GLS lamps produce light as a result of the heating effect of an electrical current. Most of the electricity goes to producing heat and a little to producing light. A fine tungsten wire is first coiled and coiled again to form the incandescent filament of the GLS lamp.

GLS LAMPS

GLS lamps produce light as a result of the heating effect of an electrical current. Most of the electricity goes to producing heat and a little to producing light. A fine tungsten wire is first coiled and coiled again to form the incandescent filament of the GLS lamp. The coiled coil arrangement reduces filament cooling and increases the light output by allowing the filament to operate at a higher temperature. The light output covers the visible spectrum, giving a warm white to yellow light with a colour rendering quality classified as fairly good. The efficacy of the GLS lamp is 14 lumens per watt over its intended lifespan of 1000 hours.

Key Fact

Energy efficiency

- Over the next few years the government will phase out GLS lamps.
- So that we will have to use more energy efficient lamps such as Compact Fluorescent Lamps (CFLs).

The filament lamp in its simplest form is a purely functional light source which is unchallenged on the domestic market despite the manufacture of more efficient lamps. One factor which may have contributed to its popularity is that lamp designers have been able to modify the glass envelope of the lamp to give a very pleasing decorative appearance, as shown in Fig. 4.82.

TUNGSTEN HALOGEN DICHROIC REFLECTOR MINIATURE SPOT LAMPS

Tungsten Halogen Dichroic Reflector Miniature Spot Lamps such as the one shown in Fig. 4.83 are extremely popular in the lighting schemes of the new millennium. Their small size and bright white illumination makes them very popular in both commercial and domestic installations. They are available as a 12 volt bi-pin package in 20, 35 and 50 watts and as a 230 volt bayonet type cap (called a GU10 or GZ10 cap) in 20, 35 and 50 watts. However, their efficacy is only 20 lumens per watt over a 2000-hour lifespan.

Definition

Discharge lamps do not produce light by means of an incandescent filament but by the excitation of a gas or metallic vapour contained within a glass envelope.

DISCHARGE LAMPS

Discharge lamps do not produce light by means of an incandescent filament but by the excitation of a gas or metallic vapour contained within a glass envelope. A voltage applied to two terminals or electrodes sealed into the end of a glass tube containing a gas or metallic vapour will excite the contents and produce light directly. Fluorescent tubes and CFLs operate on this principle.

FIGURE 4.83
Tungsten Halogen Dichroic Reflector Lamp.

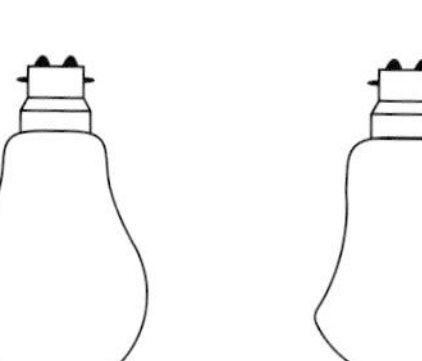
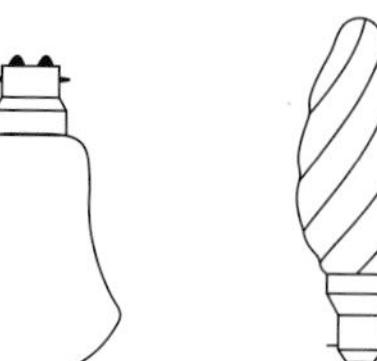

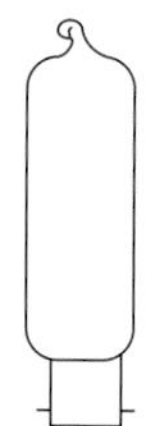

FIGURE 4.82
Some decorative GLS lamp shapes.

Fluorescent tube

A **fluorescent lamp** is a linear arc tube, internally coated with a fluorescent powder, containing a low-pressure mercury vapour discharge. The lamp construction is shown in Fig. 4.84 and the characteristics of the variously coloured tubes are given in Table 4.9.

Definition

A *fluorescent lamp* is a linear arc tube, internally coated with a fluorescent powder, containing a low-pressure mercury vapour discharge.

Passing a current through the cathodes of the tube causes them to become hot and produce a cloud of electrons which ionize the gas in the immediate vicinity of the cathodes. This ionization then spreads to the whole length of the tube, producing invisible ultraviolet rays and some blue light. The fluorescent powder on the inside of the tube is sensitive to ultraviolet rays and converts this radiation into visible light. The fluorescent powder on the inside of the tube can be mixed to give light of almost any desired colour or grade of white light. Some mixes have their maximum light output in the yellow–green region of the spectrum giving maximum efficacy but poor colour rendering. Other mixes give better colour rendering at the cost of reduced lumen output as can be seen from Table 4.9. The lamp has many domestic, industrial and commercial applications. Its efficacy varies between 30 and 70 lumens per watt depending upon the colour rendering qualities of the tube.

Definition

CFLs are miniature fluorescent lamps designed to replace ordinary GLS lamps.

Energy-efficient lamps

CFLs are miniature or compact fluorescent lamps designed to replace ordinary GLS lamps. They are available in a variety of shapes and sizes so that they can be fitted into existing light fittings. Fig. 4.85 shows three typical

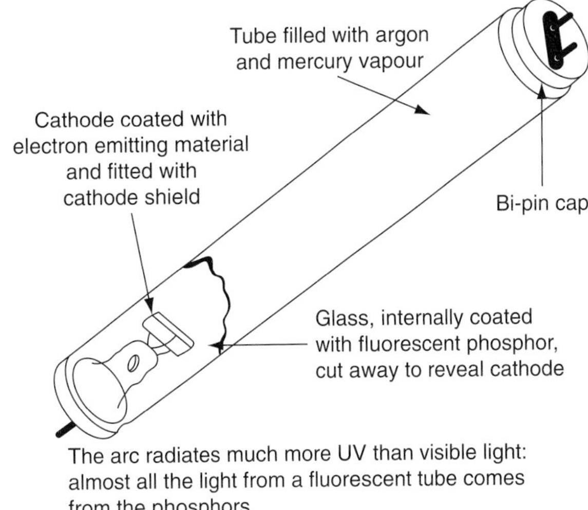

FIGURE 4.84
Fluorescent lamp construction.

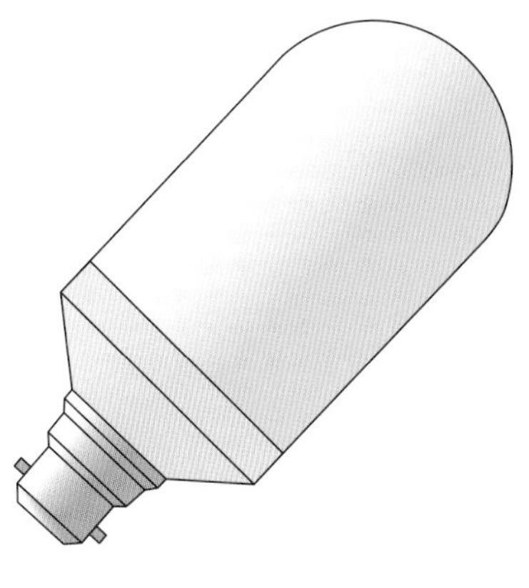

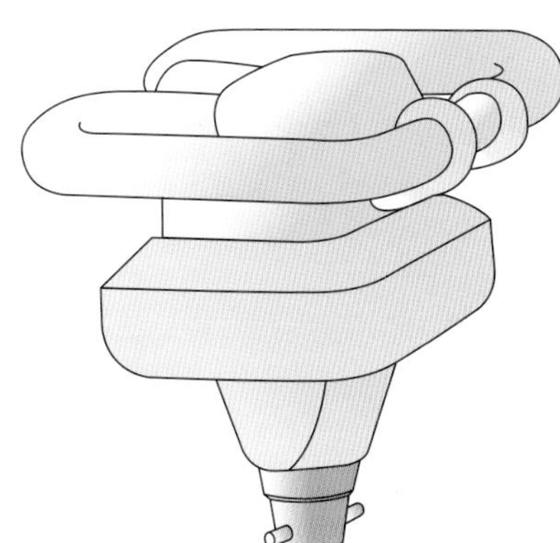

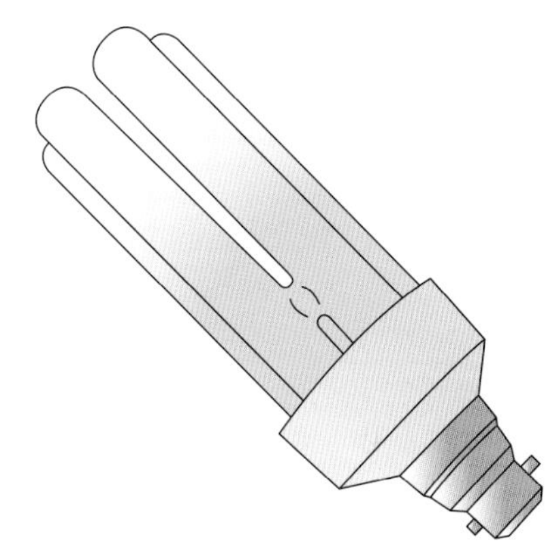

FIGURE 4.85
Energy-efficient lamps. (CFLs)

shapes. The 'stick' type gives most of their light output radially while the flat 'double D' type give most of their light output above and below.

Energy-efficient lamps use electricity much more efficiently than an equivalent GLS lamp. For example, a 20 watt energy efficient lamp will give the same light output as a 100 watt GLS lamp. An 11 watt energy efficient lamp is equivalent to a 60 watt GLS lamp. Energy-efficient lamps also have a lifespan of about eight times longer than a GLS lamp and so, they do use energy very efficiently.

However, energy-efficient lamps are expensive to purchase and they do take a few minutes to attain full brilliance after switching on. They cannot always be controlled by a dimmer switch and are unsuitable for incorporating in an automatic presence detector because they are usually not switched on long enough to be worthwhile, but energy efficient lamps are excellent for outside security lighting which is left on for several hours each night.

The electrical contractor, in discussion with a customer, must balance the advantages and disadvantages of energy-efficient lamps compared to other sources of illumination for each individual installation.

Check your Understanding

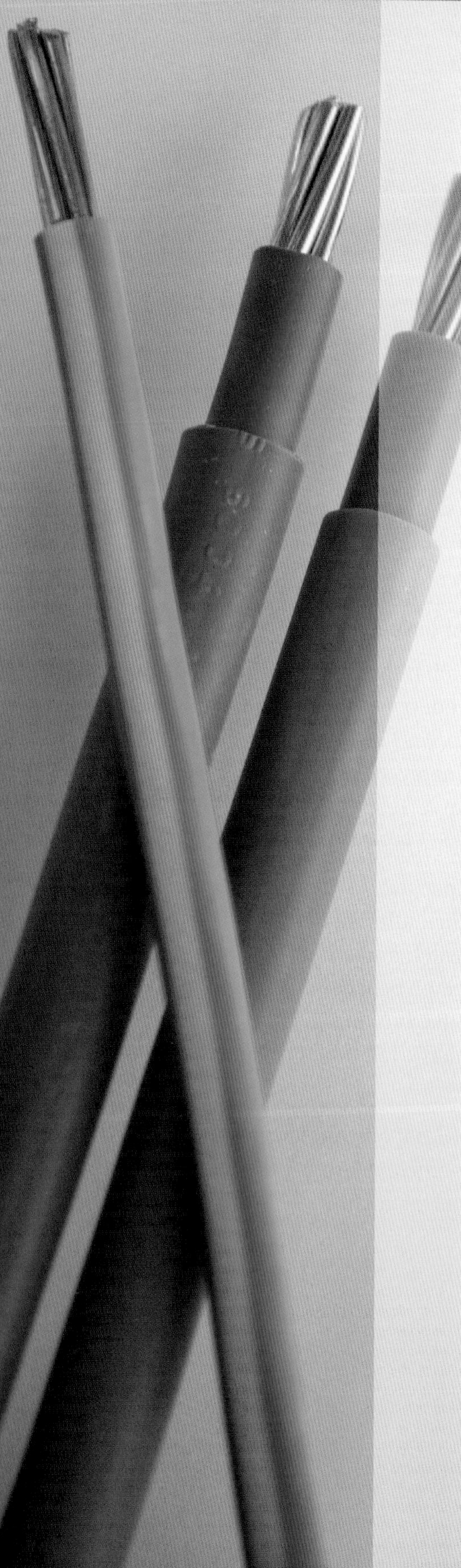

When you have completed these questions, check out your answers at the back of the book.

Note: More than one multiple choice answer may be correct.

1. An off-peak storage radiator when connected to a 230V supply was found to take a current of 12A. The resistance of the element will therefore be:
   a. 52.2 mΩ
   b. 19.2 Ω
   c. 2760 Ω
   d. 2843 Ω.

2. Calculate the resistance of 200 m of 1.5 mm^2 copper cable if the resistivity of copper is taken as 17.5×10^{-9} m.
   a. 1.3 Ω
   b. 2.3 Ω
   c. 233 Ω
   d. 1312 Ω.

3. Resistors of 3 Ω and 6 Ω are connected in series. Their combined resistor will therefore be:
   a. 0.5 Ω
   b. 2.0 Ω
   c. 9.0 Ω
   d. 18.0 Ω.

4. Resistors of 3 Ω and 6 Ω are connected in parallel. Their combined resistance will therefore be:
   a. 0.5 Ω
   b. 2.0 Ω
   c. 9.0 Ω
   d. 18.0 Ω.

5. The maximum value of the 230 V mains supply is:
   a. 207.2 V
   b. 230.0 V
   c. 325.3 V
   d. 400.0 V.

6. Calculate the reactance of a 150 μF capacitor connected to the 50 Hz mains supply:
   a. 3.14 Ω
   b. 21.2 Ω
   c. 18.8 Ω
   d. 471.3 Ω.

7. An electronic circuit resistor is colour coded green, blue, brown, gold. It has a value of:
    a. 56 Ω ± 10%
    b. 65 Ω ± 5%
    c. 560 Ω ± 5%
    d. 650 Ω ± 10%.

8. An electronic device which will allow current to flow through it in one direction only is a:
    a. light dependent resistor (LDR)
    b. light-emitting diode (LED)
    c. semiconductor diode
    d. thermistor.

9. An electronic device whose resistance varies with temperature is a:
    a. light dependent resistor (LDR)
    b. light-emitting diode (LED)
    c. semiconductor diode
    d. thermistor.

10. An electronic device which emits red, green, or yellow light when a current of about 10 mA flows through it is:
    a. light dependent resistor (LDR)
    b. light-emitting diode (LED)
    c. semiconductor diode
    d. thermistor.

11. An electronic device whose resistance changes as a result of light energy falling upon it is a:
    a. light dependent resistor (LDR)
    b. light-emitting diode (LED)
    c. semiconductor diode
    d. thermistor.

12. A street lamp has a luminous intensity of 2000 cd and is suspended 5 m above the ground. The illuminance on the pavement below the lamp will be:
    a. 40 lx
    b. 80 lx
    c. 400 lx
    d. 800 lx.

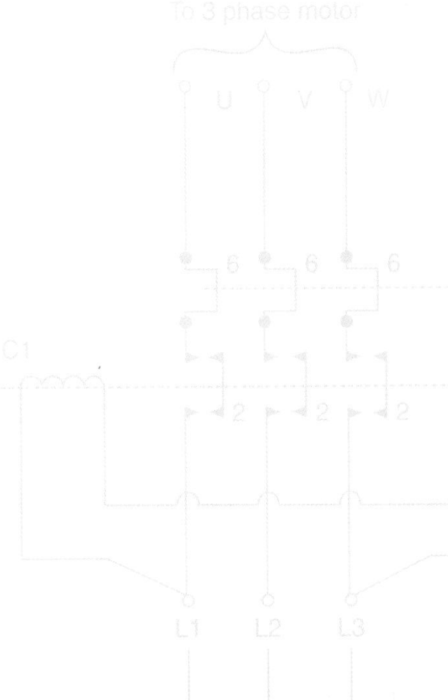

13. The method most frequently used when designing an interior lighting scheme is the:
    a. cosine law
    b. inverse square law
    c. lumen method
    d. reflective index method.
14. Identify the most energy efficient lamps in the following list:
    a. CFLs
    b. fluorescent tubes
    c. GLS lamps
    d. tungsten halogen spots.
15. Briefly state Ohm's law and the three formulae which can be derived from his Law.
16. Briefly describe the meaning of magnetic flux.
17. Sketch the magnetic field lines around a bar magnet, a horseshoe magnet and a current carrying conductor.
18. Briefly describe the meaning of inductance and mutual inductance and give one practical example of each.
19. Use a labelled sketch to describe the construction of a variable capacitor and a waxed paper capacitor. Give one practical example for the use of each.
20. An inductor and a resistor are connected in series to an a.c. supply. Sketch the circuit diagram, the phasor diagram and state the relationship between the circuit current and voltage.
21. A capacitor and resistor are connected in series to an a.c. supply. Sketch the circuit diagram, the phasor diagram and state the relationship between the circuit current and voltage.
22. Sketch and label the graphical symbols for some of the electronic components you have seen at work.

23. Use bullet points and maybe a sketch to describe how to work out the value of a resistor using the resistor colour code.
24. Use a sketch to describe the pin identification of a dual in line IC.
25. Sketch waveforms to describe how a domestic lighting dimmer switch works.
26. Define the meaning of rectification.
27. Sketch waveforms to describe the meaning of smoothing for a full-wave rectified circuit.
28. Sketch the circuit diagram for a battery charger circuit. Label the components and say what each one does.
29. Briefly describe the working principle of a GLS lamp. State one practical use for a GLS lamp. Why is the Government putting pressure on lamp manufacturers that will prevent GLS lamps being made after 2012?
30. Briefly describe how a fluorescent tube and CFLs work. Why are these the Government's preferred choice of lamp for the future?

CH 5

Electricity supply systems, protection and earthing

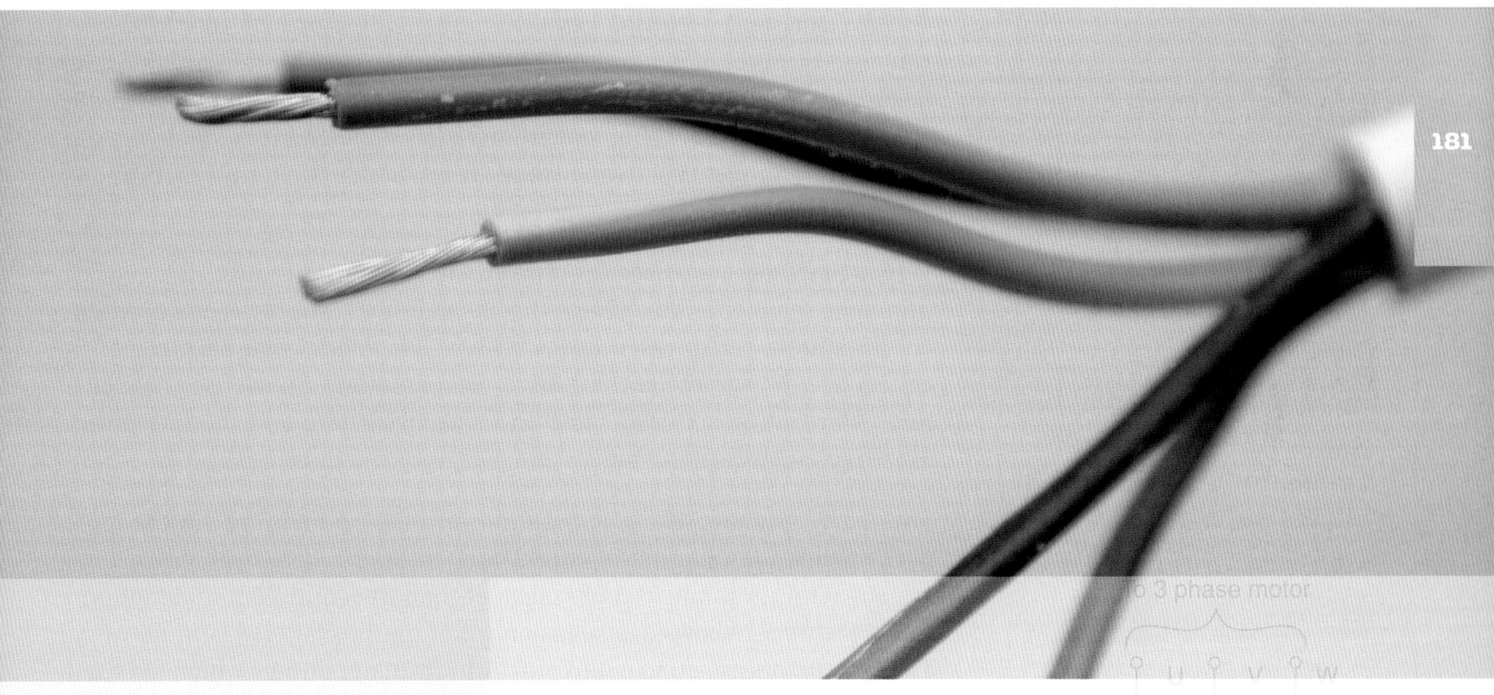

Unit 1 – Application of health and safety and electrical principles – Outcome 5

Underpinning knowledge: when you have completed this chapter you should be able to:

- describe electricity supply systems
- describe industrial distribution systems
- describe the operation of a transformer
- state transformer losses
- describe the need for earthing and protective systems

Electricity supply systems

The generation of electricity in most modern power stations is at 25 kV, and this voltage is then transformed to 400 kV for transmission. Virtually all the generators of electricity throughout the world are three-phase synchronous generators. The generator consists of a prime mover and a magnetic field excitor. The magnetic field is produced electrically by passing a direct current (d.c.) through a winding on an iron core, which rotates inside three-phase windings on the stator of the machine. The magnetic field is rotated by means of a prime mover which may be a steam turbine, water turbine, gas turbine or wind turbine.

FIGURE 5.1
Suspension tower.

The generators in modern power stations are rated between 500 and 1000 MW. A 2000 MW station might contain four 500 MW sets, three 660 MW sets and a 20 MW gas turbine generator or two 1000 MW sets. Having a number of generator sets in a single power station provides the flexibility required for seasonal variations in the load and for maintenance of equipment. When generators are connected to a single system they must rotate at exactly the same speed, hence the term synchronous generator.

Very high voltages are used for transmission systems because, as a general principle, the higher the voltage the cheaper is the supply. Since power in an a.c. system is expressed as $P = VI \cos \theta$, it follows that an increase in voltage will reduce the current for a given amount of power. A lower current will result in reduced cable and switchgear size and the line power losses, given by the equation $P = I^2R$, will also be reduced.

Key Fact

National Grid

The National Grid is a network of some 5000 miles of overhead and underground power cables.

The 132 kV grid and 400 kV supergrid transmission lines are, for the most part, steel-cored aluminium conductors suspended on steel lattice towers, since this is about 16 times cheaper than the equivalent underground cable. Figure 5.1 shows a suspension tower on the National Grid network. The conductors are attached to porcelain insulator strings which are fixed to the cross-members of the tower as shown in Fig. 5.2. Three conductors comprise a single circuit of a three-phase system so that towers with six arms carry two separate circuits.

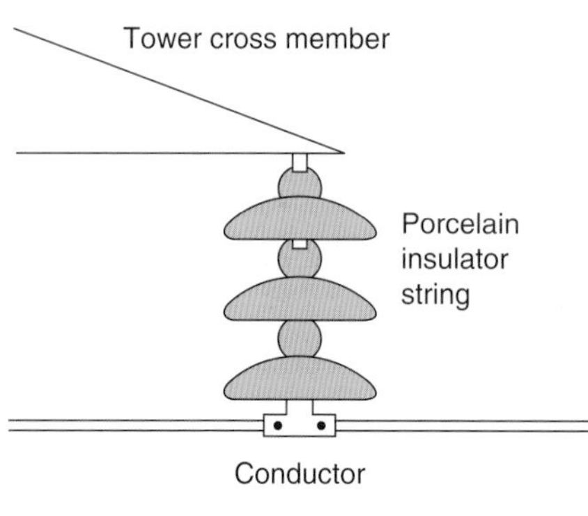

FIGURE 5.2
Steel lattice tower cable supports.

Primary distribution to consumers is from 11 kV substations, which for the most part are fed from 33 kV substations, but direct transformation between 132 and 11 kV is becoming common policy in city areas where over 100 MW can be economically distributed at 11 kV from one site. Figure 5.3 shows a block diagram indicating the voltages at the various stages of the transmission and distribution system and Fig. 5.4 shows a simplified diagram of the transmission and distribution of electricity to the consumer.

Distribution systems at 11 kV may be ring or radial systems but a ring system offers a greater security of supply. The maintenance of a secure supply is an important consideration for any electrical engineer or supply authority because electricity plays a vital part in an industrial society, and a loss of

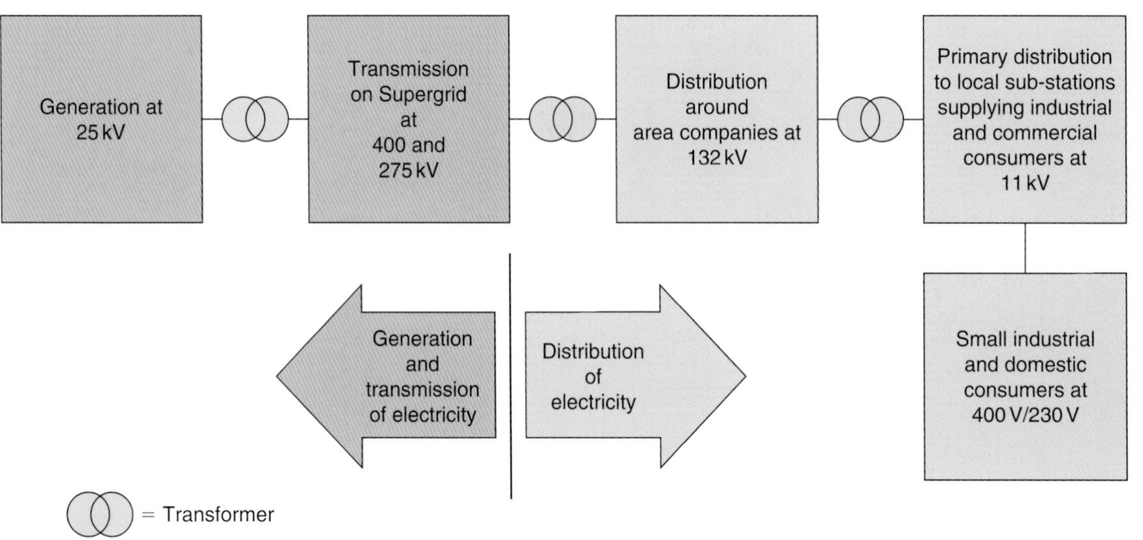

FIGURE 5.3
Generation, transmission and distribution of electrical energy.

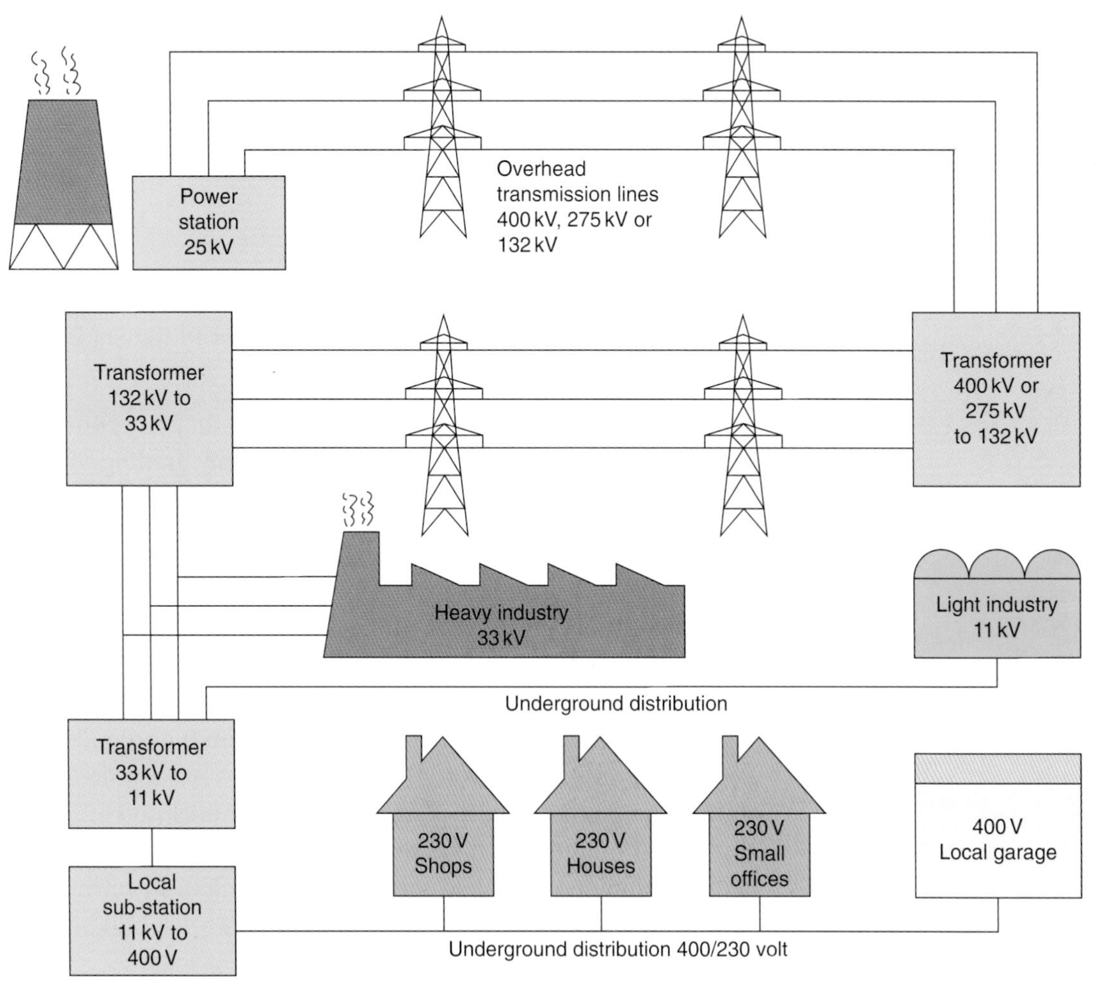

FIGURE 5.4
Simplified diagram of distribution of electricity from power station to consumer.

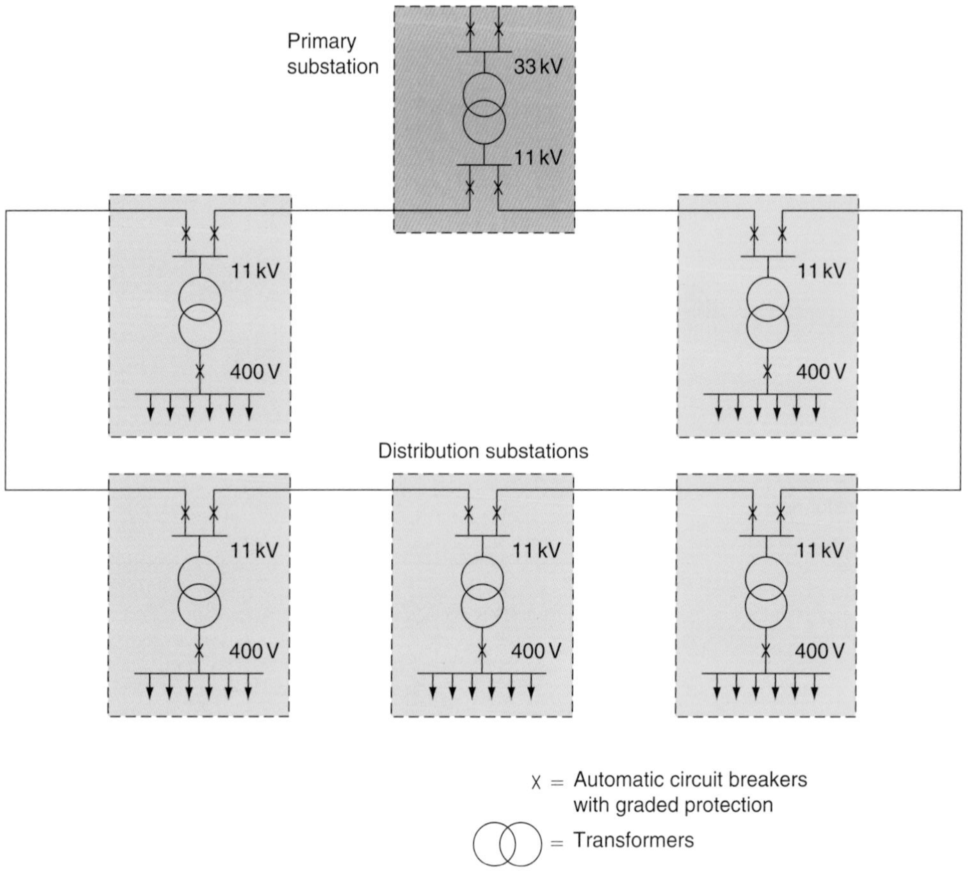

FIGURE 5.5
High-voltage ring main distribution.

supply may cause inconvenience, financial loss or danger to the consumer or the public.

The principle employed with a ring system is that any consumer's substation is fed from two directions, and by carefully grading the overload and cable protection equipment a fault can be disconnected without loss of supply to other consumers.

High-voltage distribution to primary substations is used by the electricity boards to supply small industrial, commercial and domestic consumers. This distribution method is also suitable for large industrial consumers where 11 kV substations, as shown in Fig. 5.5 may be strategically placed at load centres around the factory site. Regulation 9 of the Electricity Supply Regulations and Regulation 31 of the Factories Act require that these substations be protected by 2.44 m high fences or enclosed in some other way so that no unauthorized person may gain access to the potentially dangerous equipment required for 11 kV distribution. In towns and cities the substation equipment is usually enclosed in a brick building, as shown in Fig. 5.6.

The final connections to plant, distribution boards, commercial or domestic loads are usually by simple underground radial feeders at 400V/230V.

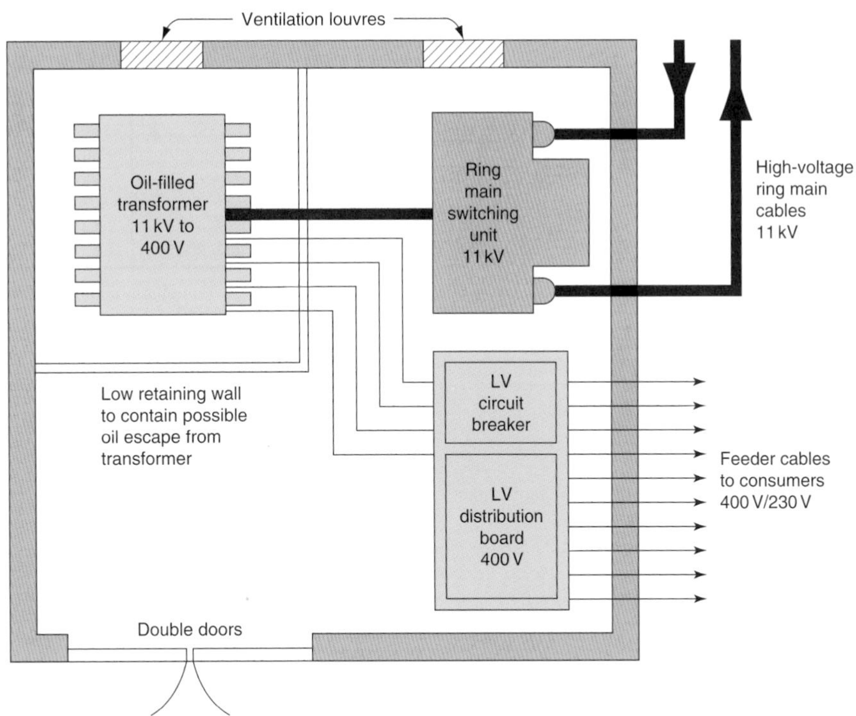

FIGURE 5.6
Typical sub-station layout.

These outgoing circuits are usually protected by circuit breakers in a distribution board.

The 400V/230V is derived from the 11 kV/400V sub-station transformer by connecting the secondary winding in star as shown in Fig. 5.7. The star point is earthed to an earth electrode sunk into the ground below the sub-station, and from this point is taken the fourth conductor, the neutral. Loads connected between phases are fed at 400V, and those fed between one phase and neutral at 230V. A three-phase 400V supply is used for supplying small industrial and commercial loads such as garages, schools and blocks of flats. A single-phase 230V supply is usually provided for individual domestic consumers.

Example

Use a suitable diagram to show how a 400 V three-phase, four-wire supply may be obtained from an 11 kV delta-connected transformer. Assuming that the three-phase four-wire supply feeds a small factory, show how the following loads must be connected:

(a) A three-phase 400 V motor.
(b) A single-phase 400 V welder.
(c) A lighting load made up of discharge lamps arranged in a way which reduces the stroboscopic effect.
(d) State why 'balancing' of loads is desirable.
(e) State the advantages of using a three-phase four-wire supply to industrial premises instead of a single-phase supply.

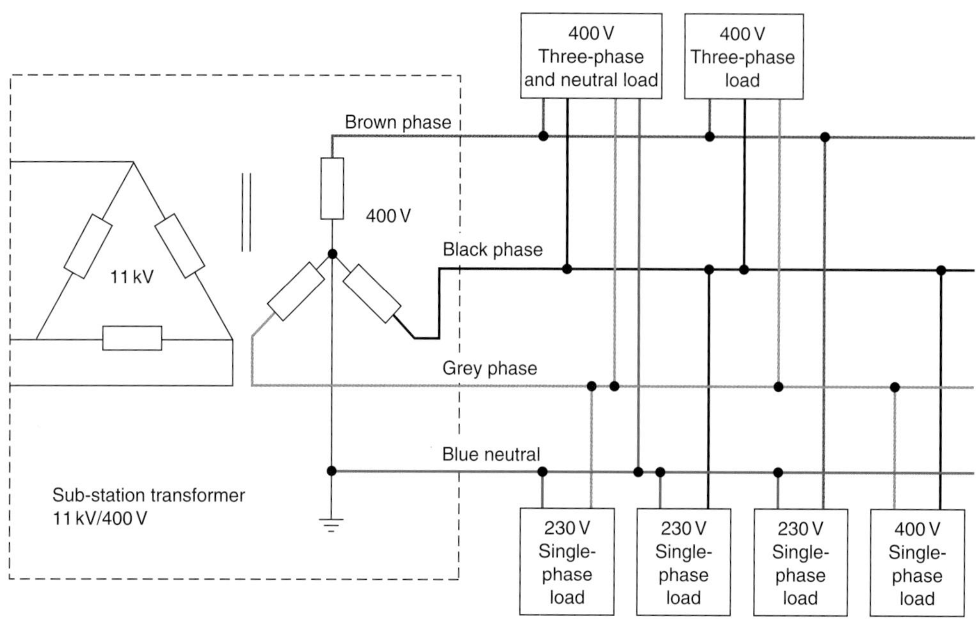

FIGURE 5.7
Three-phase four-wire distribution.

Three-phase load

Figure 5.7 shows the connections of the 11 kV to 400 V supply and the method of connecting a 400 V three-phase load such as a motor and a 400 V single-phase load such as a welder.

Reducing stroboscopic effect

The stroboscopic effect may be reduced by equally dividing the lighting load across the three phases of the supply. For example, if the lighting load were made up of 18 luminaires, then 6 luminaires should be connected to the brown phase and neutral, 6 to the black phase and neutral and 6 to the grey phase and neutral.

Balancing three-phase loads

A three-phase load such as a motor has equally balanced phases since the resistance of each phase winding will be the same. Therefore the current taken by each phase will be equal. When connecting single-phase loads to a three-phase supply, care should be taken to distribute the single-phase loads equally across the three phases so that each phase carries approximately the same current. Equally distributing the single-phase loads across the three-phase supply is known as 'balancing' the load. A lighting load of 18 luminaires would be 'balanced' if six luminaires were connected to each of the three phases.

Advantages of a three-phase four-wire supply

A three-phase four-wire supply gives a consumer the choice of a 400 V three-phase supply and a 230 V single-phase supply. Many industrial

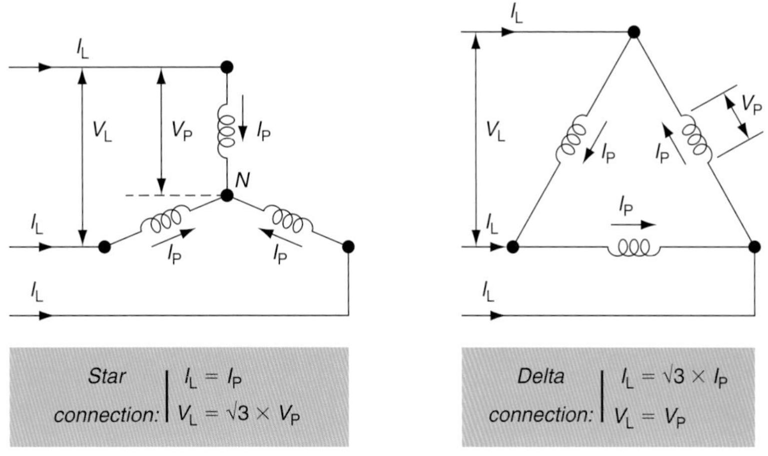

FIGURE 5.8
Star and delta connections.

Key Fact

Balancing single-phase loads

Connect single-phase loads equally across a three-phase supply, so that each phase carries approximately the same current.

loads such as motors require a three-phase 400 V supply, while the lighting load in a factory, as in a house, will be 230 V. Industrial loads usually demand more power than a domestic load, and more power can be supplied by a 400 V three-phase supply than is possible with a 230 V single-phase supply for a given size of cable since power = $VI \cos \theta$ (watts).

THREE-PHASE A.C.

A three-phase voltage is generated in exactly the same way as a single-phase a.c. voltage. For a three-phase voltage three separate windings, each separated by 120°, are rotated in a magnetic field. The generated voltage will be three identical sinusoidal waveforms each separated by 120°.

Star and delta connections

The three-phase windings may be star connected or delta connected as shown in Fig. 5.8. The important relationship between phase and line currents and voltages is also shown. The square root of 3 ($\sqrt{3}$) is simply a constant for three-phase circuits, and has a value of 1.732. The delta connection is used for electrical power transmission because only three conductors are required. Delta connection is also used to connect the windings of most three-phase motors because the phase windings are perfectly balanced and, therefore, do not require a neutral connection.

Making a star connection has the advantage that two voltages become available – a line voltage between any two phases, and a phase voltage between line and neutral which is connected to the star point.

In any star-connected system currents flow along the lines (I_L), through the load and return by the neutral conductor connected to the star point. In a *balanced* three-phase system all currents have the same value and when they are added up by phasor addition, we find the resultant current is zero. Therefore, no current flows in the neutral and the star point is at zero volts. The star point of the distribution transformer is earthed because earth is also at zero potential. A star-connected system is also called a three-phase four-wire system and allows us to connect single-phase loads to a three-phase system.

Three-phase power

We know from our single-phase theory in Chapter 4 that power can be found from the following formula:

$$\text{Power} = VI \cos \phi \text{ (W)}$$

In any balanced three-phase system, the total power is equal to three times the power in any one phase.

$$\therefore \text{ Total three-phase power} = 3\, V_P I_P \cos \phi \text{ (W)} \qquad (5.1)$$

Now for a star connection,

$$V_P = V_L/\sqrt{3} \quad \text{and} \quad I_L = I_P \qquad (5.2)$$

Substituting Equation (5.2) into Equation (5.1), we have

$$\text{Total three-phase power} = \sqrt{3}\, V_L L_L \cos \phi \text{ (W)}$$

Now consider a delta connection:

$$V_P = V_L \quad \text{and} \quad I_P = I_L/\sqrt{3} \qquad (5.3)$$

Substituting Equation (5.3) into Equation (5.1) we have, for any balanced three-phase load,

$$\text{Total three-phase power} = \sqrt{3}\, V_L L_L \cos \phi \text{ (W)}$$

Example 1

A balanced star-connected three-phase load of 10 Ω per phase is supplied from a 400 V 50 Hz mains supply at unity power factor. Calculate (a) the phase voltage, (b) the line current and (c) the total power consumed.

For a star connection,

$$V_L = \sqrt{3}\,V_P \text{ and } I_L = I_P$$

For (a),

$$V_P = V_L/\sqrt{3}\text{ (V)}$$

$$V_P = \frac{400\text{ V}}{1.732} = 230.9\text{ V}$$

For (b),

$$I_L = I_P = V_P/R_P\text{ (A)}$$

$$I_L = I_P = \frac{230.9\text{ V}}{10\ \Omega} = 23.09\text{ A}$$

For (c),

$$\text{Power} = \sqrt{3}\ V_L\ I_L \cos\phi\text{ (W)}$$

$$\therefore\ \text{Power} = 1.732 \times 400\text{ V} \times 23.09\text{ A} \times 1 = 16\text{ kW}$$

Example 2

A 20 kW 400 V balanced delta-connected load has a power factor of 0.8. Calculate (a) the line current and (b) the phase current.

We have that:

$$\text{Three-phase power} = \sqrt{3}\ V_L\ I_L \cos\phi\text{ (W)}$$

For (a),

$$I_L = \frac{\text{Power}}{\sqrt{3}\ V_L \cos\phi}\text{ (A)}$$

$$\therefore\ I_L = \frac{20\,000\text{ W}}{1.732 \times 400\text{ V} \times 0.8}$$

$$I_L = 36.08\text{ (A)}$$

For delta connection,

$$I_L = \sqrt{3}\ I_P\text{ (A)}$$

Thus, for (b),

$$I_P = \frac{I_L}{\sqrt{3}}\text{ (A)}$$

$$\therefore\ I_P = \frac{36.08\text{ A}}{1.732} = 20.83\text{ A}$$

Example 3

Three identical loads each having a resistance of 30 Ω and inductive reactance of 40 Ω are connected first in star and then in delta to a 400 V three-phase supply. Calculate the phase currents and line currents for each connection.

For each load,

$$Z = \sqrt{R^2 + X_L^2}\ (\Omega)\ \text{(from Chapter 4)}$$

$$\therefore Z = \sqrt{30^2 + 40^2}$$

$$Z = \sqrt{2500}$$

$$Z = 50\ \Omega$$

For star connection,

$$V_L = \sqrt{3}\ V_P \quad \text{and} \quad I_L = I_P$$

$$V_P = V_L/\sqrt{3}\ \text{(V)}$$

$$\therefore V_P = \frac{400\ \text{V}}{1.732} = 230.9\ \text{V}$$

$$I_P = V_P/Z_P\ \text{(A)}$$

$$\therefore I_P = \frac{230.9\ \text{V}}{50\ \Omega} = 4.62\ \text{A}$$

$$I_P = I_L$$

therefore phase and line currents are both equal to 4.62 A.

For delta, connection,

$$V_L = V_P \quad \text{and} \quad I_L = \sqrt{3}\ I_P$$

$$V_L = V_P = 400\ \text{V}$$

$$I_P = V_P/Z_P\ \text{(A)}$$

$$\therefore I = \frac{400\ \text{V}}{50\ \Omega} = 8\ \text{A}$$

$$I_L = \sqrt{3}\ I_P\ \text{(A)}$$

$$\therefore I_L = 1.732 \times 8\ \text{A} = 13.86\ \text{A}$$

INDUSTRIAL DISTRIBUTION SYSTEMS

In domestic installations the final circuits for lights, sockets, cookers, immersion heating, etc. are connected to separate fuseways in the consumer's unit mounted at the service position.

In commercial or industrial installations a three-phase 400 V supply must be distributed to appropriate equipment in addition to supplying single-phase 230 V loads such as lighting. It is now common practice to establish industrial estates speculatively, with the intention of encouraging local industry to use individual units. This presents the electrical contractor with an additional problem. The use and electrical demand of a single industrial unit are

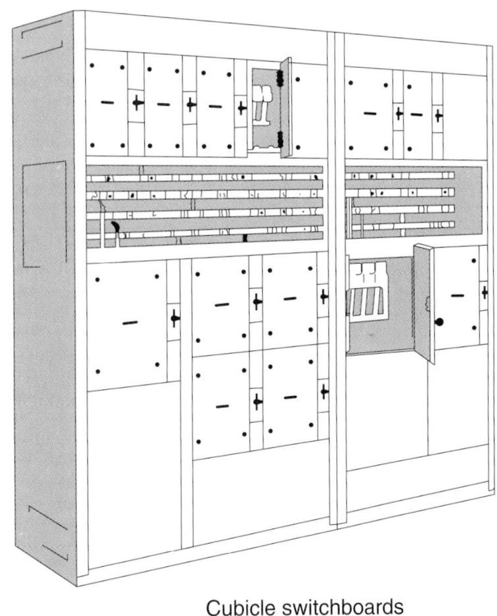

FIGURE 5.9
Industrial consumer's service position equipment.

often unknown and the electrical supply equipment will need to be flexible in order to meet a changing demand due to expansion or change of use.

Busbar chambers incorporated into cubicle switchboards or on-site assemblies of switchboards are to be found at the incoming service position of commercial and industrial consumers, since this has proved to provide the flexibility required by these consumers. This is shown in Fig. 5.9.

Distribution fuse boards, which may incorporate circuit breakers, are wired by sub-main cables from the service position to load centres in other parts of the building, thereby keeping the length of cable to the final circuit as short as possible. This is shown in Fig. 5.10.

When high-rise buildings such as multi-storey flats have to be wired, it is usual to provide a three-phase four-wire rising main. This may comprise vertical busbars running from top to bottom at some central point in the building. Each floor or individual flat is then connected to the busbar to provide the consumer's supply. When individual dwellings receive a single-phase supply the electrical contractor must balance the load across the three phases. Fig. 5.11 shows a rising main system. The rising main must incorporate fire barriers to prevent the spread of fire throughout the building (Regulations 527.1.2, 527.2.1 and 527.2.4).

Industrial wiring systems are constructed robustly so that they can withstand some minor mechanical damage and vibration, and have the adaptability to respond to the changing needs of an industrial environment. IEE Regulations 522.6 and 522.8.

Cables in trunking with conduit drops or, SWA or Mineral Insulated (MI) cables laid on cable tray, provide a flexible, adaptable electrical installation.

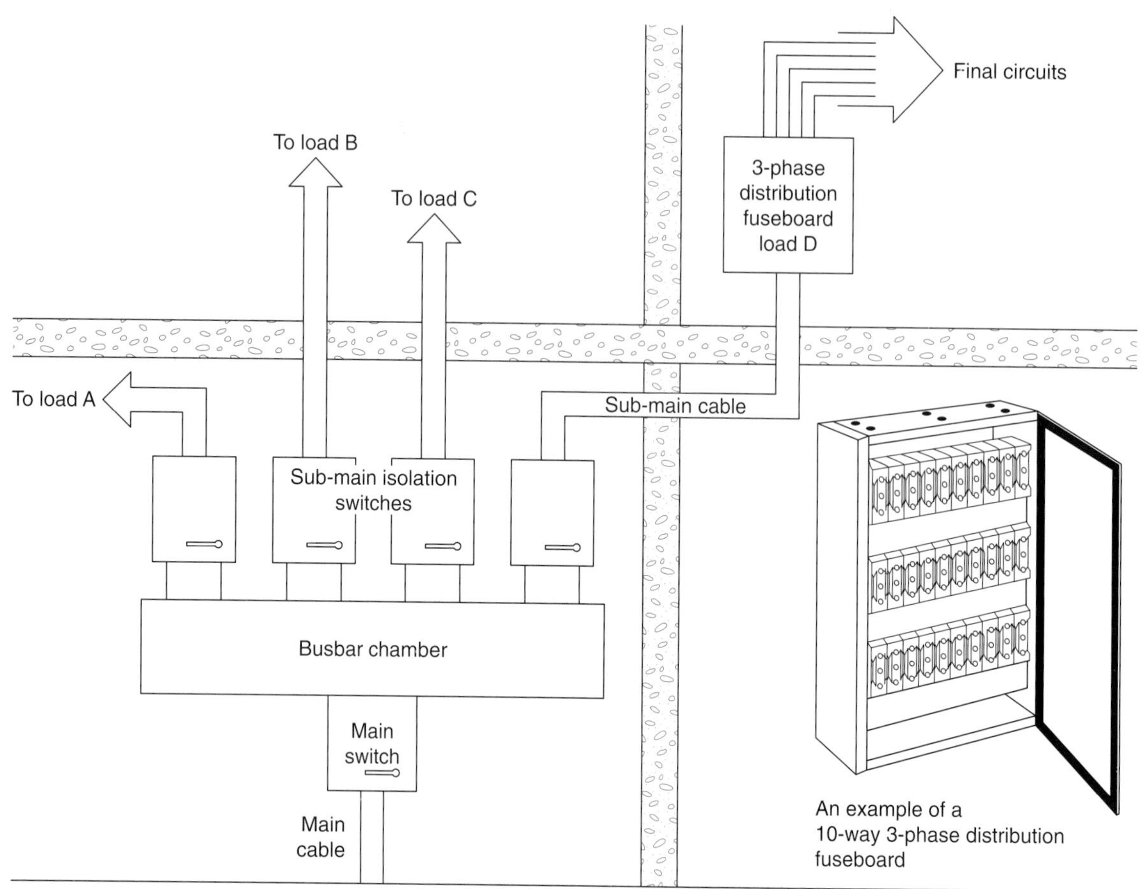

FIGURE 5.10
Typical distribution in commercial or industrial building.

Definition

Most cables can be considered to be constructed in three parts: the *conductor*, which must be of a suitable cross-section to carry the load current; the *insulation*, which has a colour or number code for identification and the *outer sheath*, which may contain some means of providing protection from mechanical damage.

Compare this flexible, adaptable type of installation to the less easily adaptable fixed wiring of domestic installations where cables are buried in the finishing plaster of the walls.

Cables

Most cables can be considered to be constructed in three parts: the **conductor**, which must be of a suitable cross-section to carry the load current; the **insulation**, which has a colour or number code for identification and the **outer sheath**, which may contain some means of providing protection from mechanical damage.

The conductors of a cable are made of either copper or aluminium and may be stranded or solid. Solid conductors are only used in fixed wiring installations and may be shaped in larger cables. Stranded conductors are more flexible and conductor sizes from 4.0 to 25 mm^2 contain seven strands. A 10 mm^2 conductor, for example, has seven 1.35 mm diameter strands which collectively make up the 10 mm^2 cross-sectional area of the cable. Conductors above 25 mm^2 have more than seven strands, depending upon the size of the cable. Flexible cords have multiple strands of very

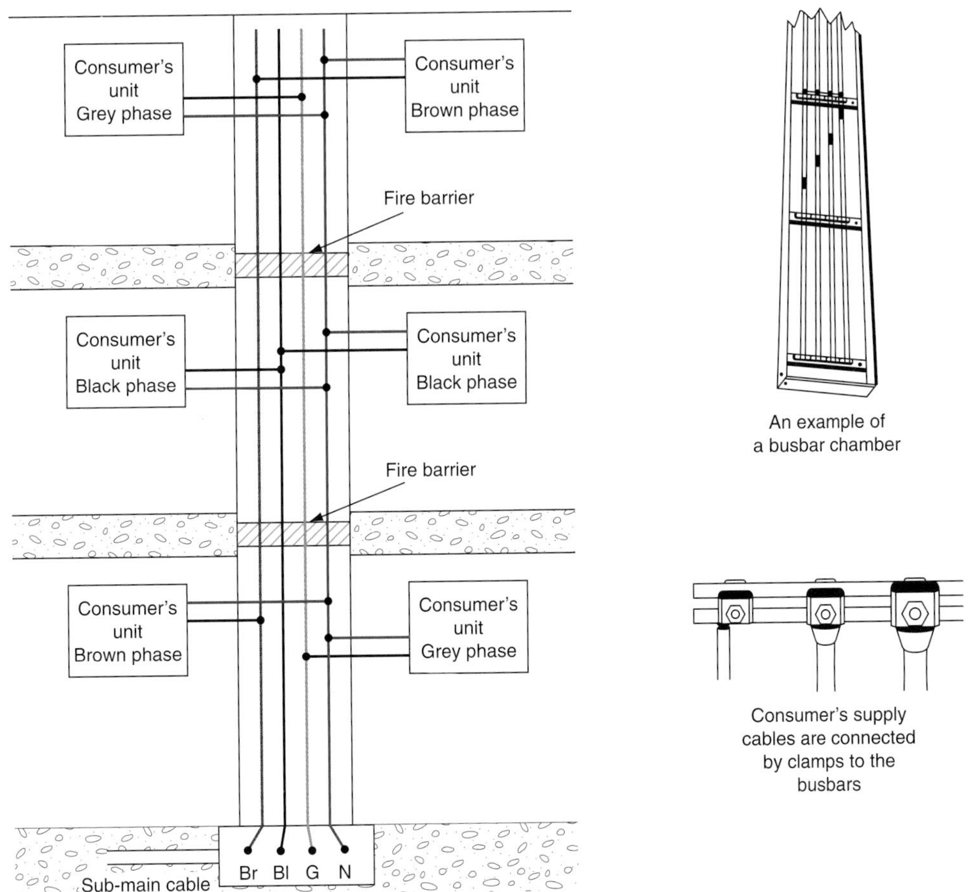

FIGURE 5.11
Busbar rising main system.

fine wire, as fine as one strand of human hair. This gives the cable its very flexible quality.

NEW WIRING COLOURS

Twenty-eight years ago the United Kingdom agreed to adopt the European colour code for flexible cords, that is, brown for live or line conductor, blue for the neutral conductor and green combined with yellow for earth conductors. However, no similar harmonization was proposed for non-flexible cables used for fixed wiring. These were to remain as red for live or line conductor, black for the neutral conductor and green combined with yellow for earth conductors.

On the 31 March 2004 the IEE published Amendment No. 2 to BS 7671: 2001 which specified new cable core colours for all fixed wiring in UK electrical installations. These new core colours will 'harmonize' the United Kingdom with the practice in mainland Europe.

FIXED CABLE CORE COLOURS UP TO 2006

Single-phase supplies – red line conductors, black neutral conductors and green combined with yellow for earth conductors.

Three-phase supplies – red, yellow and blue line conductors, black neutral conductors and green combined with yellow for earth conductors.

These core colours must *not* be used after 31 March 2006.

NEW (HARMONIZED) FIXED CABLE CORE COLOURS

Single-phase supplies – brown line conductors, blue neutral conductors and green combined with yellow for earth conductors (just like the existing flexible cords).

Three-phase supplies – brown, black and grey line conductors, blue neutral conductors and green combined with yellow for earth conductors.

These are the cable core colours to be used from 31 March 2004 onwards.

Extensions or alterations to existing ***single-phase*** installations do not require marking at the interface between the old and new fixed wiring colours. However, a warning notice must be fixed at the consumer unit or distribution fuse board which states:

> *Caution* – This installation has wiring colours to two versions of BS 7671. Great care should be taken before undertaking extensions, alterations or repair that all conductors are correctly identified.

Alterations to ***three-phase*** installations must be marked at the interface L1, L2, L3 for the lines and N for the neutral. Both new and old cables must be marked. These markings are preferred to coloured tape and a caution notice is again required at the distribution board (see Appendix 7 of the IEE Regulations).

PVC INSULATED AND SHEATHED CABLES

Domestic and commercial installations use this cable, which may be clipped direct to a surface, sunk in plaster or installed in conduit or trunking. It is the simplest and least expensive cable. Figure 5.12 shows a sketch of a twin and earth cable.

The conductors are covered with a colour-coded PVC insulation and then contained singly or with others in a PVC outer sheath.

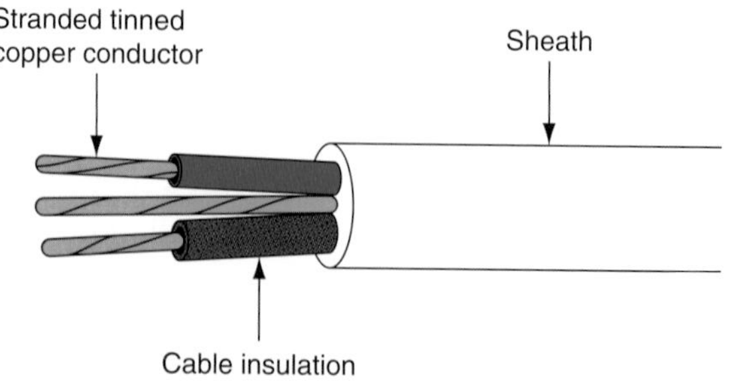

FIGURE 5.12
A twin and earth PVC insulated and sheathed cable.

Definition

PVC insulated steel wire armour cables are used for wiring underground between buildings, for main supplies to dwellings, rising sub-mains and industrial installations. They are used where some mechanical protection of the cable conductors is required.

Definition

An *MI cable* has a seamless copper sheath which makes it waterproof and fire and corrosion-resistant. These characteristics often make it the only cable choice for hazardous or high-temperature installations.

Safety first

PVC cables

- At low temperatures PVC cable insulation can become brittle when handled.
- PVC cables must not be installed when the ambient temperature is 0°C (IEE Reg 522.1.2).

PVC/SWA CABLE

PVC insulated steel wire armour cables are used for wiring underground between buildings, for main supplies to dwellings, rising sub-mains and industrial installations. They are used where some mechanical protection of the cable conductors is required.

The conductors are covered with colour-coded PVC insulation and then contained either singly or with others in a PVC sheath (see Fig. 5.13). Around this sheath is placed an armour protection of steel wires twisted along the length of the cable, and a final PVC sheath covering the steel wires protects them from corrosion. The armour sheath also provides the circuit protective conductor (CPC) and the cable is simply terminated using a compression gland.

MI CABLE

An **MI cable** has a seamless copper sheath which makes it waterproof and fire and corrosion-resistant. These characteristics often make it the only cable choice for hazardous or high-temperature installations such as oil refineries and chemical works, boiler-houses and furnaces, petrol pump and fire alarm installations.

The cable has a small overall diameter when compared to alternative cables and may be supplied as bare copper or with a PVC oversheath. It is colour-coded orange for general electrical wiring, white for emergency lighting or red for fire alarm wiring. The copper outer sheath provides the CPC, and the cable is terminated with a pot and sealed with compound and a compression gland (see Fig. 5.14).

The copper conductors are embedded in a white powder, magnesium oxide, which is non-ageing and non-combustible, but which is hygroscopic, which means that it readily absorbs moisture from the surrounding air, unless adequately terminated. The termination of an MI cable is a

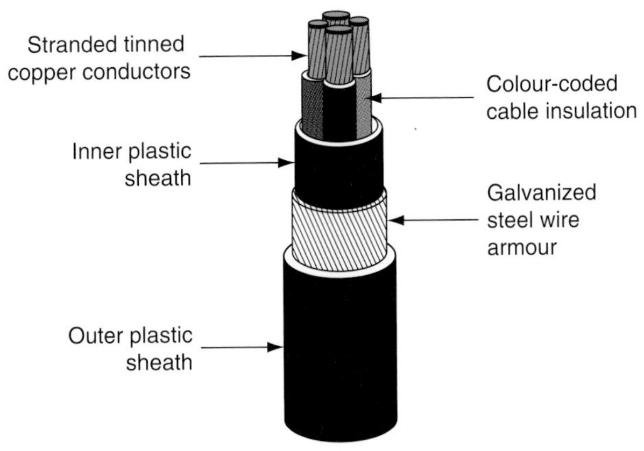

FIGURE 5.13
A four-core PVC/SWA cable.

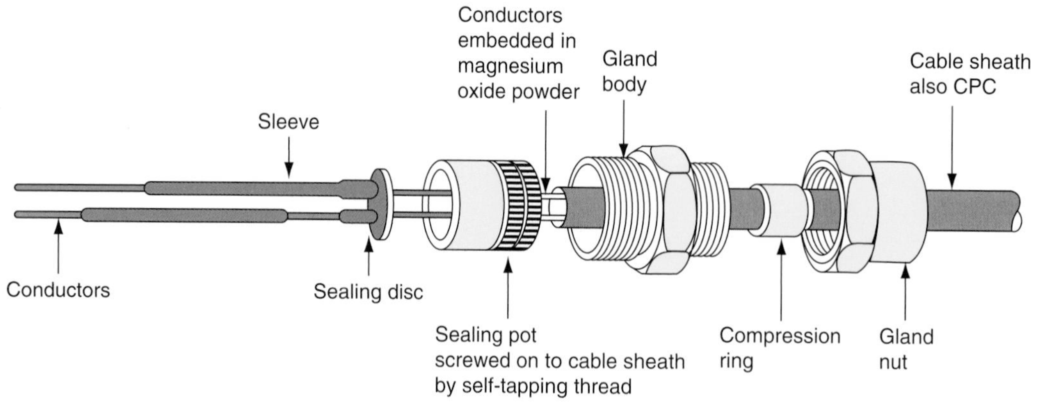

FIGURE 5.14
MI cable with terminating seal and gland.

complicated process requiring the electrician to demonstrate a high level of practical skill and expertise for the termination to be successful.

FP 200 CABLE

FP 200 cable is similar in appearance to an MI cable in that it is a circular tube, or the shape of a pencil, and is available with a red or white sheath. However, it is much simpler to use and terminate than an MI cable.

The cable is available with either solid or stranded conductors that are insulated with 'insudite' a fire resistant insulation material. The conductors are then screened by wrapping an aluminium tape around the insulated conductors, that is, between the insulated conductors and the outer sheath. This aluminium tape screen is applied metal side down and in contact with the bare CPC.

The sheath is circular and made of a robust thermoplastic low smoke, zero halogen material.

Definition

FP 200 cable is similar in appearance to an MI cable in that it is a circular tube, or the shape of a pencil, and is available with a red or white sheath. However, it is much simpler to use and terminate than an MI cable.

FP 200 is available in 2, 3, 4, 7, 12 and 19 cores with a conductor size range from 1.0mm to 4.0mm. The core colours are: two core, brown and blue; three core, brown, black and grey and four core, black, red, yellow and blue.

The cable is as easy to use as a PVC insulated and sheathed cable. No special terminations are required, the cable may be terminated through a grommet into a knock out box or terminated through a simple compression gland.

The cable is a fire resistant cable, primarily intended for use in fire alarms and emergency lighting installations or it may be embedded in plaster.

Definition

Low smoke and fume cables give off very low smoke and fumes if they are burned in a burning building. Most standard cable types are available as LSF cables.

LSF CABLES

Low smoke and fume (LSF) cables give off very low smoke and fumes if they are burned in a burning building. Most standard cable types are available as LSF cables.

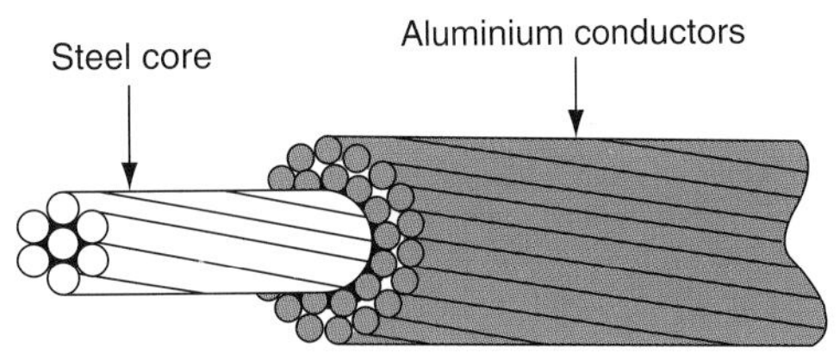

FIGURE 5.15
132 kV overhead cable construction.

HIGH-VOLTAGE POWER CABLES

The cables used for high-voltage power distribution require termination and installation expertise beyond the normal experience of a contracting electrician. The regulations covering high-voltage distribution are beyond the scope of the IEE Regulations for electrical installations. Operating at voltages in excess of 33 kV and delivering thousands of kilowatts, these cables are either suspended out of reach on pylons or buried in the ground in carefully constructed trenches.

HIGH-VOLTAGE OVERHEAD CABLES

Suspended from cable towers or pylons, overhead cables must be light, flexible and strong.

The cable is constructed of stranded aluminium conductors formed around a core of steel stranded conductors (see Fig. 5.15). The aluminium conductors carry the current and the steel core provides the tensile strength required to suspend the cable between pylons. The cable is not insulated since it is placed out of reach and insulation would only add to the weight of the cable.

HIGH-VOLTAGE UNDERGROUND CABLES – PILCSWA

Paper insulated lead covered steel wire armour (PILCSWA) cables are only used in systems above 11 kV. Very high-voltage cables are only buried underground in special circumstances when overhead cables would be unsuitable, for example, because they might spoil a view of natural beauty. Underground cables are very expensive because they are much more complicated to manufacture than overhead cables. In transporting vast quantities of power, heat is generated within the cable. This heat is removed by passing oil through the cable to expansion points, where the oil is cooled. The system is similar to the water cooling of an internal combustion engine. Figure 5.16 shows a typical high-voltage cable construction.

Definition

Paper insulated lead covered steel wire armour cables are only used in systems above 11 kV. Very high-voltage cables are only buried underground in special circumstances when overhead cables would be unsuitable, for example, because they might spoil a view of natural beauty.

The conductors may be aluminium or copper, solid or stranded. They are insulated with oil-impregnated brown paper wrapped in layers around the conductors. The oil ducts allow the oil to flow through the cable, removing excess heat. The whole cable within the lead sheath is saturated with oil, which is a good insulator. The lead sheath keeps the oil in and moisture

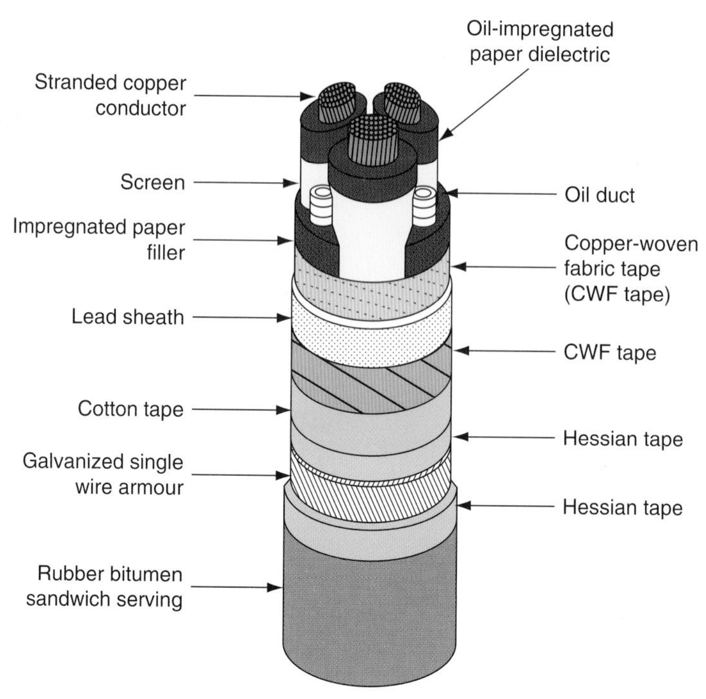

FIGURE 5.16
132 kV underground cable construction.

out of the cable, and this is supported by the copper-woven fabric tape. The cable is protected by steel wire armouring, which has bitumen or PVC serving over it to protect the armour sheath from corrosion. The termination and installation of these cables is a very specialized job, undertaken by the supply authorities only.

Installing cables

The final choice of a wiring system must rest with those designing the installation and those ordering the work, but whatever system is employed, good workmanship by competent persons and the use of proper materials is essential for compliance with the IEE Regulation 134.1.1. The necessary skills can be acquired by an electrical trainee who has the correct attitude and dedication to his craft.

PVC insulated and sheathed wiring systems are used extensively for lighting and socket installations in domestic dwellings. Mechanical damage to the cable caused by impact, abrasion, penetration, compression or tension must be minimized during installation (Regulation 522.6.1). The cables are generally fixed using plastic clips incorporating a masonry nail, which means the cables can be fixed to wood, plaster or brick with almost equal ease. Cables should be run horizontally or vertically, not diagonally, down a wall. All kinks should be removed so that the cable is run straight and neatly between clips fixed at equal distances providing adequate support for the cable so that it does not become damaged by its own weight (Regulation

522.8.4 and Table 4A of the *On Site Guide*). Where cables are bent, the radius of the bend should not cause the conductors to be damaged (Regulation 522.8.3 and Table 4E of the *On Site Guide*).

Terminations or joints in the cable may be made in ceiling roses, junction boxes, or behind sockets or switches, provided that they are enclosed in a non-ignitable material, are properly insulated and are mechanically and electrically secure (IEE Regulation 526). All joints must be accessible for inspection testing and maintenance when the installation is completed (IEE Regulation 526.3).

Try This

Definition

What do we mean by a 'competent person'?

Where PVC insulated and sheathed cables are concealed in walls, floors or partitions, they must be provided with a box incorporating an earth terminal at each outlet position. PVC cables do not react chemically with plaster, as do some cables, and consequently PVC cables may be buried under plaster. Further protection by channel or conduit is only necessary outside of designated zones if mechanical protection from nails or screws is required, or to protect them from the plasterer's trowel. However, Regulation 522.6.6 now tells us that where PVC cables are to be embedded in a wall or partition at a depth of less than 50 mm they should be run along one of the permitted routes. To identify the most probable cable routes, Regulation 522.6.6 tells us that outside a zone formed by a 150 mm border all around a wall edge, cables can only be run horizontally or vertically to a point or accessory unless they are contained in a substantial earthed enclosure, such as a conduit, which can withstand nail penetration. This is shown in Fig. 14.22 in Chapter 14 of *Basic Electrical Installation Work*, 5th edition.

Where the accessory or cable is fixed to a wall which is less than 100 mm thick, protection must also be extended to the reverse side of the wall if a position can be determined.

Where none of this protection can be complied with and the installation is to be used by ordinary people, then the cable must be given additional protection with a 30 mA residual current device (RCD) (IEE Regulation 522.6.7).

Try This

Definitions

What do we mean by an 'ordinary person'. Perhaps you could write a definition in the margin.

Key Fact

Conduit

Burrs must be removed from the cut ends of conduit so that the cable sheath will not become damaged when drawn into the conduit (IEE Regulation 522.8.1).

Where cables and wiring systems pass through walls, floors and ceilings the hole should be made good with incombustible material such as mortar or plaster to prevent the spread of fire (Regulation 527.2.1). Cables passing through metal boxes should be bushed with a rubber grommet to prevent abrasion of the cable. Holes drilled in floor joists through which cables are run should be 50 mm below the top or 50 mm above the bottom of the joist to prevent damage to the cable by nail penetration (Regulation 522.6.5). PVC cables should not be installed when the surrounding temperature is below 0°C or when the cable temperature has been below 0°C for the previous 24 hours because the insulation becomes brittle at low temperatures and may be damaged during installation (Regulation 522.1.2).

CONDUIT INSTALLATIONS

Definition

A *conduit* is a tube, channel or pipe in which insulated conductors are contained.

A **conduit** is a tube, channel or pipe in which insulated conductors are contained. The conduit, in effect, replaces the PVC outer sheath of a cable, providing mechanical protection for the insulated conductors. A conduit installation can be rewired easily or altered at any time, and this flexibility, coupled with mechanical protection, makes conduit installations popular for commercial and industrial applications. There are three types of conduit used in electrical installation work: steel, PVC and flexible.

Steel conduit

Steel conduits are made to a specification defined by BS 4568 and are either heavy gauge welded or solid drawn. Heavy gauge is made from a sheet of steel welded along the seam to form a tube and is used for most electrical installation work. Solid drawn conduit is a seamless tube which is much more expensive and only used for special gas-tight, explosion-proof or flame-proof installations.

Conduit is supplied in 3.75 m lengths and typical sizes are 16, 20, 25 and 32 mm. Conduit tubing and fittings are supplied in a black enamel finish for internal use or hot galvanized finish for use on external or damp installations. A wide range of fittings is available and the conduit is fixed using saddles or pipe hooks, as shown in Fig. 5.17.

Metal conduits are threaded with stocks and dies and bent using special bending machines. The metal conduit is also utilized as the CPC and, therefore, all connections must be screwed up tightly and all burrs removed so that cables will not be damaged as they are drawn into the conduit. Metal conduits containing a.c. circuits must contain phase and neutral conductors in the same conduit to prevent eddy currents flowing, which would result in the metal conduit becoming hot (Regulations 521.5.2, 522.8.1 and 522.8.11).

PVC conduit

PVC conduit used on typical electrical installations is heavy gauge standard impact tube manufactured to BS 4607. The conduit size and range of fittings are the same as those available for metal conduit. PVC conduit is most often joined by placing the end of the conduit into the appropriate fitting and fixing with a PVC solvent adhesive. PVC conduit can be bent by hand using a

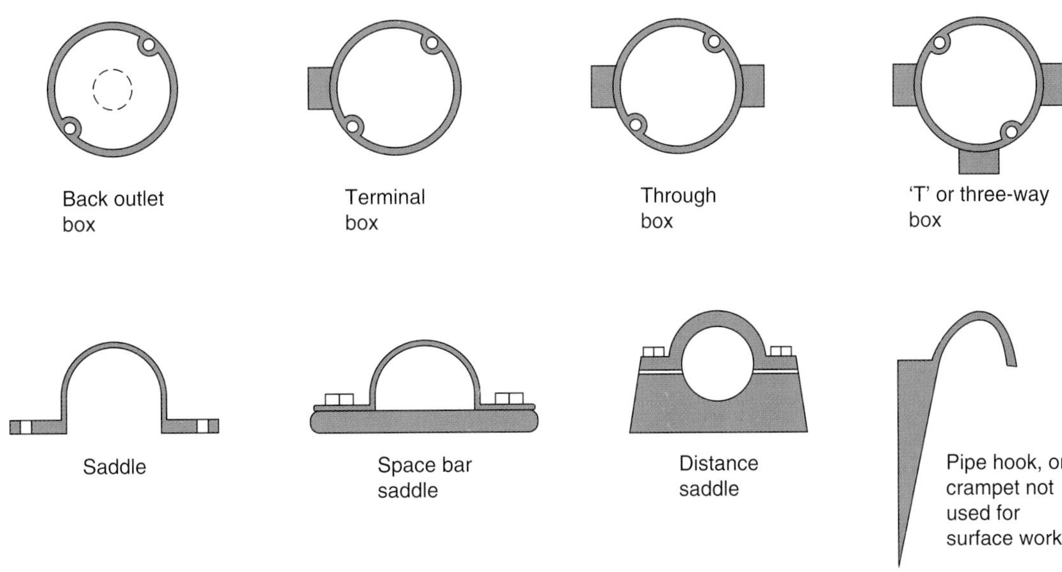

FIGURE 5.17
Conduit fittings and saddles.

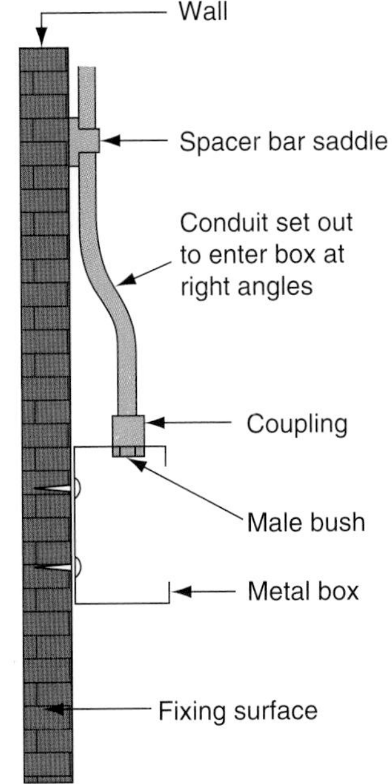

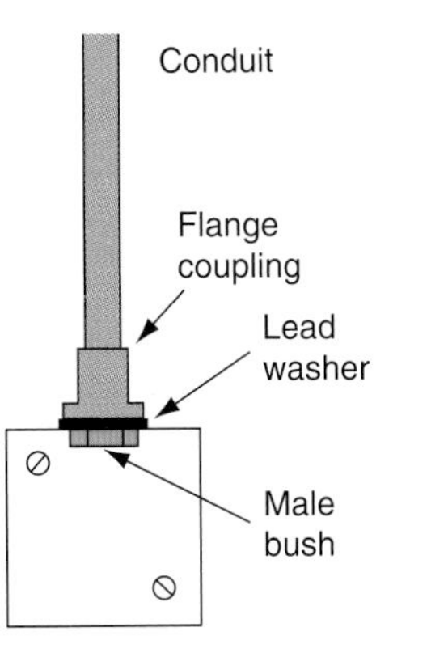

FIGURE 5.18
Terminating conduits.

bending spring of the same diameter as the inside of the conduit. The spring is pushed into the conduit to the point of the intended bend and the conduit then bent over the knee. The spring ensures that the conduit keeps its circular shape. In cold weather, a little warmth applied to the point of the intended bend often helps to achieve a more successful bend.

The advantages of a PVC conduit system are that it may be installed much more quickly than steel conduit and is non-corrosive, but it does not have the mechanical strength of steel conduit. Since PVC conduit is an insulator it cannot be used as the CPC and a separate earth conductor must be run to every outlet. It is not suitable for installations subjected to temperatures below 25°C or above 60°C. Where luminaires are suspended from PVC conduit boxes, precautions must be taken to ensure that the lamp does not raise the box temperature or that the mass of the luminaire supported by each box does not exceed the maximum recommended by the manufacturer (IEE Regulations 522.1 and 522.2). PVC conduit also expands much more than metal conduit and so long runs require an expansion coupling to allow for conduit movement and help to prevent distortion during temperature changes.

All conduit installations must be erected first before any wiring is installed (IEE Regulation 522.8.2). The radius of all bends in conduit must not cause the cables to suffer damage, and therefore the minimum radius of bends given in Table 4E of the *On Site Guide* applies (IEE Regulation 522.8.3). All conduits should terminate in a box or fitting and meet the boxes or fittings at right angles, as shown in Fig. 5.18. Any unused conduit box entries should be blanked off and all boxes covered with a box lid, fitting or accessory to provide complete enclosure of the conduit system. Conduit runs should be separate from other services, unless intentionally bonded, to

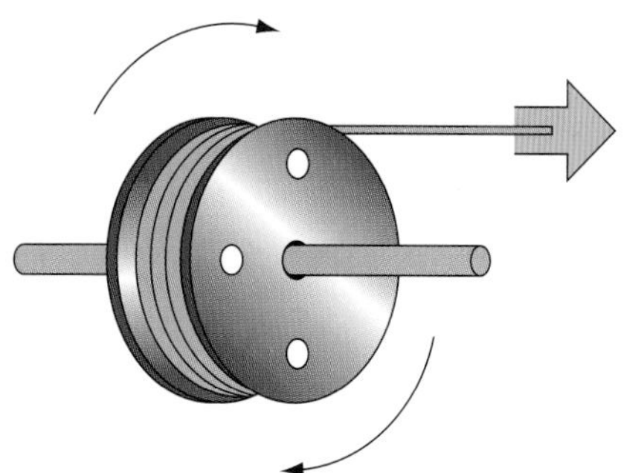

Cables *run off* will not twist, a short length of conduit can be used as an axle for the cable drum

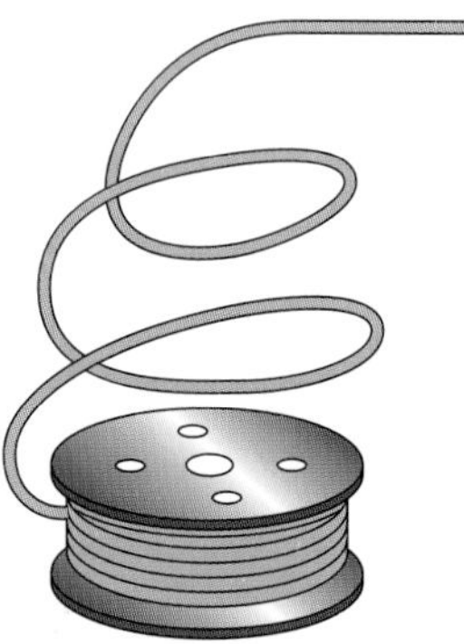

Cables allowed to *spiral off* a drum will become twisted

FIGURE 5.19
Running off cable from a drum.

Definition

Flexible conduit is made of interlinked metal spirals often covered with a PVC sleeving.

Definition

Single PVC insulated conductors are usually drawn into the installed conduit to complete the installation.

prevent arcing occurring from a faulty circuit within the conduit, which might cause the pipe of another service to become punctured.

When drawing cables into conduit they must first be *run off* the cable drum. That is, the drum must be rotated as shown in Fig. 5.19 and not allowed to *spiral off*, which will cause the cable to twist.

Cables should be fed into the conduit in a manner which prevents any cable crossing over and becoming twisted inside the conduit. The cable insulation must not be damaged on the metal edges of the draw-in box. Cables can be pulled in on a draw wire if the run is a long one. The draw wire itself may be drawn in on a fish tape, which is a thin spring steel or plastic tape.

A limit must be placed on the number of bends between boxes in a conduit run and the number of cables which may be drawn into a conduit to prevent the cables being strained during wiring. Appendix 5 of the *On Site Guide* gives a guide to the cable capacities of conduits and trunking.

Flexible conduit

Flexible conduit is made of interlinked metal spirals often covered with a PVC sleeving. The tubing must not be relied upon to provide a continuous earth path and, consequently, a separate CPC must be run either inside or outside the flexible tube (Regulation 543.2.1).

Flexible conduit is used for the final connection to motors so that the vibrations of the motor are not transmitted throughout the electrical installation and to allow for modifications to be made to the final motor position and drive belt adjustments.

Conduit capacities

Single PVC insulated conductors are usually drawn into the installed conduit to complete the installation. Having decided upon the type, size and number of cables required for a final circuit, it is then necessary to select the appropriate size of conduit to accommodate those cables.

The tables in Appendix 5 of the *On Site Guide* describe a 'factor system' for determining the size of conduit required to enclose a number of conductors. The tables are shown in Tables 5.1 and 5.2. The method is as follows:

- Identify the cable factor for the particular size of conductor. (This is given in Table 5A for straight conduit runs and Table 5C for cables run in conduits which incorporate bends, see Table 5.1.)
- Multiply the cable factor by the number of conductors, to give the sum of the cable factors.
- Identify the appropriate part of the conduit factor table given by the length of run and number of bends. (For straight runs of conduit less

Table 5.1 Conduit Cable Factors. Adapted from the *IEE On Site Guide* by kind permission of the Institution of Electrical Engineers.

TABLE 5C

Cable factors for use in conduit in long straight runs over 3 m, or runs of any length incorporating bends

Type of conductor	Conductor cross-sectional area (mm^2)	Cable factor
Solid or stranded	1	16
	1.5	22
	2.5	30
	4	43
	6	58
	10	105
	16	145
	25	217

The inner radius of a conduit bend should be not less than 2.5 times the outside diameter of the conduit.

than 3 m in length, the conduit factors are given in Table 5B. For conduit runs in excess of 3 m or incorporating bends, the conduit factors are given in Table 5D, see Table 5.2.)

- The correct size of conduit to accommodate the cables is that conduit which has a factor equal to or greater than the sum of the cable factors.

Example 1

Six 2.5 mm^2 PVC insulated cables are to be run in a conduit containing two bends between boxes 10 m apart. Determine the minimum size of conduit to contain these cables.

From Table 5C, shown in Table 5.1

The factor for one 2.5 mm^2 cable $= 30$

The sum of the cable factors $= 6 \times 30$

$= 180$

From Table 5D shown in Table 5.2, a 25 mm conduit, 10 m long and containing two bends, has a factor of 260. A 20 mm conduit containing two bends only has a factor of 141 which is less than 180, the sum of the cable factors and, therefore, 25 mm conduit is the minimum size to contain these cables.

Example 2

Ten 1.0 mm^2 PVC insulated cables are to be drawn into a plastic conduit which is 6 m long between boxes and contains one bend. A 4.0 mm PVC insulated CPC is also included. Determine the minimum size of conduit to contain these conductors.

Table 5.2 Conduit Cable Factors for Bends and Long Straight Runs. Adapted from the *IEE On Site Guide* by kind permission of the Institution of Electrical Engineers.
TABLE 5D
Cable factors for runs incorporating bends and long straight runs

	Conduit diameter (mm)																			
Length of run (m)	16	20	25	32	16	20	25	32	16	20	25	32	16	20	25	32	16	20	25	32
	Straight				One bend				Two bends				Three bends				Four bends			
1	Covered by				188	303	543	947	177	286	514	900	158	256	463	818	130	213	388	692
1.5					182	294	528	923	167	270	487	857	143	233	422	750	111	182	333	600
2	Tables				177	286	514	900	158	256	463	818	130	213	388	692	97	159	292	529
2.5					171	278	500	878	150	244	442	783	120	196	358	643	86	141	260	474
3	A and B				167	270	487	857	143	233	422	750	111	182	333	600				
3.5	179	290	521	911	162	263	475	837	136	222	404	720	103	169	311	563				
4	177	286	514	900	158	256	463	818	130	213	388	692	97	159	292	529				
4.5	174	282	507	889	154	250	452	800	125	204	373	667	91	149	275	500				
5	171	278	500	878	150	244	442	783	120	196	358	643	86	141	260	474				
6	167	270	487	857	143	233	422	750	111	182	333	600								
7	162	263	475	837	136	222	404	720	103	169	311	563								
8	158	256	463	818	130	213	388	692	97	159	292	529								
9	154	250	452	800	125	204	373	667	91	149	275	500								
10	150	244	442	783	120	196	358	643	86	141	260	474								

Additional factors: For 38 mm diameter use ...... 1.4 × (32 mm factor)
For 50 mm diameter use ...... 2.6 × (32 mm factor)
For 63 mm diameter use ...... 4.2 × (32 mm factor)

From Table 5.1

The factor for one 1.0 mm cable = 16
The factor for one 4.0 mm cable = 43.
The sum of the cable factors = (10 × 16) + (1 × 43) = 203.

From Table 5.2, a 20 mm conduit, 6 m long and containing one bend, has a factor of 233. A 16 mm conduit containing one bend only has a factor of 143 which is less than 203, the sum of the cable factors and, therefore, 20 mm conduit is the minimum size to contain these cables.

Definition

A *trunking* is an enclosure provided for the protection of cables which is normally square or rectangular in cross-section, having one removable side. Trunking may be thought of as a more accessible conduit system.

TRUNKING INSTALLATIONS

A **trunking** is an enclosure provided for the protection of cables which is normally square or rectangular in cross-section, having one removable side. Trunking may be thought of as a more accessible conduit system and for industrial and commercial installations it is replacing the larger conduit sizes. A trunking system can have great flexibility when used in conjunction with conduit; the trunking forms the background or framework for the installation, with conduits running from the trunking to the point controlling the current using apparatus. When an alteration or extension is required it is easy to drill a hole in the side of the trunking and run a conduit to the new point. The new wiring can then be drawn through the new conduit and the existing trunking to the supply point.

Trunking is supplied in 3 m lengths and various cross-sections are measured in millimetres from 50 × 50 up to 300 × 150. Most trunking is available in either steel or plastic.

Metallic trunking

Definition

Metallic trunking is formed from mild steel sheet, coated with grey or silver enamel paint for internal use or a hot-dipped galvanized coating where damp conditions might be encountered.

Metallic trunking is formed from mild steel sheet, coated with grey or silver enamel paint for internal use or a hot-dipped galvanized coating where damp conditions might be encountered and made to a specification defined by BS EN 50085. A wide range of accessories is available, such as 45° bends, 90° bends, tee and four-way junctions, for speedy on-site assembly. Alternatively, bends may be fabricated in lengths of trunking, as shown in Fig. 5.20. This may be necessary or more convenient if a bend or set is non-standard, but it does take more time to fabricate bends than merely to bolt on standard accessories.

When fabricating bends the trunking should be supported with wooden blocks for sawing and filing, in order to prevent the sheet-steel vibrating or becoming deformed. Fish plates must be made and riveted or bolted to the trunking to form a solid and secure bend. When manufactured bends are used, the continuity of the earth path must be ensured across the joint by making all fixing screw connections very tight, or fitting a separate copper strap between the trunking and the standard bend. If an earth continuity test on the trunking is found to be unsatisfactory, an insulated CPC must be installed inside the trunking. The size of the protective conductor will be determined by the largest cable contained in the trunking, as described by Table 54.7 of the IEE Regulations. If the circuit conductors are less than 16 mm^2, then a 16 mm^2 CPC will be required.

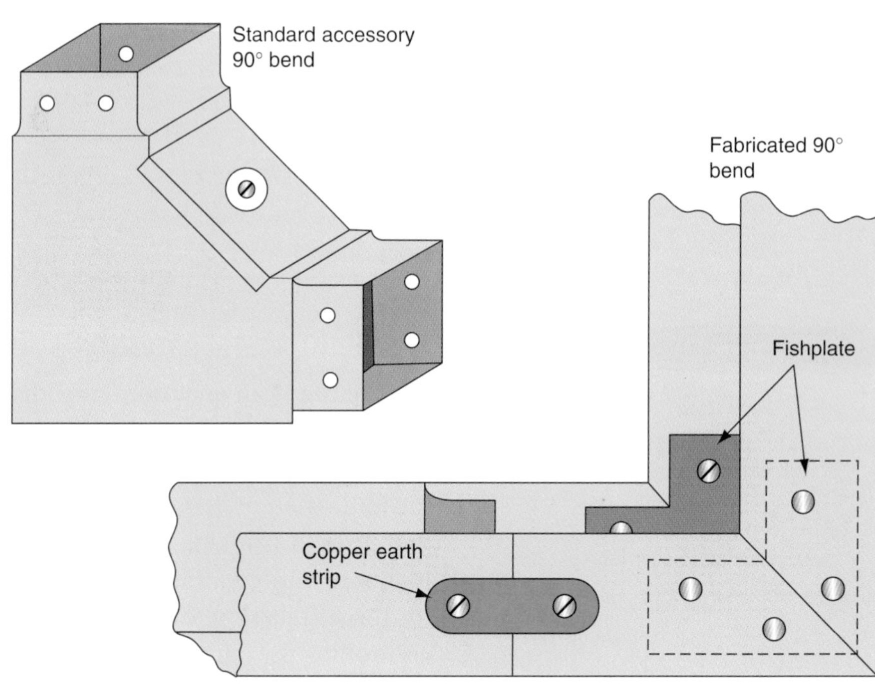

FIGURE 5.20
Alternative trunking bends.

Non-metallic trunking

Trunking and trunking accessories are also available in high-impact PVC. The accessories are usually secured to the lengths of trunking with a PVC solvent adhesive. PVC trunking, like PVC conduit, is easy to install and is non-corrosive. A separate CPC will need to be installed and non-metallic trunking may require more frequent fixings because it is less rigid than metallic trunking. All trunking fixings should use round-headed screws to prevent damage to cables since the thin sheet construction makes it impossible to countersink screw heads.

Definition

Mini-trunking is very small PVC trunking, ideal for surface wiring in domestic and commercial installations such as offices.

Mini-trunking

Mini-trunking is very small PVC trunking, ideal for surface wiring in domestic and commercial installations such as offices. The trunking has a cross-section of 16 × 16 mm, 25 × 16 mm, 38 × 16 mm or 38 × 25 mm and is ideal for switch drops or for housing auxiliary circuits such as telephone or audio equipment wiring. The modern square look in switches and sockets is complemented by the mini-trunking which is very easy to install (see Fig. 5.21).

Definition

A *trunking manufactured from PVC or steel* and in the shape of a skirting board is frequently used in commercial buildings such as hospitals, laboratories and offices.

Skirting trunking

A **trunking manufactured from PVC or steel** and in the shape of a skirting board is frequently used in commercial buildings such as hospitals, laboratories and offices. The trunking is fitted around the walls of a room and contains the wiring for socket outlets and telephone points which are mounted on the lid, as shown in Fig. 5.21.

Key Fact

Fire Safety

Where the wiring system passes through elements of building construction such as floors and walls, any damage must be made good (IEE REG 527.2).

Where any trunking passes through walls, partitions, ceilings or floors, short lengths of lid should be fitted so that the remainder of the lid may be removed later without difficulty. Any damage to the structure of the buildings must be made good with mortar, plaster or concrete in order to prevent the spread of fire. Fire barriers must be fitted inside the trunking every 5 m, or at every floor level or room dividing wall, if this is a shorter distance, as shown in Fig. 5.22(a).

Where trunking is installed vertically, the installed conductors must be supported so that the maximum unsupported length of non-sheathed cable does not exceed 5 m. Figure 5.22(b) shows cables woven through insulated pin supports, which is one method of supporting vertical cables.

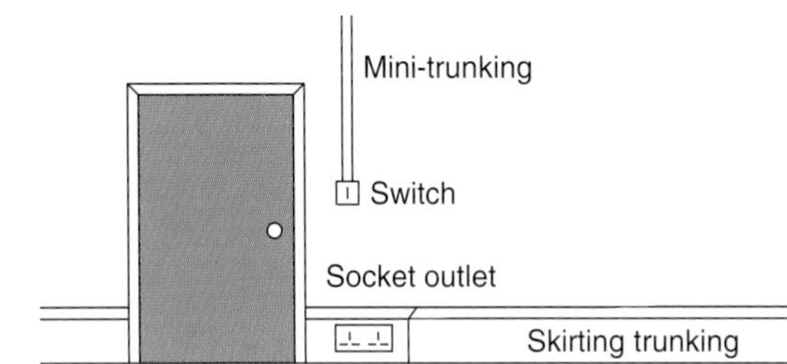

FIGURE 5.21
Typical installation of skirting trunking and mini-trunking.

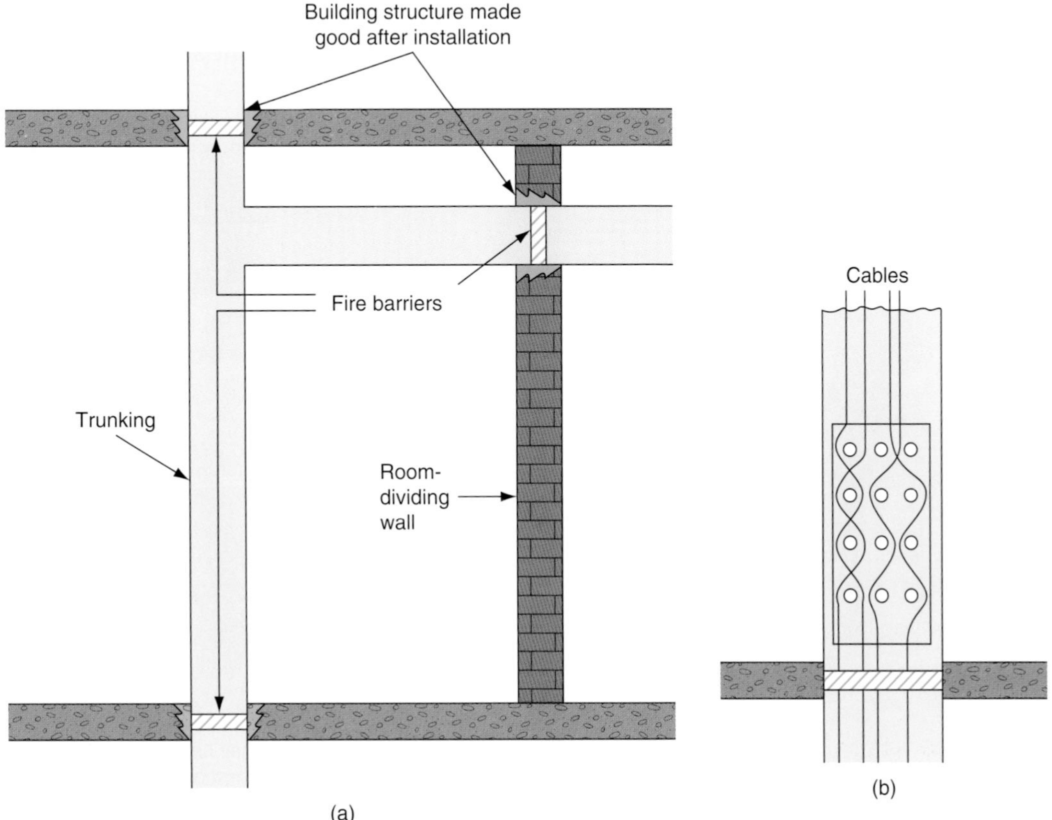

FIGURE 5.22
Installation of trunking.

PVC insulated cables are usually drawn into an erected conduit installation or laid into an erected trunking installation. Table 5D of the *On Site Guide* only gives factors for conduits up to 32 mm in diameter, which would indicate that conduits larger than this are not in frequent or common use. Where a cable enclosure greater than 32 mm is required because of the number or size of the conductors, it is generally more economical and convenient to use trunking.

Trunking capacities

Definition

The ratio of the space occupied by all the cables in a conduit or trunking to the whole space enclosed by the conduit or trunking is known as the *space factor*.

The ratio of the space occupied by all the cables in a conduit or trunking to the whole space enclosed by the conduit or trunking is known as the **space factor**. Where sizes and types of cable and trunking are not covered by the tables in Appendix 5 of the *On Site Guide* a space factor of 45% must not be exceeded. This means that the cables must not fill more than 45% of the space enclosed by the trunking. The tables of Appendix 5 take this factor into account.

To calculate the size of trunking required to enclose a number of cables:

- Identify the cable factor for the particular size of conductor. See Table 5.3.
- Multiply the cable factor by the number of conductors to give the sum of the cable factors.
- Consider the factors for trunking and shown in Table 5.4. The correct size of trunking to accommodate the cables is that trunking which has a factor equal to or greater than the sum of the cable factors.

Table 5.3 Trunking, Cable Factors. Adapted from the IEE *On Site Guide* by kind permission of the Institution of Electrical Engineers.
TABLE 5E
Cable factors for trunking

Type of conductor	Conductor cross-sectional area (mm^2)	PVC, BS 6004 Cable factor	Thermosetting BS 7211 Cable factor
Solid	1.5	8.0	8.6
	2.5	11.9	11.9
Stranded	1.5	8.6	9.6
	2.5	12.6	13.9
	4	16.6	18.1
	6	21.2	22.9
	10	35.3	36.3
	16	47.8	50.3
	25	73.9	75.4

Note: (i) These factors are for metal trunking and may be optimistic for plastic trunking where the cross-sectional area available may be significantly reduced from the nominal by the thickness of the wall material. (ii) The provision of spare space is advisable; however, any circuits added at a later date must take into account grouping. Appendix 4, BS 7671.

Example

Calculate the minimum size of trunking required to accommodate the following single-core PVC cables:

20 × 1.5 mm solid conductors
20 × 2.5 mm solid conductors
21× 4.0 mm stranded conductors
16 × 6.0 mm stranded conductors

From Table 5.3, the cable factors are:

for 1.5 mm solid cable	–	8.0
for 2.5 mm solid cable	–	11.9
for 4.0 mm stranded cable	–	16.6
for 6.0 mm stranded cable	–	21.2

The sum of the cable terms is:

$$(20 \times 8.0) + (20 \times 11.9) + (21 \times 16.6) + (16 \times 21.2) = 1085.8.$$

From Table 5.4, 75 × 38 mm trunking has a factor of 1146 and, therefore, the minimum size of trunking to accommodate these cables is 75 × 38 mm, although a larger size, say 75 × 50 mm would be equally acceptable if this was more readily available as a standard stock item.

SEGREGATION OF CIRCUITS

Where an installation comprises a mixture of low-voltage and very low-voltage circuits such as mains lighting and power, fire alarm and telecommunication circuits, they must be separated or *segregated* to prevent electrical contact (IEE Regulation 528.1).

For the purpose of these regulations various circuits are identified by one of two bands as follows:

Band I telephone, radio, bell, call and intruder alarm circuits, emergency circuits for fire alarm and emergency lighting.

Band II mains voltage circuits.

When Band I circuits are insulated to the same voltage as Band II circuits, they may be drawn into the same compartment.

When trunking contains rigidly fixed metal barriers along its length, the same trunking may be used to enclose cables of the separate Bands without further precautions, provided that each Band is separated by a barrier, as shown in Fig. 5.23.

Multi-compartment PVC trunking cannot provide band segregations since there is no metal screen between the Bands. This can only be provided in PVC trunking if screened cables are drawn into the trunking.

Table 5.4 Trunking Cable Factors. Adapted from the *IEE On Site Guide* by kind permission of the Institution of Electrical Engineers. Factors for trunking

Dimensions of trunking (mm × mm)	Factor	Dimensions of trunking (mm × mm)	Factor
50 × 38	767	200 × 100	8572
50 × 50	1037	200 × 150	13001
75 × 25	738	200 × 200	17429
75 × 38	1146	225 × 38	3474
75 × 50	1555	225 × 50	4671
75 × 75	2371	225 × 75	7167
100 × 25	993	225 × 100	9662
100 × 38	1542	225 × 150	14652
100 × 50	2091	225 × 200	19643
100 × 75	3189	225 × 225	22138
100 × 100	4252	300 × 38	4648
150 × 38	2999	300 × 50	6251
150 × 50	3091	300 × 75	9590
150 × 75	4743	300 × 100	12929
150 × 100	6394	300 × 150	19607
150 × 150	9697	300 × 200	26285
200 × 38	3082	300 × 225	29624
200 × 50	4145	300 × 300	39428
200 × 75	6359		

Space factor – 45% with trunking thickness taken into account

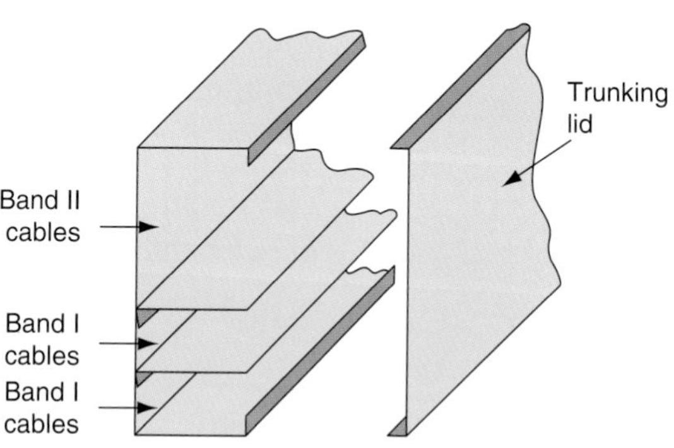

FIGURE 5.23
Segregation of cables in trunking.

CABLE TRAY INSTALLATIONS

Cable tray is a sheet-steel channel with multiple holes. The most common finish is hot-dipped galvanized but PVC-coated tray is also available. It is used extensively on large industrial and commercial installations for supporting MI and SWA cables which are laid on the cable tray and secured with cable ties through the tray holes.

Definition

Cable tray is a sheet-steel channel with multiple holes. The most common finish is hot-dipped galvanized but PVC-coated tray is also available. It is used extensively on large industrial and commercial installations for supporting MI and SWA cables which are laid on the cable tray and secured with cable ties through the tray holes.

Cable tray should be adequately supported during installation by brackets which are appropriate for the particular installation. The tray should be bolted to the brackets with round-headed bolts and nuts, with the round head inside the tray so that cables drawn along the tray are not damaged.

The tray is supplied in standard widths from 50 to 900 mm, and a wide range of bends, tees and reducers is available. Figure 5.24 shows a factory-made 90° bend at B. The tray can also be bent using a cable tray bending machine to create bends such as that shown at A in Fig. 5.24. The installed tray should be securely bolted with round-headed bolts where lengths or accessories are attached, so that there is a continuous earth path which may be bonded to an electrical earth. The whole tray should provide a firm support for the cables and therefore the tray fixings must be capable of supporting the weight of both the tray and cables.

PVC/SWA CABLE INSTALLATIONS

Steel wire armoured PVC insulated cables are now extensively used on industrial installations and often laid on cable tray. This type of installation has the advantage of flexibility, allowing modifications to be made speedily as the need arises. The cable has a steel wire armouring giving mechanical protection and permitting it to be laid directly in the ground or in ducts, or it may be fixed directly or laid on a cable tray. Figure 5.13 shows a PVC/SWA cable.

Definition

Steel wire armoured PVC insulated cables are now extensively used on industrial installations and often laid on cable tray.

It should be remembered that when several cables are grouped together the current rating will be reduced according to the correction factors given in Appendix 4. Table 4C1 of the IEE Regulations and Table 6C of the *On Site Guide.*

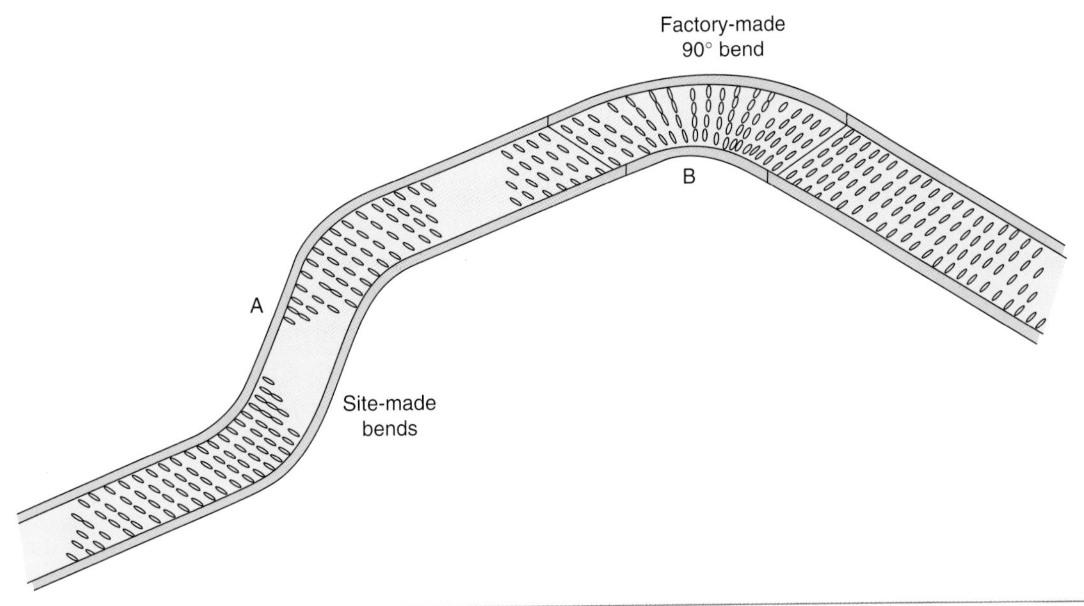

FIGURE 5.24
Cable tray with bends.

The cable is easy to handle during installation, is pliable and may be bent to a radius of eight times the cable diameter. The PVC insulation would be damaged if installed in ambient temperatures over 70°C or below 0°C, but once installed the cable can operate at low temperatures.

The cable is terminated with a simple gland which compresses a compression ring onto the steel wire armouring to provide the earth continuity between the switchgear and the cable.

MI CABLE INSTALLATIONS

MI cables are available for general wiring as:

- light-duty MI cables for voltages up to 600V and sizes from 1.0 to 10 mm^2 and
- heavy-duty MI cables for voltages up to 1000V and sizes from 1.0 to 150 mm^2.

Figure 5.14 shows an MI cable and termination. The cables are available with bare sheaths or with a PVC oversheath. The cable sheath provides sufficient mechanical protection for all but the most severe situations, where it may be necessary to fit a steel sheath or conduit over the cable to give extra protection, particularly near floor level in some industrial situations.

The cable may be laid directly in the ground, in ducts, on cable tray or clipped directly to a structure. It is not affected by water, oil or the cutting fluids used in engineering and can withstand very high temperature or even fire. The cable diameter is small in relation to its current carrying capacity and it should last indefinitely if correctly installed because it is made from inorganic materials. These characteristics make the cable ideal for Band I emergency circuits, boiler-houses, furnaces, petrol stations and chemical plant installations.

The cable is supplied in coils and should be run off during installation and not spiralled off, as described in Fig. 5.19 for conduit. The cable can be work-hardened if over-handled or over-manipulated. This makes the copper outer sheath stiff and may result in fracture. The outer sheath of the cable must not be penetrated, otherwise moisture will enter the magnesium oxide insulation and lower its resistance. To reduce the risk of damage to the outer sheath during installation, cables should be straightened and formed by hammering with a hide hammer or a block of wood and a steel hammer. When bending MI cables the radius of the bend should not cause the cable to become damaged and clips should provide adequate support (Regulations 522.8.5).

The cable must be prepared for termination by removing the outer copper sheath to reveal the copper conductors. This can be achieved by using a rotary stripper tool or, if only a few cables are to be terminated, the outer sheath can be removed with side cutters, peeling off the cable in a similar way to peeling the skin from a piece of fruit with a knife. When enough conductor has been revealed, the outer sheath must be cut off square to facilitate the fitting of the sealing pot, and this can be done with a ringing tool. All excess magnesium oxide powder must be wiped from the conductors with a clean cloth. This is to prevent moisture from penetrating the seal by capillary action.

Cable ends must be terminated with a special seal to prevent the entry of moisture. Figure 5.14 shows a brass screw-on seal and gland assembly, which allows termination of the MI cables to standard switchgear and conduit fittings. The sealing pot is filled with a sealing compound, which is pressed in from one side only to prevent air pockets forming, and the pot closed by crimping home the sealing disc. Such an assembly is suitable for working temperatures up to 105°C. Other compounds or powdered glass can increase the working temperature up to 250°C.

The conductors are not identified during the manufacturing process and so it is necessary to identify them after the ends have been sealed. A simple continuity or polarity test, as described in Chapter 8 of this book can identify the conductors which are then sleeved or identified with coloured markers.

Connection of MI cables can be made directly to motors, but to absorb the vibrations a 360° loop should be made in the cable just before the termination. If excessive vibration is to be expected the MI cable should be terminated in a conduit through-box and the final connection made by flexible conduit.

Copper MI cables may develop a green incrustation or patina on the surface, even when exposed to normal atmospheres. This is not harmful and should not be removed. However, if the cable is exposed to an environment which might encourage corrosion, an MI cable with an overall PVC sheath should be used.

Transformers

You have seen so far in this chapter that:

- electricity is generated as an alternating a.c. power supply,
- it is generated at 11 kV and then transformed up to 132 or 400 kV for transmission and distribution on the National Grid network,
- it is then transformed down for local underground distribution at 11 kV and then,
- further reduced to 400V and 230V for industrial, commercial and domestic consumers.

All of this is only possible because of one piece of electrical equipment, the transformer.

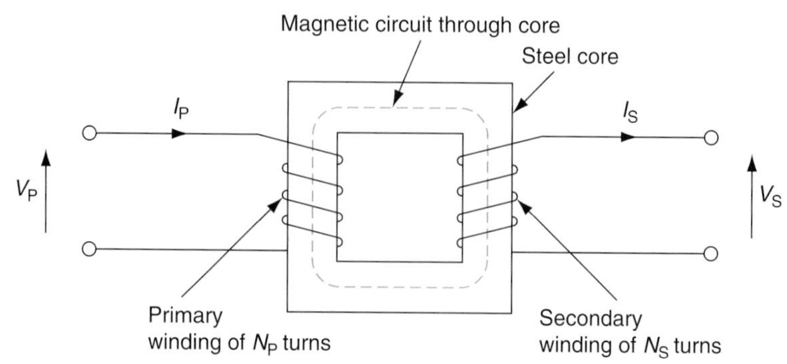

FIGURE 5.25
A simple transformer.

Definition

A *transformer* is an electrical machine which is used to change the value of an alternating voltage. They vary in size from miniature units used in electronics to huge power transformers used in power stations.

A **transformer** is an electrical machine which is used to change the value of an alternating voltage. They vary in size from miniature units used in electronics to huge power transformers used in power stations. A transformer will only work when an alternating voltage is connected. It will not normally work from a d.c. supply such as a battery.

A transformer, as shown in Fig. 5.25, consists of two coils, called the primary and secondary coils, or windings, which are insulated from each other and wound onto the same steel or iron core.

An alternating voltage applied to the primary winding produces an alternating current, which sets up an alternating magnetic flux throughout the core. This magnetic flux induces an electromotive force (emf) in the secondary winding, as described by Faraday's law, which says that when a conductor is cut by a magnetic field, an emf is induced in that conductor. Since both windings are linked by the same magnetic flux, the induced emf per turn will be the same for both windings. Therefore, the emf in both windings is proportional to the number of turns. In symbols,

$$\frac{V_P}{N_P} = \frac{V_S}{N_S} \tag{5.4}$$

Most practical power transformers have a very high efficiency, and for an ideal transformer having 100% efficiency the primary power is equal to the secondary power:

$$\text{Primary power} = \text{Secondary power}$$

and, since

$$\text{Power} = \text{Voltage} \times \text{Current} \tag{5.5}$$

then,

$$V_P \times I_P = V_S \times I_S$$

Combining Equations (5.4) and (5.5), we have

$$\frac{V_P}{V_S} = \frac{N_P}{N_S} = \frac{I_S}{I_P}$$

Example

A 230 V to 12V bell transformer is constructed with 800 turns on the primary winding. Calculate the number of secondary turns and the primary and secondary currents when the transformer supplies a 12V 12W alarm bell.

Collecting the information given in the question into a usable form, we have:

$$V_P = 230\text{ V}$$
$$V_S = 12\text{ V}$$
$$N_P = 800$$

Power = 12W

Information required: N_S, I_S and I_P

Secondary turns,

$$N_S = \frac{N_P V_S}{V_P}$$
$$\therefore N_S = \frac{800 \times 12\text{ V}}{230\text{ V}} = 42\text{ turns}$$

Secondary current,

$$I_S = \frac{\text{Power}}{V_S}$$
$$\therefore I_S = \frac{12\text{ W}}{12\text{ V}} = 1\text{ A}$$

Primary current,

$$I_P = \frac{I_S \times V_S}{V_P}$$
$$\therefore I_P = \frac{1\text{ A} \times 12\text{ V}}{230\text{ V}} = 0.052\text{ A}$$

Transformer rating

Definition

Transformers are *rated* in kVA (kilovolt-amps) rather than power in watts because the output current and power factor will be affected by the load connected to the transformer.

Transformers are **rated** in kVA (kilovolt-amps) rather than power in watts because the output current and power factor will be affected by the load connected to the transformer.

The kVA rating of a transformer tells us the current that can be delivered for a known voltage. Thus a 5kVA transformer can deliver 12.5A at 400V.

Definition

As they have no moving parts causing frictional losses, most transformers have a very high efficiency, usually better than 90%. However, the losses which do occur in a transformer can be grouped under two general headings: *copper losses* and *iron losses*.

Definition

Copper losses occur because of the small internal resistance of the windings.

Definition

Iron losses are made up of hysteresis loss and eddy current loss. The hysteresis loss depends upon the type of iron used to construct the core and consequently core materials are carefully chosen.

Definition

Eddy currents are circulating currents created in the core material by the changing magnetic flux. These are reduced by building up the core of thin slices or laminations of iron and insulating the separate laminations from each other.

Transformer losses

As they have no moving parts causing frictional losses, most transformers have a very high efficiency, usually better than 90%. However, the losses which do occur in a transformer can be grouped under two general headings: **copper losses** and **iron losses**.

Copper losses occur because of the small internal resistance of the windings. They are proportional to the load, increasing as the load increases because copper loss is an 'I^2R' loss.

Iron losses are made up of *hysteresis loss* and *eddy current loss*. The hysteresis loss depends upon the type of iron used to construct the core and consequently core materials are carefully chosen. We looked at magnetic hysteresis loops at Fig. 4.21 in the last chapter. Transformers will only operate on an alternating supply. Thus, the current which establishes the core flux is constantly changing from positive to negative. Each time there is a current reversal, the magnetic flux reverses and it is this build-up and collapse of magnetic flux in the core material which accounts for the hysteresis loss.

Eddy currents are circulating currents created in the core material by the changing magnetic flux. These are reduced by building up the core of thin slices or laminations of iron and insulating the separate laminations from each other. The iron loss is a constant loss consuming the same power from no load to full load.

Transformer efficiency

The efficiency of any machine is determined by the losses incurred by the machine in normal operation. The efficiency of rotating machines is usually in the region of 50–60% because they incur windage and friction losses; but the transformer has no moving parts so, therefore, these losses do not occur. However, the efficiency of a transformer can be calculated in the same way as for any other machine. The efficiency of a machine is generally given by:

$$\eta = \frac{\text{Output power}}{\text{Input power}}$$

(η is the Greek letter 'eta'). However, the input to the transformer must supply the output plus any losses which occur within the transformer. We can therefore say:

$$\text{Input power} = \text{Output power} + \text{Losses}$$

Rewriting the basic formula, we have:

$$\eta = \frac{\text{Output power}}{\text{Output power} + \text{Losses}}$$

Example

A 100 kVA power transformer feeds a load operating at a power factor of 0.8. Find the efficiency of the transformer if the combined iron and copper loss at this load is 1 kW.

$$\text{Output power} = \text{kVA} \times \text{p.f}$$

$$\therefore \text{Output power} = 100\ \text{kVA} \times 0.8$$

$$\text{Output power} = 80\ \text{kW}$$

$$\eta = \frac{\text{Output power}}{\text{Output power} + \text{losses}}$$

$$\eta = \frac{80\,\text{kW}}{80\,\text{kW} + 1\,\text{kW}}$$

$$\eta = 0.987$$

or, multiplying by 100 to give a percentage, the transformer has an efficiency of 98.7%.

Transformer construction

Transformers are constructed in a way which reduces the losses to a minimum. The core is usually made of silicon–iron laminations, because at fixed low frequencies silicon–iron has a small hysteresis loss and the laminations reduce the eddy current loss. The primary and secondary windings are wound close to each other on the same limb. If the windings are spread over two limbs, there will usually be half of each winding on each limb, as shown in Fig. 5.26.

AUTO-TRANSFORMERS

Transformers having a separate primary and secondary winding, as shown in Fig. 5.26, are called double-wound transformers, but it is possible to construct a transformer which has only one winding which is common to the primary and secondary circuits. The secondary voltage is supplied by

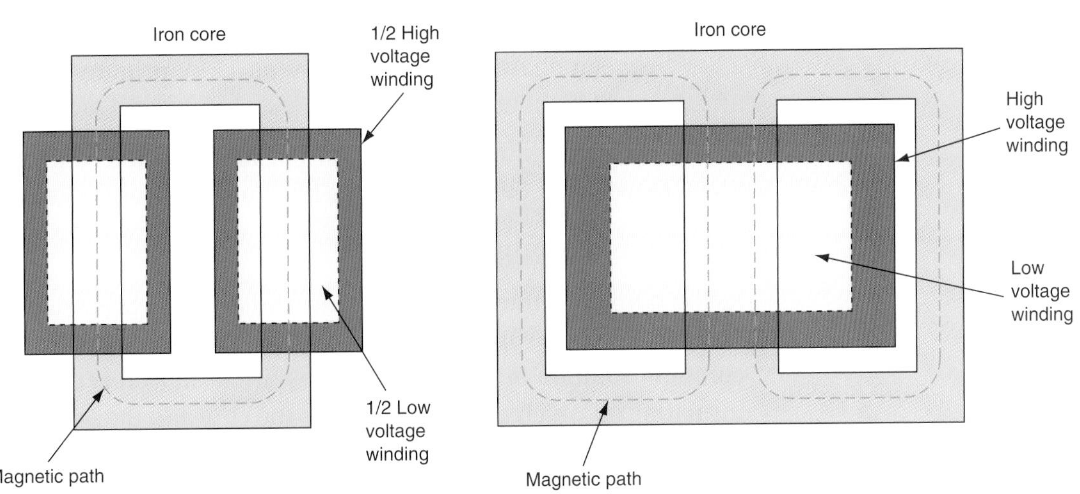

FIGURE 5.26
Transformer construction.

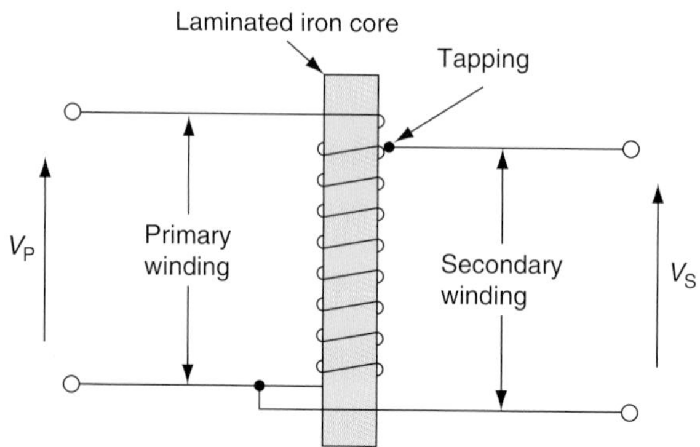

FIGURE 5.27
An auto-transformer.

means of a 'tapping' on the primary winding. An arrangement such as this is called an auto-transformer.

The auto-transformer is much cheaper and lighter than a double-wound transformer because less copper and iron are used in its construction. However, the primary and secondary windings are not electrically separate and a short circuit on the upper part of the winding shown in Fig. 5.27 would result in the primary voltage appearing across the secondary terminals. For this reason auto-transformers are mostly used where only a small difference is required between the primary and secondary voltages. When installing transformers, the regulations of Section 555 must be complied with, in addition to any other regulations relevant to the particular installation.

THREE-PHASE TRANSFORMERS

Most of the transformers used in industrial applications are designed for three-phase operation. In the double-wound type construction, as shown in Fig. 5.25, three separate single-phase transformers are wound onto a common laminated silicon–steel core to form the three-phase transformer. The primary and secondary windings may be either star or delta connected but in distribution transformers the primary is usually connected in delta and the secondary in star. This has the advantage of providing two secondary voltages, typically 400 V between phases and 230 V between phase and neutral from an 11 kV primary voltage. The coil arrangement is shown in Fig. 5.28.

The construction of the three-phase transformer is the same as the single-phase transformer, but because of the larger size the core is often cooled by oil.

OIL-IMMERSED CORE

As the rating of a transformer increases so does the problem of dissipating the heat generated in the core. In power distribution transformers the most common solution is to house the transformer in a steel casing containing insulating oil which completely covers the core and the windings. The oil is a coolant and an insulating medium for the core. On load the transformer heats up and establishes circulating convection currents in the oil which flows through the external tubes. Air passing over the tubes carries the heat away and cools the transformer. Figure 5.29 shows the construction of

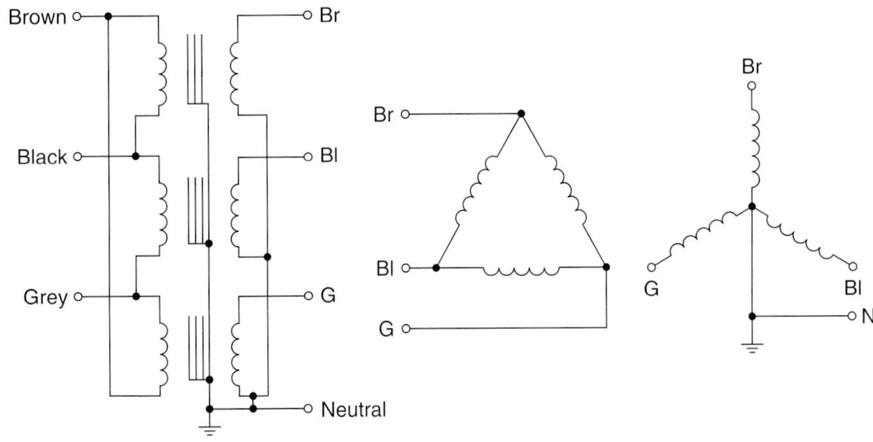

FIGURE 5.28
Delta-star connected three-phase transformer.

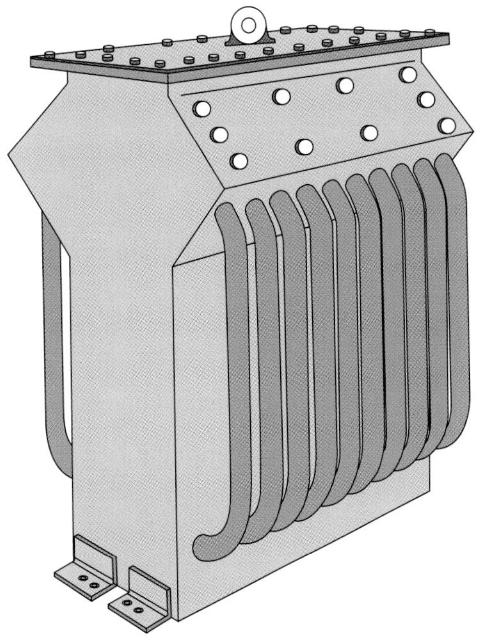

FIGURE 5.29
Typical oil-filled power transformer.

a typical oil-filled transformer, and the arrangement is typical of a distribution transformer used in a sub-station.

TAP CHANGING

Under load conditions the secondary voltage of a transformer may fall and become less than that permitted by the Regulations. The tolerance permitted by the Regulations in 2008 is the open-circuit voltage plus 10% or minus 6%. However, it is proposed that these tolerance levels will be adjusted to ±10% of the declared nominal voltage. Because the voltage of a transformer is proportional to the number of turns, one solution is to vary the number of turns on either the primary or secondary winding, to achieve the desired voltage. This process is called tap changing, and most distribution transformers are fitted with a tap changing switch on the high-voltage winding so that the number of turns can be varied. These switches are always off load devices and, therefore, the transformer must be isolated before the tap changing operation is carried out. The switch is usually padlocked to prevent unauthorized operation.

Earthing and protection systems

We know from the earlier chapters in this book that using electricity is one of the causes of accidents in the workplace. Using electricity is a hazard because it has the potential, the possibility, to cause harm and, therefore, the provision of protective devices in an electrical installation is fundamental to the whole concept of the safe use of electricity.

The consumer's mains equipment is fixed close to the point at which the supply cable enters the building. The IEE Regulations require the consumer's mains equipment to provide:

- Protection against electric shock (Chapter 41)
- Protection against overcurrent (Chapter 43)
- Isolation and switching (Chapter 53).

Protection against electric shock, both 'Basic Protection' and 'Fault Protection' is provided by placing live parts out of reach in suitable enclosures, by insulation and by earthing and bonding metalwork for the purpose of safety. Also by providing fuses or circuit breakers so that the supply is automatically disconnected under fault conditions.

Protection against overcurrent is achieved by providing a device which will automatically disconnect the supply before the overcurrent can cause a rise in temperature that would damage the installation. A fuse or miniature circuit board (MCB) would meet these requirements.

Definition

An *isolator* is a mechanical device that is operated manually and is provided so that the whole of the installation, one circuit or one piece of equipment, may be cut off from the live supply.

An **isolator** is a mechanical device that is operated manually and is provided so that the whole of the installation, one circuit or one piece of equipment, may be cut off from the live supply.

In addition, a means of switching off for maintenance or emergency switching, must be provided. A switch may provide the means of isolation but an isolator differs from a switch in that it is intended to be opened when the circuit is not carrying current. Its purpose is to ensure the safety of those working on the circuit by making dead those parts that are live in normal service.

Definition

The switching of electrical equipment in normal service is called *'functional switching'*.

One device may provide both isolation and switching functions. The switching of electrical equipment in normal service is called '**functional switching**'.

All electrical installation circuits are controlled by switchgear that is assembled so that the individual circuits may be operated safely under normal conditions, isolated automatically under fault conditions or isolated manually for safe maintenance or repair work. These requirements are met by good workmanship carried out by competent persons using approved British Standard cables equipment and accessories.

Earthing systems and protection systems are described in detail in Chapter 7 of this book.

Try This

Definitions

Write down the meaning of

- functional switching and
- isolation switching.

Check your Understanding

When you have completed these questions, check out your answers at the back of the book.
Note: more than one multiple choice answer may be correct.

1. The most suitable cable for a hazardous or high-temperature installation is:
   a. PVC insulated and sheathed cables
   b. PVC/SWA cable
   c. PILCSWA cables
   d. MI cables.

2. The part of the cable which sometimes requires some means of mechanical protection is the:
   a. conductor
   b. cable insulation
   c. outer sheath
   d. PVC/SWA.

3. The most suitable cable for wiring underground when some mechanical protection is required is:
   a. PVC insulated and sheathed cable
   b. PVC/SWA cable
   c. FP 200 cable
   d. MI cable.

4. The most suitable cable for wiring domestic installations is:
   a. PVC insulated and sheathed cable
   b. PVC/SWA cable
   c. FP 200 cable
   d. MI cable.

5. The most suitable cable for wiring a fire alarm in fire resistant cable is:
   a. PVC insulated and sheathed cable
   b. PVC/SWA cable
   c. FP 200 cable
   d. MI cable.

6. It is a tube, a channel or pipe in which insulated conductors are contained is one definition of a:
   a. flexible conduit
   b. steel conduit
   c. tray
   d. trunking.

7. It is an enclosure provided for the protection of cables that is normally square or rectangular with one removable side, is one definition of:
   a. flexible conduit
   b. steel conduit
   c. tray
   d. trunking.

8. A 12V mini-spot lamp transformer has 1400 turns on the primary winding. When connected to the 230V mains supply the number of secondary turns will be approximately:
   a. 2
   b. 73
   c. 386
   d. 4000.

9. A 12V 12W mini-spot lamp transformer is connected to the 230V mains supply. The primary current will be:
   a. 52.2mA
   b. 62.6mA
   c. 19.2A
   d. 33.12A.

10. Transformer losses are:
    a. copper loss
    b. current loss
    c. iron loss
    d. voltage loss.

11. Use a labelled sketch to describe the electricity supply system from generation, transmission and distribution to the end user. State all voltages.

12. Use a labelled sketch to show how a 230V and 400V supply is obtained from an 11kV sub-station transformer.

13. Use a labelled sketch to show a delta-star connected transformer. State the formulae for calculating the voltage and current for each type of connection.

14. Use a labelled sketch to describe the construction of a PVC/SWA cable. State one suitable application for this type of cable.

15. Use a labelled sketch to describe a trunking plus conduit installation in an industrial environment. State the method to be used to prevent the spread of fire through the fabric of the building and internally through the trunking where it passes through walls and ceilings.

16. Use a labelled sketch to describe how you would fix to a brick wall:
    a. conduit
    b. trunking.

    State the type of screws to be used in each case.

17. Use the Tables 5.1 and 5.2 in this chapter to calculate the maximum number of 1.5 mm PVC cables which can be drawn into a 20 mm conduit having one bend in a 4 m length of run.

18. Using the information given in question 17, calculate the maximum number of 2.5 mm cables.

19. Use the Tables 5.3 and 5.4 in this chapter to calculate the maximum number of solid 2.5 mm PVC cables which can be drawn into a 100 × 50 trunking.

20. List all the losses that occur in a transformer and state how the losses are reduced to a minimum.

21. Describe what we mean by a 'transformer's rating' and calculate the current which a 2 kVA site transformer can deliver at 110 V.

22. Describe the meaning of
    a. Basic Protection and
    b. Fault Protection.

    Note: There are definitions in this book and in Part 2 of the IEE Regulations.

23. Describe how an electrical installation provides protection from electric shock.
24. Describe how an electrical installation is protected against overcurrent.
25. Explain the difference between switching for isolation purposes and functional switching.

Electrical machines and motors

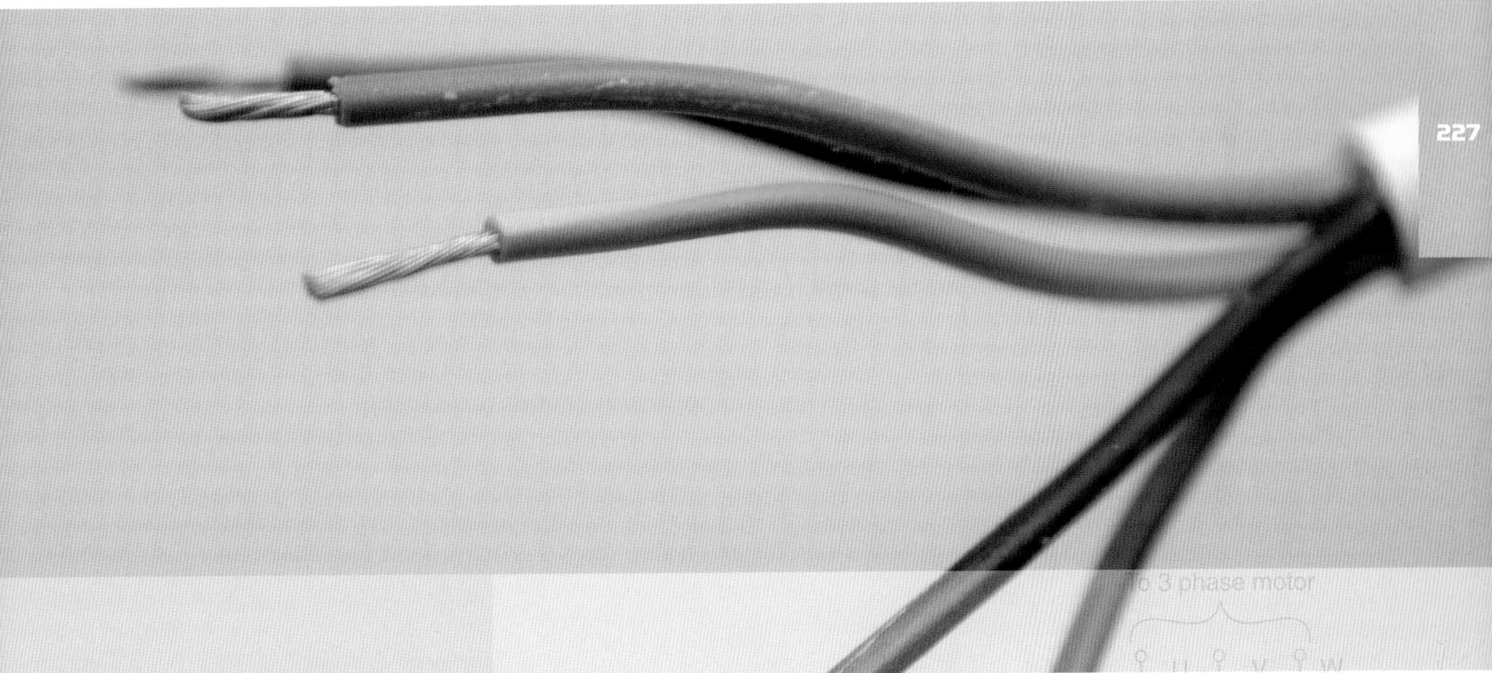

Unit 1 – Application of health and safety and electrical principles – Outcome 6

Underpinning knowledge: when you have completed this chapter you should be able to:

- describe the operation of electrical rotating machines
- identify a three-phase induction motor
- describe the production of a three-phase rotating magnetic field
- identify single-phase a.c. motors
- identify the basic principles of motor starting and speed control

Electrical machines and motors

The fundamental principles of electrical machines were laid down in *Basic Electrical Installation Work*. In this chapter we will essentially be looking at d.c. and a.c. motors, their control equipment and maintenance.

Direct current motors

If a current carrying conductor is placed into the field of a permanent magnet as shown in Fig. 6.1(c) a force F will be exerted on the conductor to push it out of the magnetic field.

To understand the force, let us consider each magnetic field acting alone. Figure 6.1(a) shows the magnetic field due to the current carrying conductor only. Figure 6.1(b) shows the magnetic field due to the permanent magnet in which is placed the conductor carrying no current. Figure 6.1(c) shows the effect of the combined magnetic fields which are distorted and, because lines of magnetic flux never cross, but behave like stretched elastic bands, always trying to find the shorter distance between a north and south pole, the force F is exerted on the conductor, pushing it out of the permanent magnetic field.

This is the basic motor principle, and the force F is dependent upon the strength of the magnetic field B, the magnitude of the current flowing in the conductor I and the length of conductor within the magnetic field l. The following equation expresses this relationship:

$$F = BlI \text{ (N)}$$

where B is in teslas, l is in metres, I is in amperes and F is in newtons.

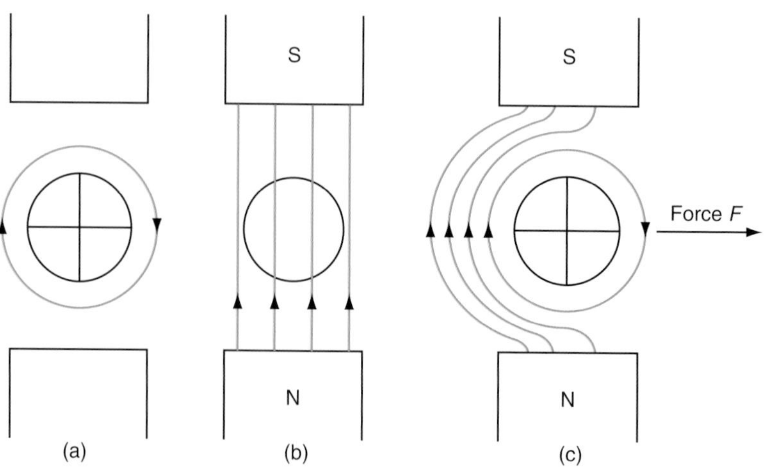

FIGURE 6.1
Force on a conductor in a magnetic field.

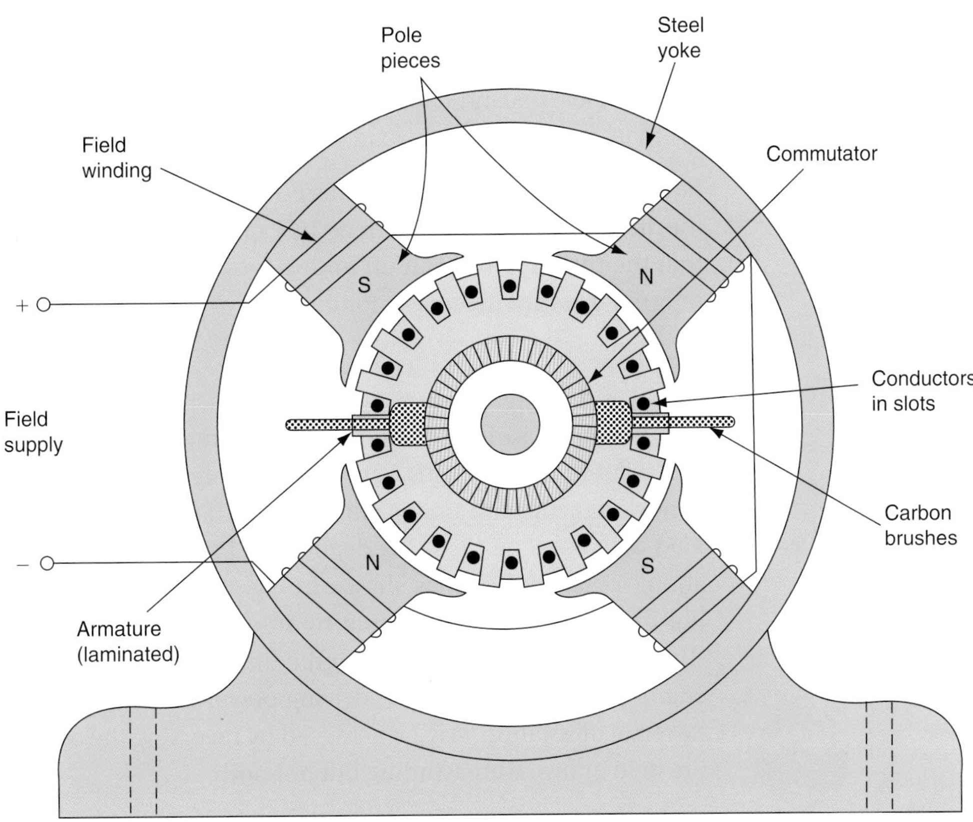

FIGURE 6.2
Showing d.c. machine construction.

Definition

Direct current motors are classified by the way in which the field and armature windings are connected, which may be in series or in parallel.

Example

A coil which is made up of a conductor some 15 m in length, lies at right angles to a magnetic field of strength 5 T. Calculate the force on the conductor when 15 A flows in the coil.

$$F = Bll \text{ (N)}$$

$$F = 5\,\text{T} \times 15\,\text{m} \times 15\,\text{A} = 1125 \text{ (N)}$$

PRACTICAL D.C. MOTORS

Practical motors are constructed as shown in Fig. 6.2. All d.c. motors contain a field winding wound on pole pieces attached to a steel yoke. The armature winding rotates between the poles and is connected to the commutator. Contact with the external circuit is made through carbon brushes rubbing on the commutator segments. **Direct current motors** are classified by the way in which the field and armature windings are connected, which may be in series or in parallel.

Series motor

The field and armature windings are connected in series and consequently share the same current. The series motor has the characteristics of a high starting torque but a speed which varies with load. Theoretically the motor would speed up to self-destruction, limited only by the windage of the rotating armature and friction, if the load were completely removed. Figure 6.3 shows

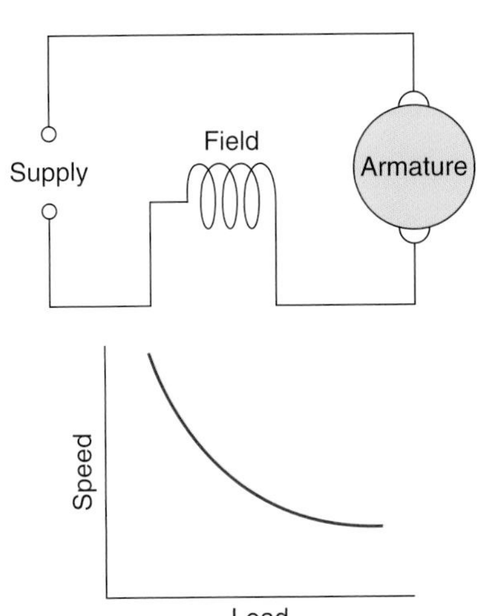

FIGURE 6.3
Series motor connections and characteristics.

Definition

Series motor: the machine will run on both a.c. or d.c. and is, therefore, sometimes referred to as a *'universal'* motor.

Key Fact

Series motor

- an electric drill motor,
- a vacuum cleaner motor,
- an electric train motor,
- are all series motors.

series motor connections and characteristics. For this reason the motor is only suitable for direct coupling to a load, except in very small motors, such as vacuum cleaners and hand drills, and is ideally suited for applications where the machine must start on load, such as electric trains, cranes and hoists.

Reversal of rotation may be achieved by reversing the connections of either the field or armature windings but not both. This characteristic means that the machine will run on both a.c. or d.c. and is, therefore, sometimes referred to as a '**universal**' motor.

Shunt motor

The field and armature windings are connected in parallel (see Fig. 6.4). Since the field winding is across the supply, the flux and motor speed are considered constant under normal conditions. In practice, however, as the load increases the field flux distorts and there is a small drop in speed of about 5% at full load, as shown in Fig. 6.4. The machine has a low starting torque and it is advisable to start with the load disconnected. The shunt motor is a very desirable d.c. motor because of its constant speed characteristics. It is used for driving power tools, such as lathes and drills. Reversal of rotation may be achieved by reversing the connections to either the field or armature winding but not both.

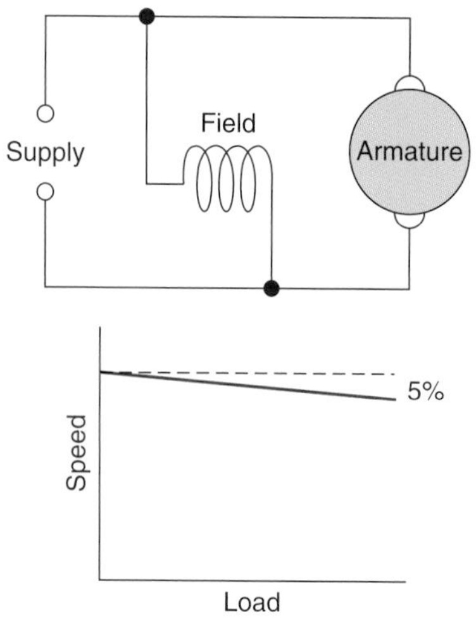

FIGURE 6.4
Shunt motor connections and characteristics.

Compound motor

The compound motor has two field windings – one in series with the armature and the other in parallel. If the field windings are connected so that the field flux acts in opposition, the machine is known as a *short shunt* and has the characteristics of a series motor. If the fields are connected so that the field flux is strengthened, the machine is known as a *long shunt* and has constant speed characteristics similar to a shunt motor. The arrangement of compound motor connections is given in Fig. 6.5. The compound motor may be designed to possess the best characteristics of both series and shunt motors, that is, good starting torque together with almost constant speed. Typical applications are for electric motors in steel rolling mills, where a constant speed is required under varying load conditions.

Definition

One of the *advantages of a d.c. machine* is the ease with which the speed may be controlled.

SPEED CONTROL OF D.C. MACHINES

One of the **advantages of a d.c. machine** is the ease with which the speed may be controlled. The speed of a d.c. motor is inversely proportional to the strength of the magnetic flux in the field winding. The magnetic flux in the field winding can be controlled by the field current and, as a result, controlling the field current will control the motor speed.

A variable resistor connected into the field circuit, as shown in Fig. 6.6 provides one method of controlling the field current and the motor speed. This method has the disadvantage that much of the input energy is dissipated in the variable resistor and an alternative, when an a.c. supply is available, is to use thyristor control.

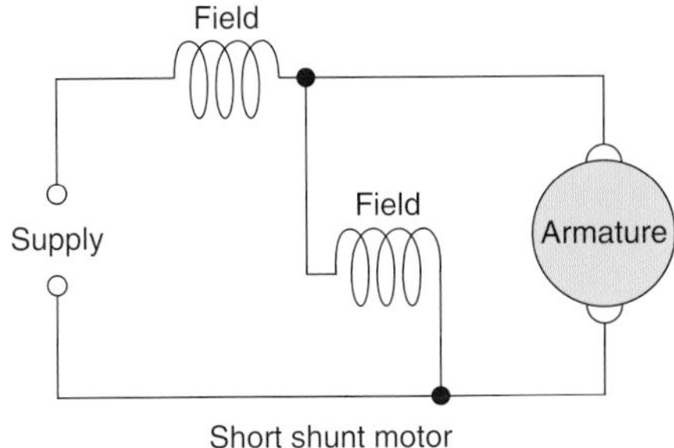

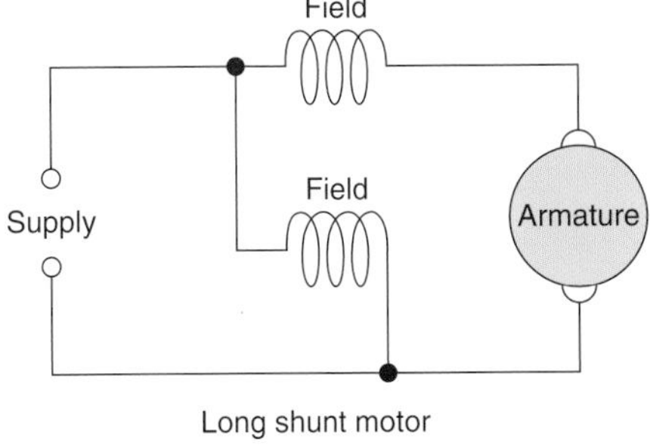

FIGURE 6.5
Compound motor connections.

Definition

If a *three-phase supply* is connected to three separate windings equally distributed around the stationary part or stator of an electrical machine, an alternating current circulates in the coils and establishes a magnetic flux.

BACK EMF AND MOTOR STARTING

When the armature conductors cut the magnetic flux of the main field, an emf is induced in the armature, as described earlier in Chapter 4 (Fig. 4.20). This induced emf is known as the back emf, since it acts in opposition to the supply voltage. During normal running, the back emf is always a little smaller than the supply voltage, and acts as a limit to the motor current. However, when the motor is first switched on, the back emf does not exist because the conductors are stationary and so a motor starter is required to limit the starting current to a safe value. This applies to all but the very smallest of motors and is achieved by connecting a resistor in series with the armature during starting, so that the resistance can be gradually reduced as the speed builds up.

The control switch of Fig. 6.7 is moved to the start position, which connects the variable resistors in series with the motor, thereby limiting the starting current. The control switch is moved progressively over the variable resistor contacts to the run position as the motor speed builds up. A practical motor starter is designed so that the control switch returns automatically to the 'off' position whenever the motor stops, so that the starting resistors are connected when the machine is once again switched on.

Three-phase a.c. motors

If a **three-phase supply** is connected to three separate windings equally distributed around the stationary part or stator of an electrical machine, an alternating current circulates in the coils and establishes a magnetic flux. The magnetic field established by the three-phase currents travels in a clockwise direction around the stator, as can be seen by considering

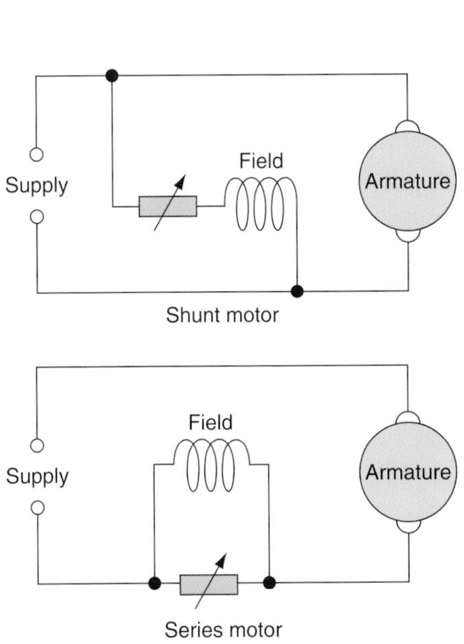

FIGURE 6.6
Speed control of a d.c. motor.

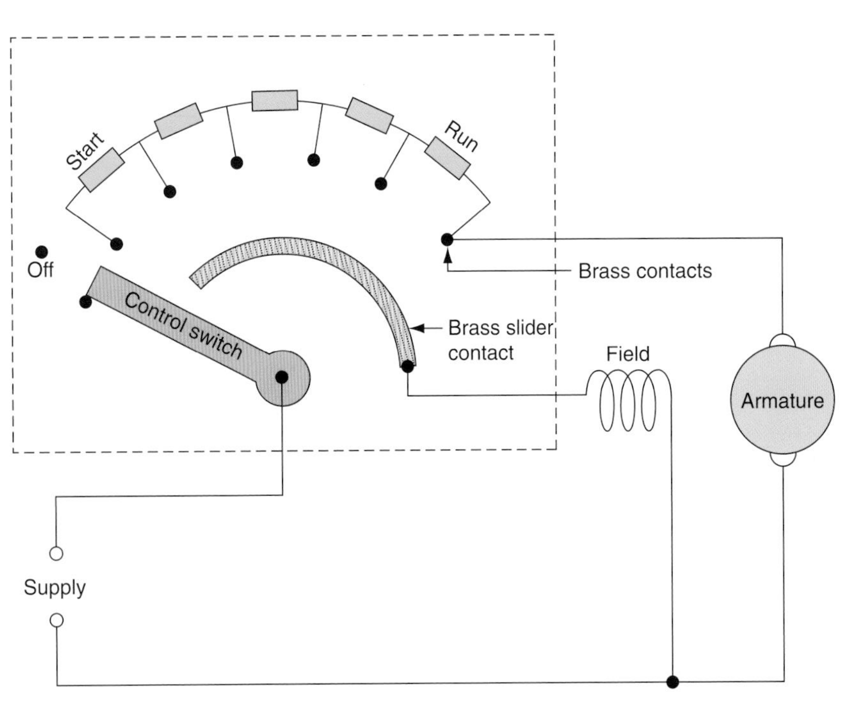

FIGURE 6.7
A d.c. motor starting.

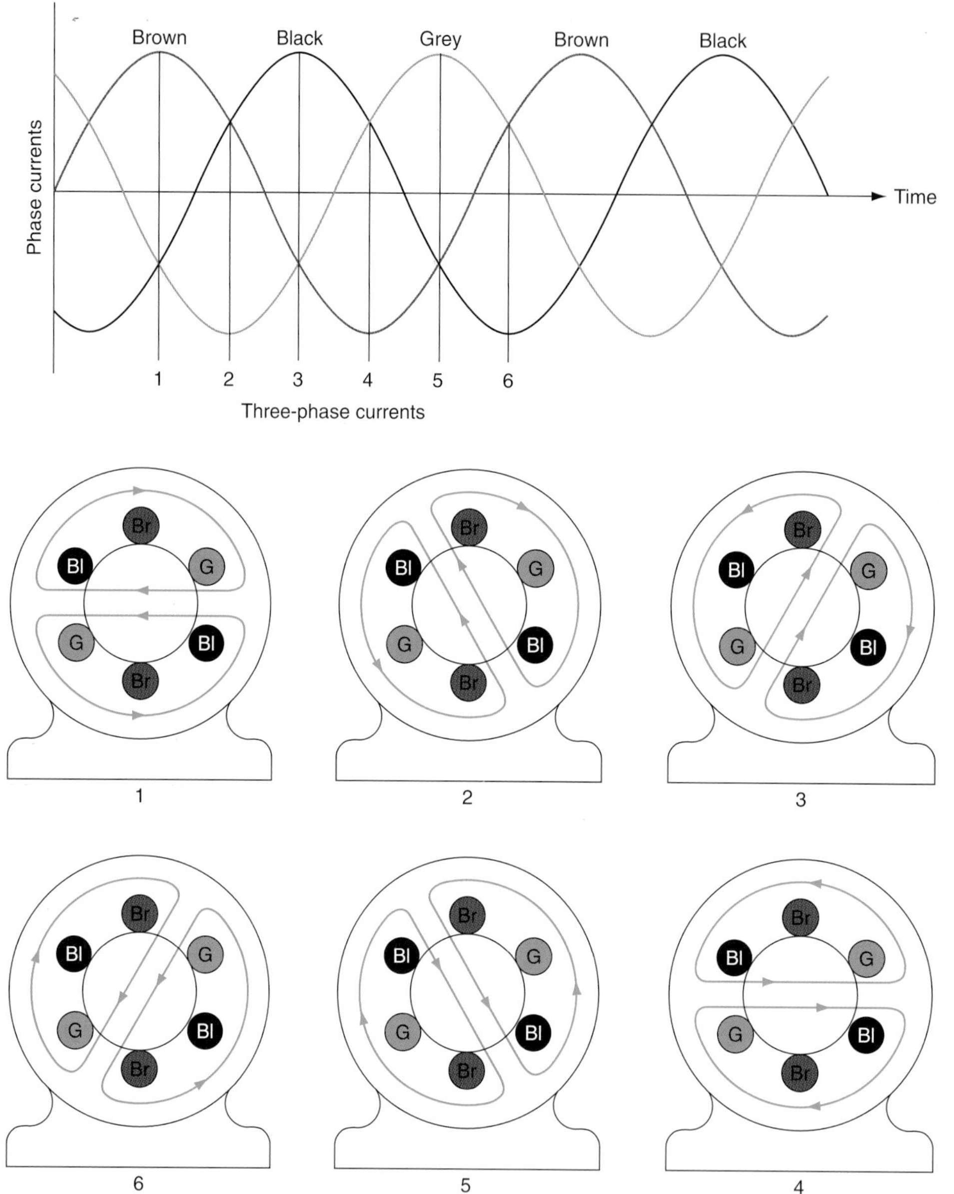

FIGURE 6.8
Distribution of magnetic flux due to three-phase currents.

the various intervals of time 1–6 shown in Fig. 6.8. The three-phase supply establishes a rotating magnetic flux which rotates at the same speed as the supply frequency. This is called synchronous speed, denoted as n_S:

$$n_s = \frac{f}{P} \text{ or } N_S = \frac{60f}{P}$$

where:

n_S is measured in revolutions per second,

N_S is measured in revolutions per minute,

f is the supply frequency measured in hertz,

P is the number of pole pairs.

Example

Calculate the synchronous speed of a four-pole machine connected to a 50 Hz mains supply.

We have

$$n_S = \frac{f}{p}\,(\text{rps})$$

A four-pole machine has two pairs of poles:

$$\therefore\ n_S = \frac{50\ \text{Hz}}{2} = 25\ \text{rps}$$

$$\text{or } N_S = \frac{60 \times 50\ \text{Hz}}{2} = 1500\ \text{rpm}$$

This rotating magnetic field is used for practical effect in the induction motor.

THREE-PHASE INDUCTION MOTOR

When a three-phase supply is connected to insulated coils set into slots in the inner surface of the stator or stationary part of an induction motor as shown in Fig. 6.9(a), a rotating magnetic flux is produced. The rotating magnetic flux cuts the conductors of the rotor and induces an emf in the rotor conductors by Faraday's law, which states that when a conductor cuts or is cut by a magnetic field, an emf is induced in that conductor, the magnitude of which is proportional to the *rate* at which the conductor cuts or is cut by the magnetic flux. This induced emf causes rotor currents to flow and establish a magnetic flux which reacts with the stator flux and causes a force to be exerted on the rotor conductors, turning the rotor as shown in Fig. 6.9(b).

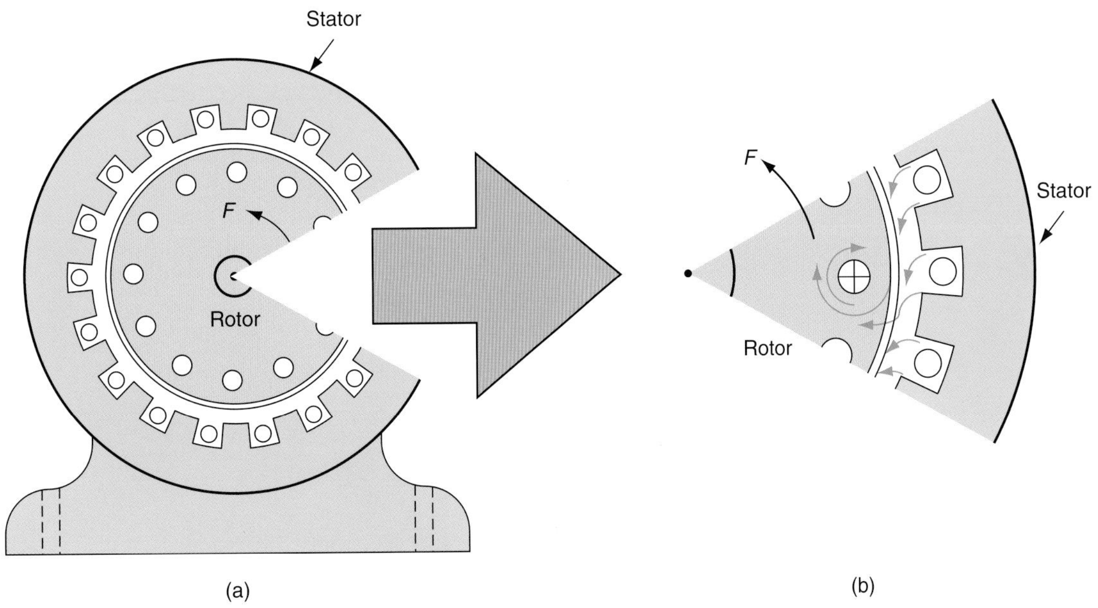

FIGURE 6.9
Segment taken out of an induction motor to show turning force: (a) construction of an induction motor; (b) production of torque by magnetic fields.

The turning force or torque experienced by the rotor is produced by inducing an emf into the rotor conductors due to the *relative* motion between the conductors and the rotating field. The torque produces rotation in the same direction as the rotating magnetic field. At switch-on, the rotor speed increases until it approaches the speed of the rotating magnetic flux, that is, the synchronous speed. The faster the rotor revolves the less will be the difference in speed between the rotor and the rotating magnetic field. By Faraday's laws, this will result in less induced emf, less rotor current and less torque on the rotor. The rotor can never run at synchronous speed because, if it did so, there would be no induced emf, no current and no torque. The induction motor is called an asynchronous motor. In practice, the rotor runs at between 2% and 5% below the synchronous speed so that a torque can be maintained on the rotor which overcomes the rotor losses and the applied load.

The difference between the rotor speed and synchronous speed is called slip; the per-unit slip, denoted *s*, is given by:

$$s = \frac{n_S - n}{n_S} = \frac{N_S - N}{N_S}$$

where

n_S = synchronous speed in revolutions per second

N_S = synchronous speed in revolutions per minute

n = rotor speed in revolutions per second

N = rotor speed in revolutions per minute.

The percentage slip is just the per-unit slip multiplied by 100.

Example

A two-pole induction motor runs at 2880 rpm when connected to the 50 Hz mains supply. Calculate the percentage slip.

The synchronous speed is given by:

$$N_S = \frac{60 \times f}{p} \text{(rpm)}$$

$$\therefore N_S = \frac{60 \times 50\ \text{Hz}}{1} = 3000\ \text{rpm}$$

Thus the per-unit slip is:

$$s = \frac{N_S - N}{N_S}$$

$$\therefore s = \frac{3000\ \text{rpm} - 2880\ \text{rpm}}{3000\ \text{rpm}}$$

$$s = 0.04$$

So the percentage slip is $0.04 \times 100 = 4\%$.

Definition

There are two types of *induction motor rotor* – the wound rotor and the cage rotor.

ROTOR CONSTRUCTION

There are two types of **induction motor rotor** – the wound rotor and the cage rotor. The cage rotor consists of a laminated cylinder of silicon steel with copper or aluminium bars slotted in holes around the circumference and short-circuited at each end of the cylinder as shown in Fig. 6.10. In small motors the rotor is cast in aluminium. Better starting and quieter running are achieved if the bars are slightly skewed. This type of rotor is extremely robust and since there are no external connections there is no need for slip-rings or brushes. A machine fitted with a cage rotor does suffer from a low starting torque and the machine must be chosen which has a higher starting torque than the load, as shown by curve (b) in Fig. 6.11. A machine with the characteristic shown by curve (a) in Fig. 6.11 would not start since the load torque is greater than the machine starting torque.

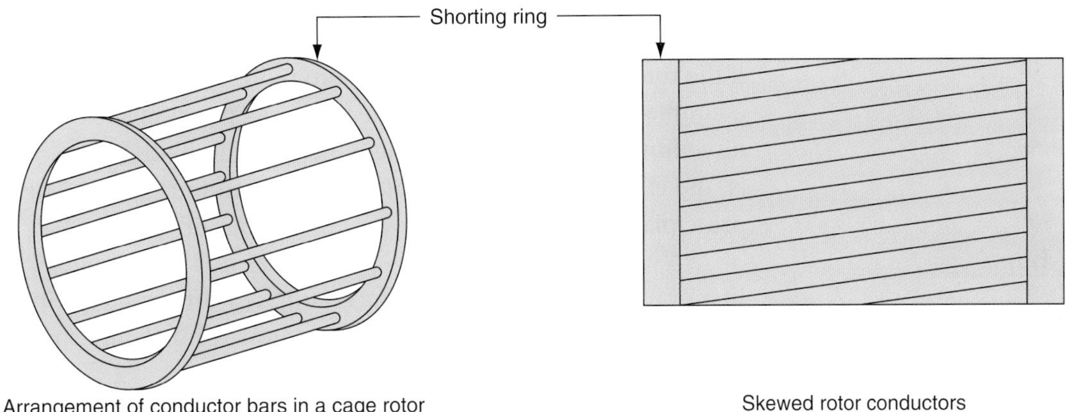

FIGURE 6.10
Construction of a cage rotor.

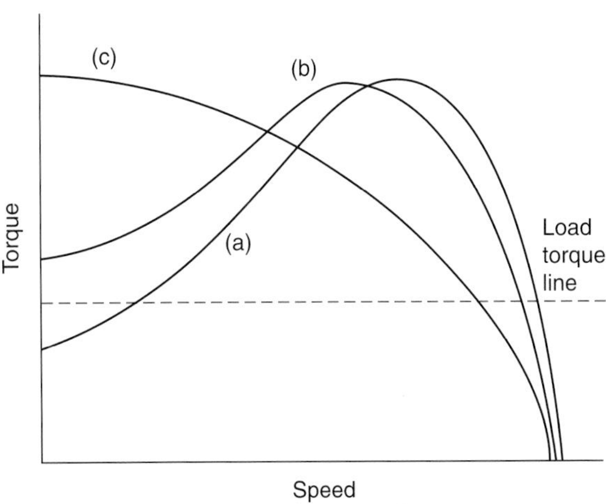

FIGURE 6.11
Various speed–torque characteristics for an induction motor.

Alternatively, the load may be connected after the motor has been run up to full speed, or extra resistance can be added to a wound rotor through slip-rings and brushes since this improves the starting torque, as shown by curve (c) in Fig. 6.11. The wound rotor consists of a laminated cylinder of silicon steel with copper coils embedded in slots around the circumference. The windings may be connected in star or delta and the end connections brought out to slip-rings mounted on the shaft. Connection by carbon brushes can then be made to an external resistance to improve starting, but once normal running speed is achieved the external resistance is short circuited. Therefore, the principle of operation for both types of rotor is the same.

The cage induction motor has a small starting torque and should be used with light loads or started with the load disconnected. The speed is almost constant at about 5% less than synchronous speed. Its applications are for constant speed machines such as fans and pumps. Reversal of rotation is achieved by reversing any two of the stator winding connections.

THREE-PHASE SYNCHRONOUS MOTOR

If the rotor of a three-phase induction motor is removed and replaced with a simple magnetic compass, the compass needle will rotate in the same direction as the rotating magnetic field set up by the stator winding. That is, the compass needle will rotate at synchronous speed. This is the basic principle of operation of the synchronous motor.

Key Fact

Cage rotor

A cage rotor induction motor

- requires very little maintenance,
- does not have carbon brushes,
- does not have a commutator,
- is almost indestructible,
- is used extensively in industry,
- is used for fans and pumps.

In a practical machine, the rotor is supplied through slip-rings with a d.c. supply which sets up an electromagnet having north and south poles.

When the supply is initially switched on, the rotor will experience a force, first in one direction and then in the other direction every cycle as the stator flux rotates around the rotor at synchronous speed. Therefore, the synchronous motor is not self-starting. However, if the rotor is rotated at or near synchronous speed, then the stator and rotor poles of opposite polarity will 'lock together' producing a turning force or torque which will cause the rotor to rotate at synchronous speed.

If the rotor is slowed down and it comes out of synchronism, then the rotor will stop because the torque will be zero. The synchronous motor can, therefore, only be run at synchronous speed, which for a 50 Hz supply will be 3000, 1500, 1000 or 750 rpm depending upon the number of poles, as discussed earlier in this chapter.

A practical synchronous machine can be brought up to synchronous speed by either running it initially as an induction motor or by driving it up to synchronous speed by another motor called a 'pony motor'. Once the rotor achieves synchronous speed the pony motor is disconnected and the load applied to the synchronous motor.

With such a complicated method of starting the synchronous motor, it is clearly not likely to find applications which require frequent stopping and starting. However, the advantage of a synchronous motor is that it runs at a constant speed and operates at a leading power factor (p.f.). It can, therefore, be used to improve a bad p.f. while driving constant speed machines such as ventilation fans and pumping compressors.

Definition

A *single-phase a.c. supply* produces a pulsating magnetic field, not the rotating magnetic field produced by a three-phase supply. All a.c. motors require a rotating field to start. Therefore, single-phase a.c. motors have two windings which are electrically separated by about 90°. The two windings are known as the start and run windings.

Definition

Once rotation is established, the *pulsating field* in the run winding is sufficient to maintain rotation and the start winding is disconnected by a centrifugal switch which operates when the motor has reached about 80% of the full load speed.

Single-phase a.c. motors

A **single-phase a.c. supply** produces a pulsating magnetic field, not the rotating magnetic field produced by a three-phase supply. All a.c. motors require a rotating field to start. Therefore, single-phase a.c. motors have two windings which are electrically separated by about 90°.The two windings are known as the start and run windings. The magnetic fields produced by currents flowing through these out-of-phase windings create the rotating field and turning force required to start the motor. Once rotation is established, the **pulsating field** in the run winding is sufficient to maintain rotation and the start winding is disconnected by a centrifugal switch which operates when the motor has reached about 80% of the full load speed.

A cage rotor is used on single-phase a.c. motors, the turning force being produced in the way described previously for three-phase induction motors and shown in Fig. 6.9. Because both windings carry currents which are out-of-phase with each other, the motor is known as a 'split-phase' motor. The phase displacement between the currents in the windings is achieved in one of two ways:

- by connecting a capacitor in series with the start winding, as shown in Fig. 6.12(a), which gives a 90° phase difference between the currents in the start and run windings;
- by designing the start winding to have a high resistance and the run winding a high inductance, once again creating a 90° phase shift between the currents in each winding, as shown in Fig. 6.12(b).

When the motor is first switched on, the centrifugal switch is closed and the magnetic fields from the two coils produce the turning force required

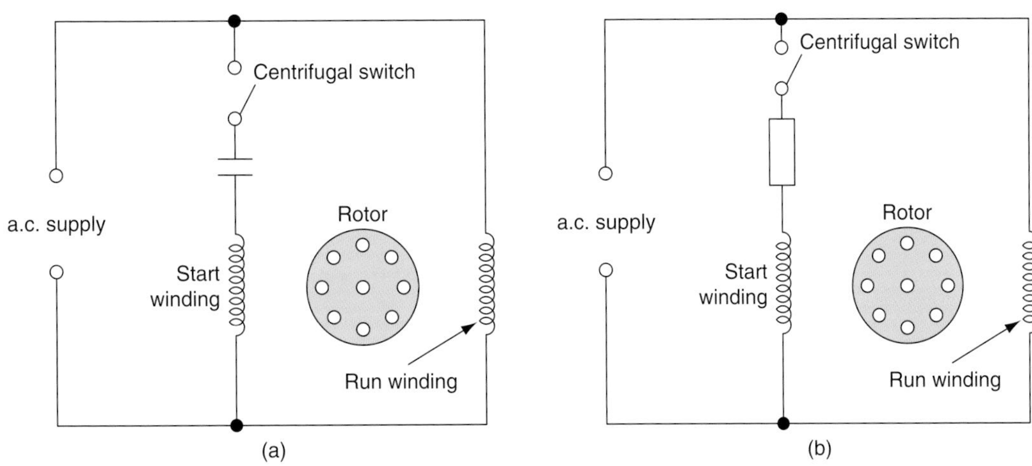

FIGURE 6.12
Circuit diagram of: (a) capacitor split-phase motors; (b) resistance split-phase motors.

to run the rotor up to full speed. When the motor reaches about 80% of full speed, the centrifugal switch clicks open and the machine continues to run on the magnetic flux created by the run winding only.

Split-phase motors are constant speed machines with a low starting torque and are used on light loads such as fans, pumps, refrigerators and washing machines. Reversal of rotation may be achieved by reversing the connections to the start or run windings, but not both.

SHADED POLE MOTORS

The shaded pole motor is a simple, robust single-phase motor, which is suitable for very small machines with a rating of less than about 50 W. Figure 6.13 shows a shaded pole motor. It has a cage rotor and the moving field is produced by enclosing one side of each stator pole in a solid copper or brass ring, called a shading ring, which displaces the magnetic field and creates an artificial phase shift.

Shaded pole motors are constant speed machines with a very low starting torque and are used on very light loads such as oven fans, record turntable motors and electric fan heaters. Reversal of rotation is theoretically possible by moving the shading rings to the opposite side of the stator pole face. However, in practice this is often not a simple process, but the motors are symmetrical and it is sometimes easier to reverse the rotor by removing the fixing bolts and reversing the whole motor.

There are more motors operating from single-phase supplies than all other types of motor added together. Most of them operate as very small motors in domestic and business machines where single-phase supplies are most common.

Definition

The purpose of the *motor starter* is not to start the machine, as the name implies, but to reduce heavy starting currents and provide overload and no-volt protection in accordance with the requirements of Regulations 552.

Motor starters

The magnetic flux generated in the stator of an induction motor rotates immediately the supply is switched on, and therefore the machine is self-starting. The purpose of the **motor starter** is not to start the machine, as the name implies, but to reduce heavy starting currents and provide overload and no-volt protection in accordance with the requirements of Regulations 552.

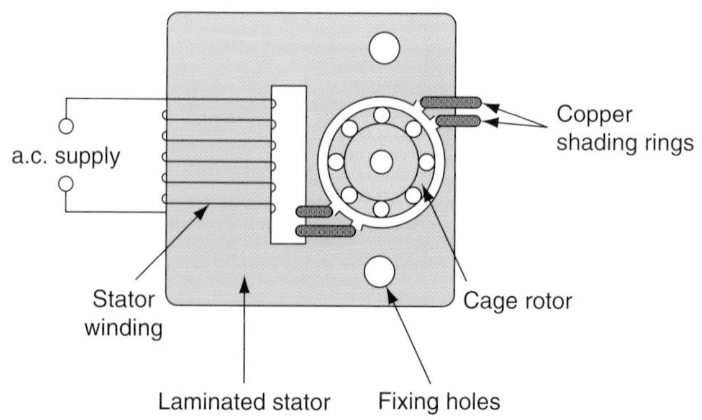

FIGURE 6.13
Shaded pole motor.

Thermal overload protection is usually provided by means of a bimetal strip bending under overload conditions and breaking the starter contactor coil circuit. This de-energizes the coil and switches off the motor under fault conditions such as overloading or single phasing. Once the motor has automatically switched off under overload conditions or because a remote stop/start button has been operated, it is an important safety feature that the motor cannot restart without the operator going through the normal start-up procedure. Therefore, no-volt protection is provided by incorporating the safety devices into the motor starter control circuit which energizes the contactor coil.

Electronic thermistors (thermal transistors) provide an alternative method of sensing if a motor is overheating (see Chapter 4 of this book). These tiny heat-sensing transistors, about the size of a matchstick head, are embedded in the motor windings to sense the internal temperature, and the thermistor either trips out the contactor coil as described above or operates an alarm.

All electric motors with a rating above 0.37 kW must be supplied from a suitable motor starter and we will now consider the more common types.

DIRECT ON LINE (D.O.L.) STARTERS

The d.o.l. starter switches the main supply directly on to the motor. Since motor starting currents can be seven or eight times greater than the running current, the d.o.l. starter is only used for small motors of less than about 5 kW rating.

When the start button is pressed current will flow from the brown line through the control circuit and contactor coil to the grey line which energizes the contactor coil and the contacts close, connecting the three-phase supply to the motor, as can be seen in Fig. 6.14. If the start button is released the control circuit is maintained by the hold on contact. If the

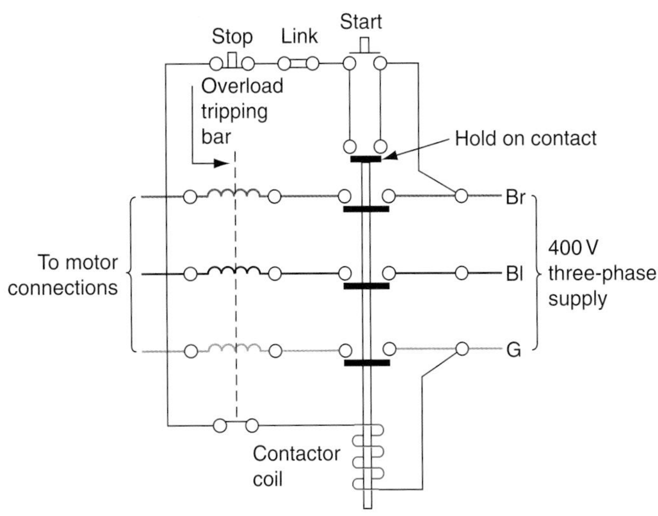

FIGURE 6.14
Three-phase d.o.l. starter.

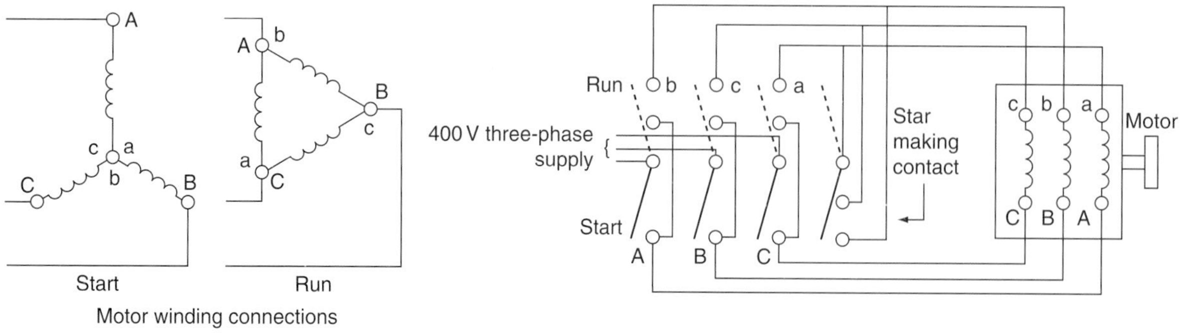

Start Run
Motor winding connections

FIGURE 6.15
Star delta starter.

stop button is pressed or the overload coils operate, the control circuit is broken and the contractor drops out, breaking the supply to the load. Once the supply is interrupted the supply to the motor can only be reconnected by pressing the start button. Therefore this type of arrangement also provides no-volt protection.

When large industrial motors have to be started, a way of reducing the excessive starting currents must be found. One method is to connect the motor to a star delta starter.

STAR DELTA STARTERS

When three loads, such as the three windings of a motor, are connected in star, the line current has only one-third of the value it has when the same load is connected in delta. A starter which can connect the motor windings in star during the initial starting period and then switch to delta connection will reduce the problems of an excessive starting current. This arrangement is shown in Fig. 6.15, where the six connections to the three stator phase windings are brought out to the starter. For starting, the motor windings are star-connected at the a–b–c end of the winding by the star making contacts. This reduces the phase voltage to about 58% of the running voltage which reduces the current and the motor's torque. Once the motor is running a double-throw switch makes the changeover from star starting to delta running, thereby achieving a minimum starting current and maximum running torque. The starter will incorporate overload and no-volt protection, but these are not shown in Fig. 6.15 in the interests of showing more clearly the principle of operation.

AUTO-TRANSFORMER STARTER

An auto-transformer motor starter provides another method of reducing the starting current by reducing the voltage during the initial starting period. Since this also reduces the starting torque, the voltage is only reduced by a sufficient amount to reduce the starting current, being permanently connected to the tapping found to be most appropriate by the installing electrician. Switching the changeover switch to the start position connects the auto-transformer windings in series with the delta-connected motor starter winding. When sufficient speed has been achieved by the

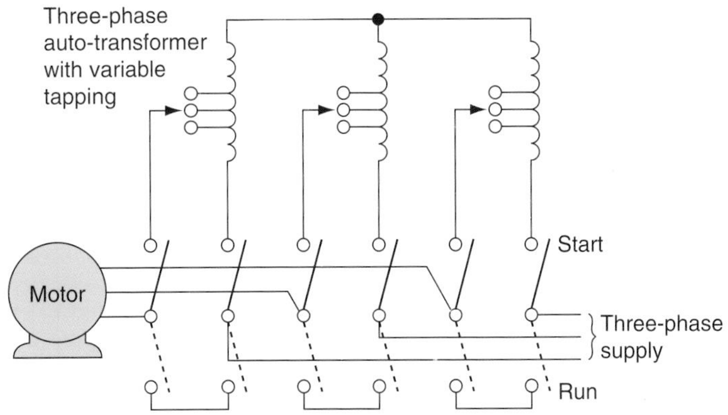

FIGURE 6.16
Auto-transformer starting.

motor the changeover switch is moved to the run connections which connect the three-phase supply directly on to the motor as shown in Fig. 6.16.

This starting method has the advantage of only requiring three connection conductors between the motor starter and the motor. The starter will incorporate overload and no-volt protection in addition to some method of preventing the motor being switched to the run position while the motor is stopped. These protective devices are not shown in Fig. 6.16 in order to show more clearly the principle of operation.

Key Fact

Shaded pole motors

- are very small motors,
- have a constant speed,
- use a cage rotor,
- are used in electric ovens, microwave ovens and fan heaters.

ROTOR RESISTANCE STARTER

When starting a machine on load a wound rotor induction motor must generally be used since this allows an external resistance to be connected to the rotor winding through slip-rings and brushes, which increases the starting torque as shown in Fig. 6.11 curve (c).

When the motor is first switched on the external rotor resistance is at a maximum. As the motor speed increases the resistance is reduced until at full speed the external resistance is completely cut out and the machine runs as a cage induction motor. The starter is provided with overload and no-volt protection and an interlock to prevent the machine being switched on with no rotor resistance connected, but these are not shown in Fig. 6.17 since the purpose of the diagram is to show the principle of operation.

Remote control of motors

When it is required to have stop/start control of a motor at a position other than the starter position, additional start buttons may be connected in parallel and additional stop buttons in series, as shown in Fig. 6.18 for the d.o.l. starter. This is the diagram shown in Fig. 6.14 with the link removed and a remote stop/start button connected. Additional stop and start facilities are often provided for the safety and convenience of the machine operator.

Installation of motors

Electric motors vibrate when running and should be connected to the electrical installation through a flexible connection. This may also make

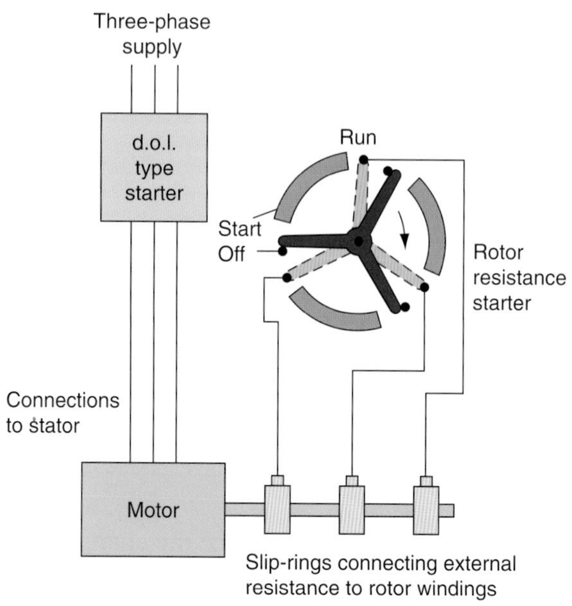

FIGURE 6.17
Rotor resistance starter for a wound rotor machine.

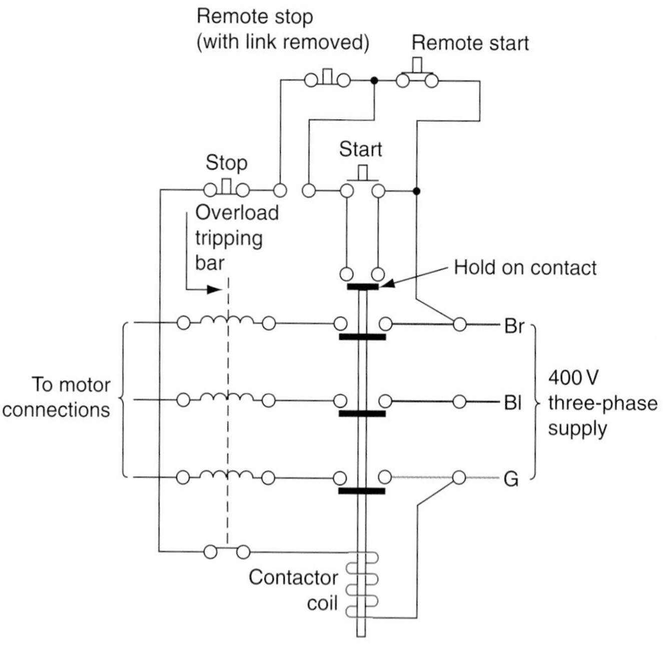

FIGURE 6.18
Remote stop/start connections to d.o.l. starter.

final adjustments of the motor position easier. Where the final connection is made with flexible conduit, the tube must not be relied upon to provide a continuous earth path and a separate circuit protective conductor (CPC) must be run either inside or outside the flexible conduit (Regulation 543.2.1).

All motors over 0.37 kW rating must be connected to the source of supply through a suitable starter which incorporates overload protection and a device which prevents dangerous restarting of the motor following a mains failure (Regulations 552.1.2 and 3).

The cables supplying the motor must be capable of carrying at least the full load current of the motor (Regulation 552.1.1) and a local means of isolation must be provided to facilitate safe mechanical maintenance (Regulations 537.2.1.2 and 537.3.1.1).

At the supply end, the motor circuit will be protected by a fuse or miniature circuit breaker (MCB). The supply protection must be capable of withstanding the motor starting current while providing adequate overcurrent protection. There must also be discrimination so that the overcurrent device in the motor starter operates first in the event of an excessive motor current.

Most motors are 'continuously rated'. This is the load at which the motor may be operated continuously without overheating.

Many standard motors have class A insulation which is suitable for operating in ambient temperatures up to about 55°C. If a class A motor is to be operated in a higher ambient temperature, the continuous rating may need to be reduced to prevent damage to the motor. The motor and its enclosure must be suitable for the installed conditions and must additionally prevent anyone coming into contact with the internal live or moving parts. Many different enclosures are used, depending upon the atmosphere in which the motor is situated. Clean air, damp conditions, dust particles in the atmosphere, chemical or explosive vapours will determine the type of motor enclosure. In high ambient temperatures it may be necessary to provide additional ventilation to keep the motor cool and prevent the lubricating oil thinning. The following motor enclosures are examples of those to be found in industry.

Safety First

Motors

- All electric motors over 0.37 kW must have a local means of isolation for safe mechanical maintenance.
- IEE Regulation 537.3.1.1

Screen protected enclosures prevent access to the internal live and moving parts by covering openings in the motor casing with metal screens of perforated metal or wire mesh. Air flow for cooling is not restricted and is usually assisted by a fan mounted internally on the machine shaft. This type of enclosure is shown in Fig. 6.19.

A duct ventilated enclosure is used when the air in the room in which the motor is situated is unsuitable for passing through the motor for cooling – for example, when the atmosphere contains dust particles or chemical vapour. In these cases the air is drawn from a clean air zone outside the room in which the machine is installed, as shown in Fig. 6.19.

A totally enclosed enclosure is one in which the air inside the machine casing has no connection with the air in the room in which it is installed, but it is not necessarily airtight. A fan on the motor shaft inside the casing circulates the air through the windings and cooling is by conduction through the motor casing. To increase the surface area and assist cooling, the casing is surrounded by fins, and an externally mounted fan can increase the flow of air over these fins. This type of enclosure is shown in Fig. 6.19.

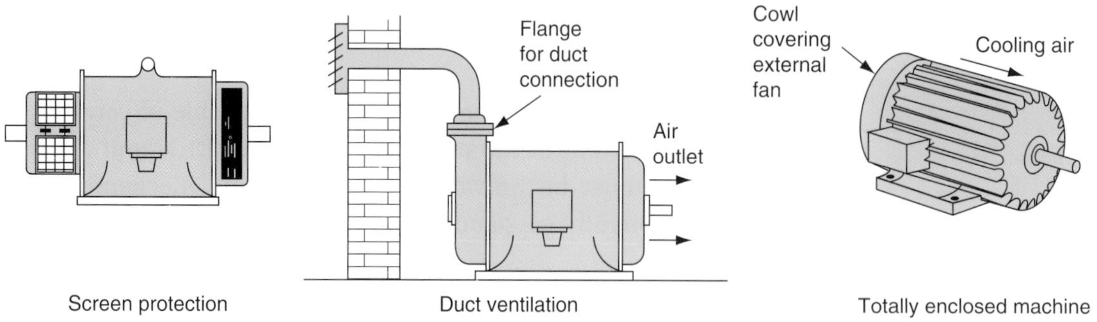

FIGURE 6.19
Motor enclosures.

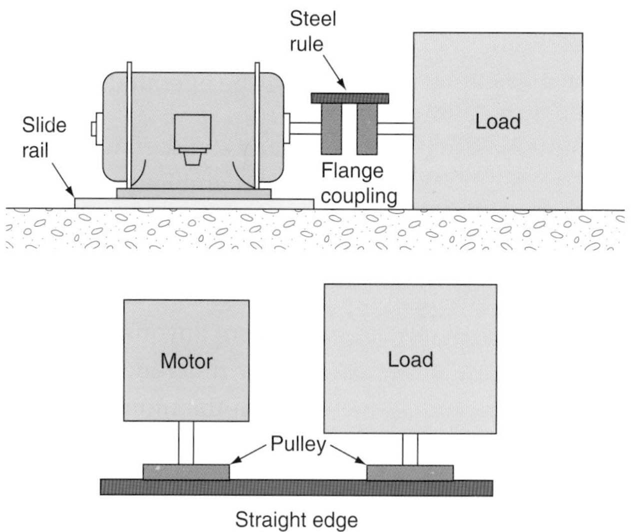

FIGURE 6.20
Pulley and flange coupling arrangement.

A flameproof enclosure requires that further modifications be made to the totally enclosed casing to prevent inflammable gases coming into contact with sparks or arcing inside the motor. To ensure that the motor meets the stringent regulations for flameproof enclosures the shaft is usually enclosed in special bearings and the motor connections contained by a wide flange junction box.

When a motor is connected to a load, either by direct coupling or by a vee belt, it is important that the shafts or pulleys are exactly in line. This is usually best achieved by placing a straight edge or steel rule across the flange coupling of a direct drive or across the flat faces of a pair of pulleys, as shown in Fig. 6.20. Since pulley belts stretch in use it is also important to have some means of adjusting the tension of the vee belt. This is usually achieved by mounting the motor on a pair of slide rails as shown in Fig. 6.21. Adjustment is carried out by loosening the motor fixing bolts, screwing in the adjusting bolts which push the motor back, and when the correct belt tension has been achieved the motor fixing bolts are tightened.

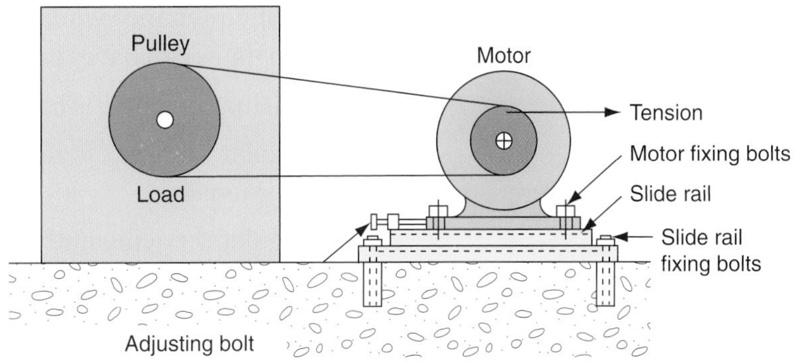

FIGURE 6.21
Vee belt adjustment of slide rail mounted motor.

Motor maintenance

All rotating machines are subject to wear, simply because they rotate. Motor fans which provide cooling also pull dust particles from the surrounding air into the motor enclosure. Bearings dry out, drive belts stretch and lubricating oils and greases require replacement at regular intervals. Industrial electric motors are often operated in a hot, dirty, dusty or corrosive environment for many years. If they are to give good and reliable service they must be suitable for the task and the conditions in which they must operate. Maintenance at regular intervals is required, in the same way that a motor car is regularly serviced.

Definition

The solid construction of the *cage rotor* used in many a.c. machines makes them almost indestructible.

The solid construction of the **cage rotor** used in many a.c. machines makes them almost indestructible, and, since there are no external connections to the rotor, the need for slip-rings and brushes is eliminated. These characteristics give cage rotor a.c. machines maximum reliability with the minimum of maintenance and make the induction motor the most widely used in industry. Often the only maintenance required with an a.c. machine is to lubricate in accordance with the manufacturer's recommendations.

However, where high torque and variable speed characteristics are required d.c. machines are often used. These require a little more maintenance than a.c. machines because the carbon brushes, rubbing on the commutator, wear down and require replacing. New brushes must be of the correct grade and may require 'bedding in' or shaping with a piece of fine abrasive cloth to the curve of the commutator.

The commutator itself should be kept clean and any irregularities smoothed out with abrasive cloth. As the commutator wears, the mica insulation between the segments must be cut back with an undercutting tool or a hacksaw blade to keep the commutator surface smooth. If the commutator has become badly worn, and a groove is evident, the armature will need to be removed from the motor, and the commutator turned in a lathe.

Motors vibrate when operating and as a result fixing bolts and connections should be checked as part of the maintenance operation. Where a motor drives a load via a pulley belt, the motor should be adjusted on the slider rails until there is about 10 mm of play in the belt.

Planning maintenance work with forethought and keeping records of work done with dates can have the following advantages:

- The maintenance is carried out when it is most convenient.
- Regular simple maintenance often results in less emergency maintenance.
- Regular servicing and adjustment maintain the plant and machines at peak efficiency.

The result of planned maintenance is often that fewer breakdowns occur, which result in loss of production time. Therefore, a planned maintenance programme must be a sensible consideration for any commercial operator.

Power-factor correction

Most electrical installations have a low p.f. because loads such as motors, transformers and discharge lighting circuits are inductive in nature and cause the current to lag behind the voltage. A capacitor has the opposite effect to an inductor, causing the current to lead the voltage. Therefore, by adding capacitance to an inductive circuit the bad p.f. can be corrected. The load current I_L is made up of an in-phase component I and a quadrature component I_Q. The p.f. can be corrected to unity when the capacitor current I_C is equal and opposite to the quadrature or reactive current I_Q of the inductive load. The quadrature or reactive current is responsible for setting up the magnetic field in an inductive circuit. Figure 6.22 shows the p.f. corrected to unity, that is when $I_Q = I_C$.

A low p.f. is considered a disadvantage because a given load takes more current at a low p.f. than it does at a high p.f. In Chapter 7 we calculated that a 1.84 kW load at unity p.f. took 8 A, but at a bad p.f. of 0.4 a current of 20 A was required to supply the same load.

The supply authorities discourage industrial consumers from operating at a bad p.f. because:

- larger cables and switchgear are necessary to supply a given load;
- larger currents give rise to greater copper losses in transmission cables and transformers;
- larger currents give rise to greater voltage drops in cables;

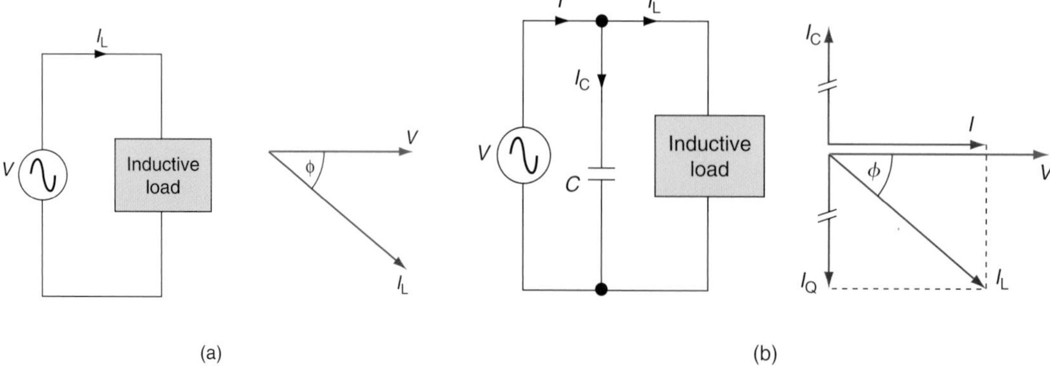

FIGURE 6.22
Power-factor correction of inductive load: (a) circuit and phasor diagram for an inductive load with low p.f.; (b) circuit and phasor diagram for (a) with capacitor correcting p.f. to unity.

- larger cables may be required on the consumer's side of the electrical installation to carry the larger currents demanded by a load operating with a bad p.f.

Bad p.f.s are corrected by connecting a capacitor either across the individual piece of equipment or across the main busbars of the installation. When individual capacitors are used they are usually of the paper dielectric type of construction (see Fig. 4.23 in Chapter 4). This is the type of capacitor used for p.f. correction in a fluorescent luminaire. When large banks of capacitors are required to correct the p.f. of a whole installation, paper dielectric capacitors are immersed in an oil tank in a similar type construction to a transformer, and connected on to the main busbars of the electrical installation by suitably insulated and mechanically protected cables.

The current to be carried by the capacitor for p.f. correction and the value of the capacitor may be calculated as shown by the following example.

Example

An 8 kW load with a p.f. of 0.7 is connected across a 400 V, 50 Hz supply. Calculate:

(a) the current taken by this load,
(b) the capacitor current required to raise the p.f. to unity,
(c) the capacitance of the capacitor required to raise the p.f. to unity.

For (a), since $P = VI\cos\phi\ (W)$

$$I = \frac{P}{V\cos\phi}\ (A)$$

$$\therefore I = \frac{8000\,W}{400\,V \times 0.7} = 28.57\ (A)$$

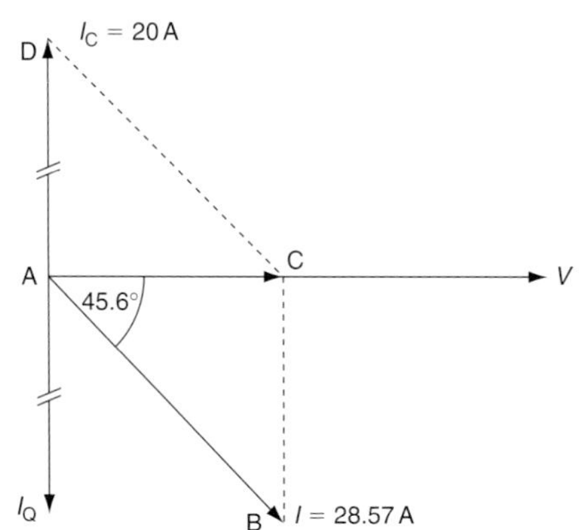

FIGURE 6.23
Phasor diagram.

This current lags the voltage by an angle of 45.6° (since $\cos^{-1} 0.7 = 45.6°$) and can therefore be drawn to scale as shown in Fig. 6.23 and represented by line AB.

For (b) at unity p.f. the current will be in phase with the voltage, represented by line AC in Fig. 6.23. To raise the load current to this value will require a capacitor current I_c which is equal and opposite to the value of the quadrature or reactive component I_Q. The value of I_Q is measured from the phasor diagram and found to be 20 A which is the value of the capacitor current required to raise the p.f. to unity and shown by line AD in Fig. 6.23.

For (c), since:

$$I_C = \frac{V}{X_C}\ (A) \quad \text{and} \quad X_C = \frac{V}{I_C}\ (\Omega)$$

$$\therefore X_C = \frac{400\,V}{20\,A} = 20\,\Omega$$

Since:

$$X_C = \frac{I}{2\pi fC}\ (\Omega) \quad \text{and} \quad C = \frac{I}{2\pi fX_C}\ (F)$$

$$\therefore C = \frac{I}{2 \times \pi \times 50\,Hz \times 20\,\Omega} = 159\ \mu F$$

A 159 μF capacitor connected in parallel with the 8 kW load would correct the p.f. to unity.

Check your Understanding

When you have completed these questions, check out the answers at the back of the book.
Note: more than one multiple choice answer may be correct.

1. The series motor finds many applications such as:
   a. constant speed steel rolling mills
   b. electric hand drills
   c. electric trains and trams
   d. vacuum cleaners.
2. The compound motor finds applications such as:
   a. constant speed steel rolling mills
   b. electric hand drills
   c. electric trains and trams
   d. vacuum cleaners.
3. The advantage of a d.c. machine is:
   a. it is easy to control the speed
   b. it runs at an almost constant speed
   c. it requires little maintenance
   d. it is practically indestructible.
4. The advantage of a cage rotor induction motor is:
   a. it is easy to control the speed
   b. it runs at an almost constant speed
   c. it requires little maintenance
   d. it is practically indestructible.
5. A cage rotor induction motor finds applications operating:
   a. electric oven fans and fan heaters
   b. pumps and fans in industry
   c. refrigerators and washing machines
   d. trains and trams.
6. A split-phase a.c. motor finds applications operating:
   a. electric oven fans and fan heaters
   b. pumps and fans in industry
   c. refrigerators and washing machines
   d. trains and trams.

7. A shaded pole motor finds applications operating:
   a. electric oven fans and fan heaters
   b. pumps and fans in industry
   c. refrigerators and washing machines
   d. trains and trams.

8. The purpose of the motor starter connected to an electric motor is:
   a. to start the motor
   b. to reduce high starting currents
   c. to provide overload protection
   d. to provide no-volt protection.

9. A capacitor connected in parallel with an electric motor provides:
   a. improved starting
   b. motor speed control
   c. power-factor correction
   d. safe mechanical maintenance.

10. Use bullet points and a sketch to describe how a three-phase a.c. supply connected to the stator winding of an induction motor causes the rotor shaft to turn around. (You might want to look at Fig. 6.9.)

11. Use a simple sketch to explain how a turning force is exerted upon the armature winding of a d.c. motor. (You might want to look at Fig. 6.1.)

12. Very briefly describe how a three-phase supply produces a rotating magnetic flux in an induction motor.

13. Use a labelled sketch to describe the construction of a cage rotor.

14. Why is a cage rotor practically indestructible?

15. Why does a wound rotor or armature winding require maintenance of the commutator and carbon brushes.

16. Briefly describe how a centrifugal switch works.

17. Briefly describe why a centrifugal switch is required in a single-phase a.c. motor.

18. List three types of single-phase a.c. motors and give one application for each type.
19. List three things that a motor starter does.
20. Which electric motors require a motor starter?
21. Section 552 of the IEE Regulations – Rotating Machines – list five Regulations in this section. Very briefly state the requirements of each Regulation.
22. Why is p.f. improvement necessary for motor circuits?

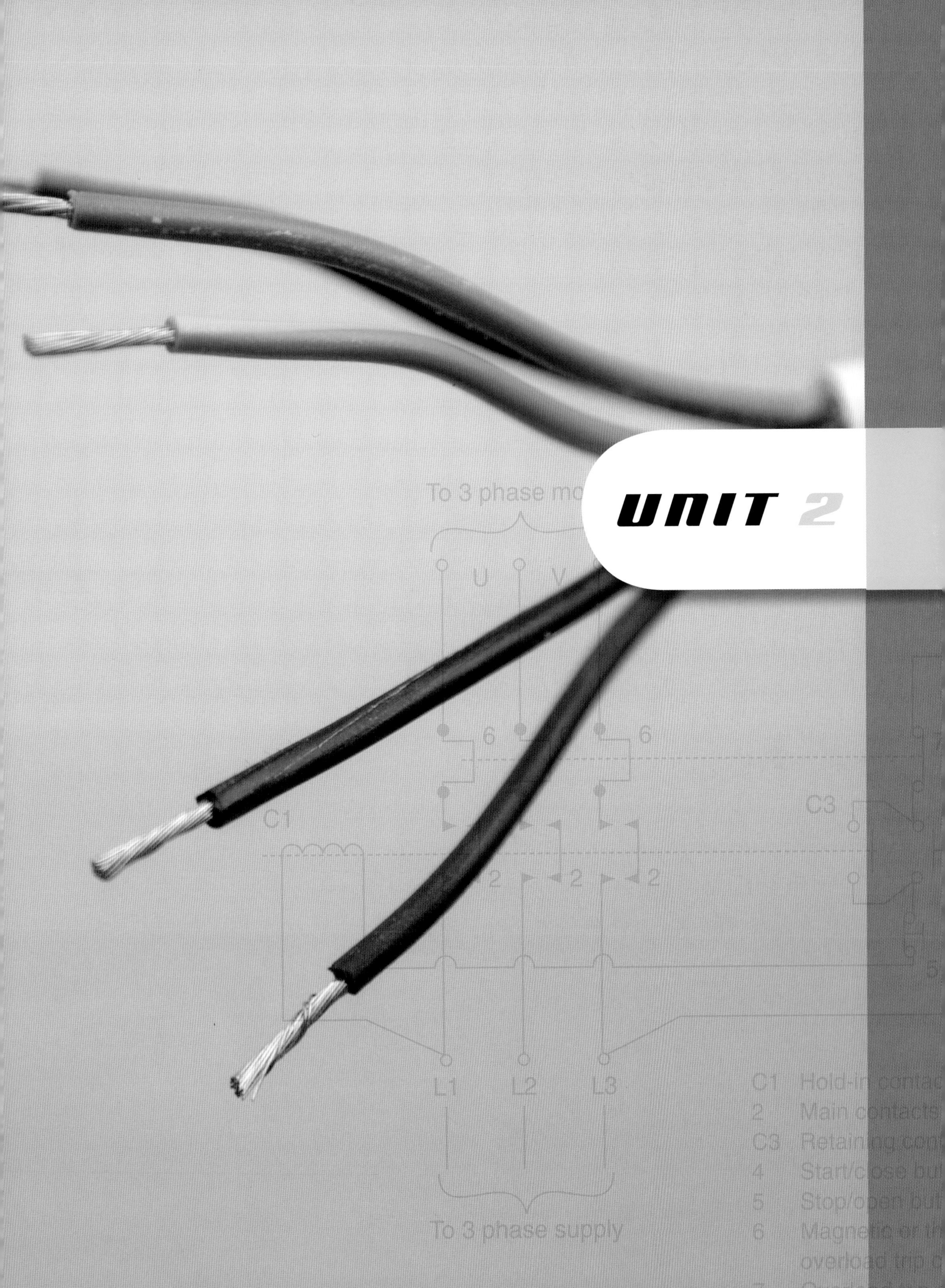
UNIT 2
U
C1
C3
L1
L2
L3
To 3 phase supply

CH 7

Safe, effective and efficient working practices

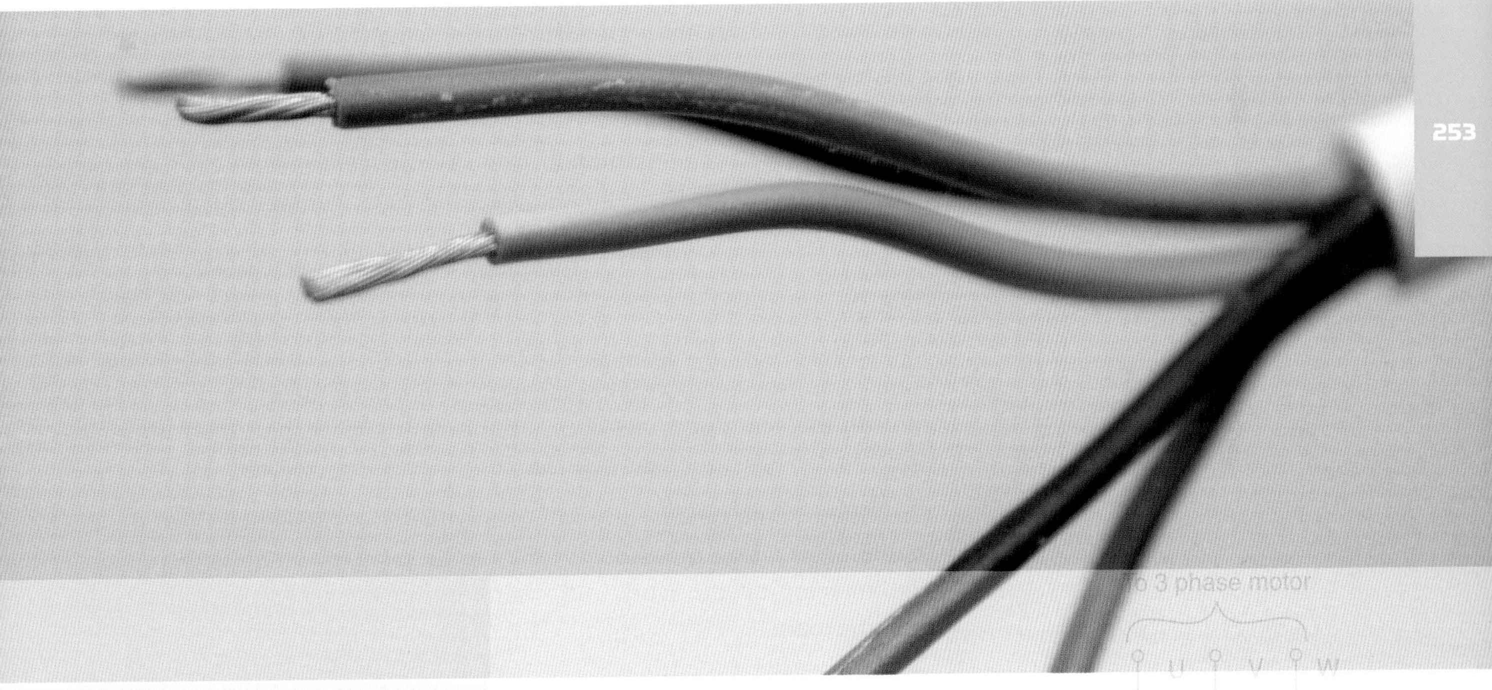

Unit 2 – Installation (buildings and structures): inspection, testing and commissioning – Outcome 1

Underpinning knowledge: when you have completed this chapter you should be able to:

- describe the procedures or activities necessary to make an area safe before work commences
- describe four installation work activities
- identify the parties involved in installation activities and their relationships
- state the management systems used to monitor contract progress
- state the systems, equipment and procedures which create a safe electrical installation
- describe the supply system earthing arrangements
- describe an installation overcurrent protection equipment
- state the factors affecting the selection of conductors

Those involved with installation activities

An electrician working for an electrical contracting company works as a part of the broader construction industry. This is a multi-million-pound industry carrying out all types of building work, from basic housing to hotels, factories, schools, shops, offices and airports. The construction industry is one of the UK's biggest employers, and carries out contracts to the value of about 10% of the UK's gross national product.

Although a major employer, the construction industry is also very fragmented. Firms vary widely in size, from the local builder employing two or three people to the big national companies employing thousands. Of the total workforce of the construction industry, 92% are employed in small firms of less than 25 people.

The yearly turnover of the construction industry is about £35 billion. Of this total sum, about 60% is spent on new building projects and the remaining 40% on maintenance, renovation or restoration of mostly housing.

In all these various construction projects the electrotechnical industries play an important role, supplying essential electrical services to meet the needs of those who will use the completed building.

The building team

The construction of a new building is a complex process which requires a team of professionals working together to produce the desired results. We can call this team of professionals the building team, and their interrelationship can be expressed as in Fig. 7.1.

The client is the person or group of people with the actual need for the building, such as a new house, office or factory. The client is responsible for financing all the work and, therefore, in effect, employs the entire building team.

The architect is the client's agent and is considered to be the leader of the building team. The architect must interpret the client's requirements and produce working drawings. During the building process the architect will supervise all aspects of the work until the building is handed over to the client.

The quantity surveyor measures the quantities of labour and material necessary to complete the building work from drawings supplied by the architect.

Specialist engineers advise the architect during the design stage. They will prepare drawings and calculations on specialist areas of work.

The clerk of works is the architect's 'on-site' representative. He or she will make sure that the contractors carry out the work in accordance with the drawings and other contract documents. They can also agree general matters directly with the building contractor as the architect's representative.

The local authority will ensure that the proposed building conforms to the relevant planning and building legislation.

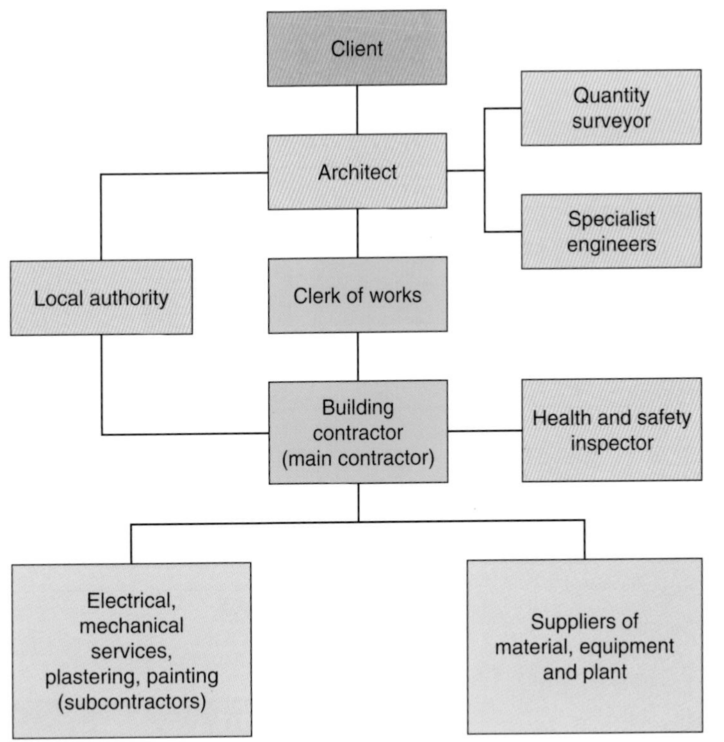

FIGURE 7.1
The building team.

The health and safety inspectors will ensure that the government's legislation concerning health and safety is fully implemented by the building contractor.

The building contractor will enter into a contract with the client to carry out the construction work in accordance with contract documents. The building contractor is usually the main contractor and he or she, in turn, may engage sub-contractors to carry out specialist services such as electrical installation, mechanical services, plastering and painting.

The electrical team

The electrical contractor is the sub-contractor responsible for the installation of electrical equipment within the building.

Electrical installation activities include:

- installing electrical equipment and systems into new sites or locations;
- installing electrical equipment and systems into buildings that are being refurbished because of change of use;
- installing electrical equipment and systems into buildings that are being extended or updated;
- replacement, repairs and maintenance of existing electrical equipment and systems.

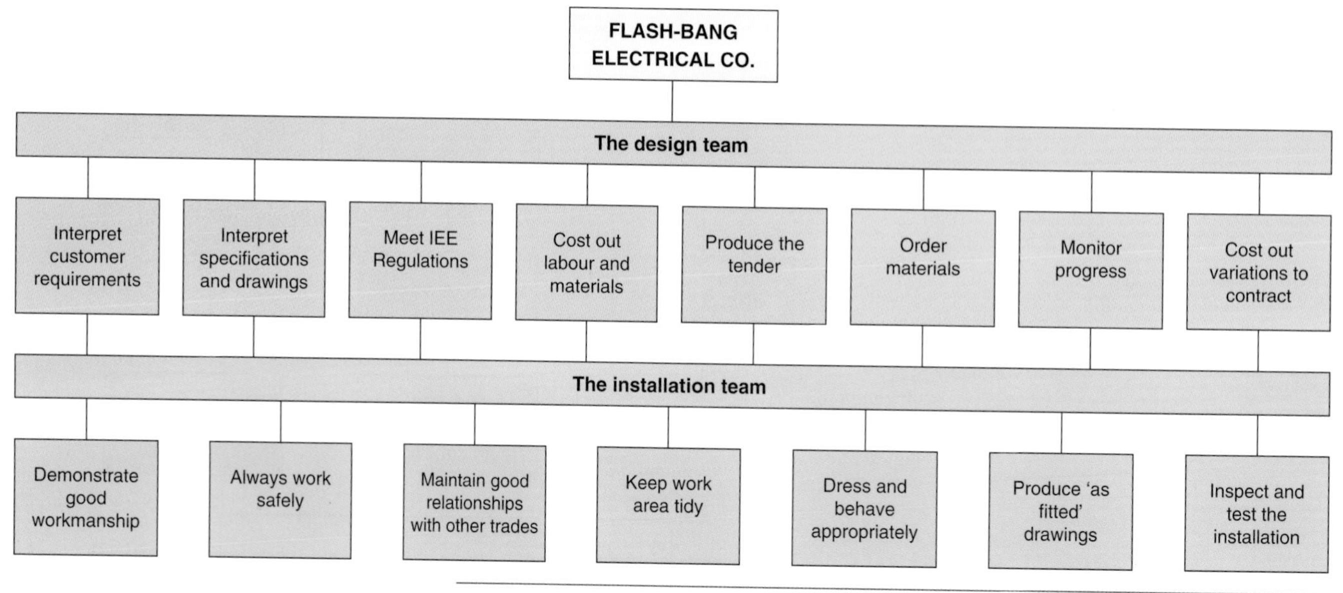

FIGURE 7.2
The electrical team.

Try This

My Team

Sketch a block diagram, similar to those shown in Figs 7.1 and 7.2, that represents the team in which you work.

An electrical contracting firm is made up of a group of individuals with varying duties and responsibilities. There is often no clear distinction between the duties of the individuals, and the responsibilities carried by an employee will vary from one employer to another. If the firm is to be successful, the individuals must work together to meet the requirements of their customers. Good customer relationships are important for the success of the firm and the continuing employment of the employee.

The customer or his representatives will probably see more of the electrician and the electrical trainee than the managing director of the firm and, therefore, the image presented by them is very important. They should always be polite and seen to be capable and in command of the situation. This gives a customer confidence in the firm's ability to meet his or her needs. The electrician and his trainee should be appropriately dressed for the job in hand, which probably means an overall of some kind. Footwear is also important, but sometimes a difficult consideration for a journeyman electrician. For example, if working in a factory, the safety regulations may insist that protective footwear be worn, but rubber boots may be most appropriate for a building site. However, neither of these would be the most suitable footwear for an electrician fixing a new light fitting in the home of the managing director!

The electrical installation in a building is often carried out alongside other trades. It makes sound sense to help other trades where possible and to develop good working relationships with other employees.

The employer has the responsibility of finding sufficient work for his employees, paying government taxes and meeting the requirements of the Health and Safety at Work Act described in Chapter 1. The rates of pay and conditions for electricians and trainees are determined by negotiation between the Joint Industry Board and the AMICUS Union, which will also represent their members in any disputes. Electricians are usually paid at a rate agreed for their grade as an electrician, approved electrician or technician electrician; movements through the grades are determined by a combination of academic achievement and practical experience.

The electrical team will consist of a group of professionals and their interrelationship can be expressed as shown in Fig. 7.2.

Designing an electrical installation

The designer of an electrical installation must ensure that the design meets the requirements of the IEE Wiring Regulations for electrical installations and any other regulations which may be relevant to a particular installation. The designer may be a professional technician or engineer whose job is to design electrical installations for a large contracting firm. In a smaller firm, the designer may also be the electrician who will carry out the installation to the customer's requirements. The **designer** of any electrical installation is the person who interprets the electrical requirements of the customer within the regulations, identifies the appropriate types of installation, the most suitable methods of protection and control and the size of cables to be used.

Definition

The *designer* of any electrical installation is the person who interprets the electrical requirements of the customer within the regulations.

A large electrical installation may require many meetings with the customer and his professional representatives in order to identify a specification of what is required. The designer can then identify the general characteristics of the electrical installation and its compatibility with other services and equipment, as indicated in Part 3 of the Regulations. The protection and safety of the installation, and of those who will use it, must be considered, with due regard to Part 4 of the Regulations. An assessment of the frequency and quality of the maintenance to be expected will give an indication of the type of installation which is most appropriate.

The size and quantity of all the materials, cables, control equipment and accessories can then be determined. This is called a '**bill of quantities**'.

Definition

The size and quantity of all the materials, cables, control equipment and accessories can then be determined. This is called a '*bill of quantities*'.

It is a common practice to ask a number of electrical contractors to tender or submit a price for work specified by the bill of quantities. The contractor must cost all the materials, assess the labour cost required to install the materials and add on profit and overhead costs in order to arrive at a final estimate for the work. The contractor tendering the lowest cost is usually, but not always, awarded the contract.

To complete the contract in the specified time the electrical contractor must use the management skills required by any business to ensure that men and materials are on site as and when they are required. If alterations or modifications are made to the electrical installation as the work proceeds which are outside the original specification, then a **variation order** must be issued so that the electrical contractor can be paid for the additional work.

The specification for the chosen wiring system will be largely determined by the building construction and the activities to be carried out in the completed building.

An industrial building, for example, will require an electrical installation which incorporates flexibility and mechanical protection. This can be achieved by a conduit, tray or trunking installation.

In a block of purpose-built flats, all the electrical connections must be accessible from one flat without intruding upon the surrounding flats. A loop-in conduit system, in which the only connections are at the light switch and outlet positions, would meet this requirement.

For a domestic electrical installation an appropriate lighting scheme and multiple socket outlets for the connection of domestic appliances, all at a reasonable cost, are important factors which can usually be met by a PVC insulated and sheathed wiring system.

The final choice of a wiring system must rest with those designing the installation and those ordering the work, but whatever system is employed, good workmanship by competent persons is essential for compliance with the regulations (IEE Regulation 134.1.1). The necessary skills can be acquired by an electrical trainee who has the correct attitude and dedication to his craft.

Legal contracts

Before work commences, some form of legal contract should be agreed between the two parties, that is, those providing the work (e.g. the subcontracting electrical company) and those asking for the work to be carried out (e.g. the main building company).

A contract is a formal document which sets out the terms of agreement between the two parties. A standard form of building contract typically contains four sections:

1. *The articles of agreement* – this names the parties, the proposed building and the date of the contract period.
2. *The contractual conditions* – this states the rights and obligations of the parties concerned, for example, whether there will be interim payments for work completed, or a penalty if work is not completed on time.
3. *The appendix* – this contains details of costings, for example, the rate to be paid for extras as daywork, who will be responsible for defects, how much of the contract tender will be retained upon completion and for how long.
4. *The supplementary agreement* – this allows the electrical contractor to recoup any value-added tax paid on materials at interim periods.

In signing the contract, the electrical contractor has agreed to carry out the work to the appropriate standards in the time stated and for the agreed cost. The other party, say the main building contractor, is agreeing to pay the price stated for that work upon completion of the installation.

If a dispute arises the contract provides written evidence of what was agreed and will form the basis for a solution.

For smaller electrical jobs, a verbal contract may be agreed, but if a dispute arises there is no written evidence of what was agreed and it then becomes a matter of one person's word against another's.

Effective and efficient management systems

Smaller electrical contracting firms will know where their employees are working and what they are doing from day to day because of the level of personal contact between the employer, employee and customer.

As a firm expands and becomes engaged on larger contracts, it becomes less likely that there is anyone in the firm with a complete knowledge of the firm's operations, and there arises an urgent need for sensible management and planning skills so that men and materials are on site when they are required and a healthy profit margin is maintained.

When the electrical contractor is told that he has been successful in tendering for a particular contract he is committed to carrying out the necessary work within the contract period. He must therefore consider:

- by what date the job must be finished;
- when the job must be started if the completion date is not to be delayed;
- how many men will be required to complete the contract;
- when certain materials will need to be ordered;
- when the supply authorities must be notified that a supply will be required;
- if it is necessary to obtain authorization from a statutory body for any work to commence.

In thinking ahead and planning the best method of completing the contract, the individual activities or jobs must be identified and consideration given to how the various jobs are interrelated. To help in this process a number of management techniques are available. In this chapter we will consider only two: bar charts and network analysis. The very preparation of a bar chart or network analysis forces the contractor to think deeply, carefully and logically about the particular contract, and it is therefore a very useful aid to the successful completion of the work.

Definition

There are many different types of bar chart used by companies but the *object of any bar chart* is to establish the sequence and timing of the various activities involved in the contract as a whole.

BAR CHARTS

There are many different types of bar chart used by companies but the object of any bar chart is to establish the sequence and timing of the various activities involved in the contract as a whole. They are a visual aid in the process of communication. In order to be useful they must be clearly understood by the people involved in the management of a contract. The chart is constructed on a rectangular basis, as shown in Fig. 7.3.

All the individual jobs or activities which make up the contract are identified and listed separately down the vertical axis on the left-hand side, and time flows from left to right along the horizontal axis. The unit of time can be chosen to suit the length of the particular contract, but for most practical purposes either days or weeks are used.

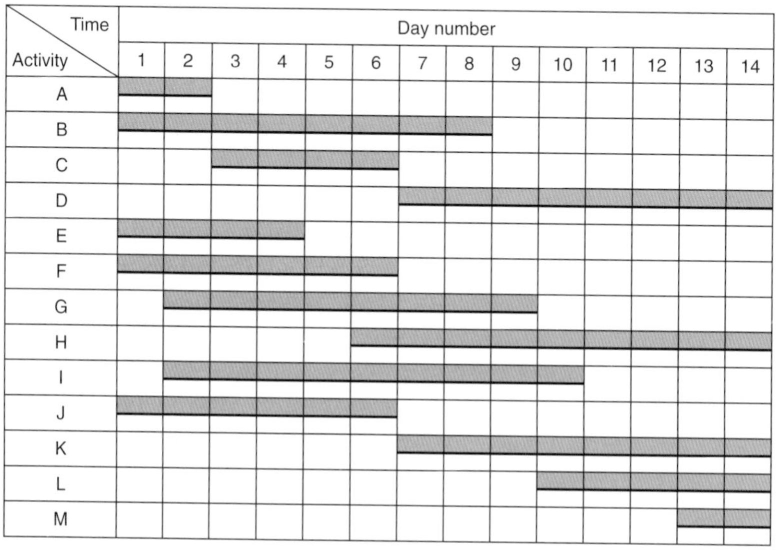

(a) A simple bar chart or schedule of work

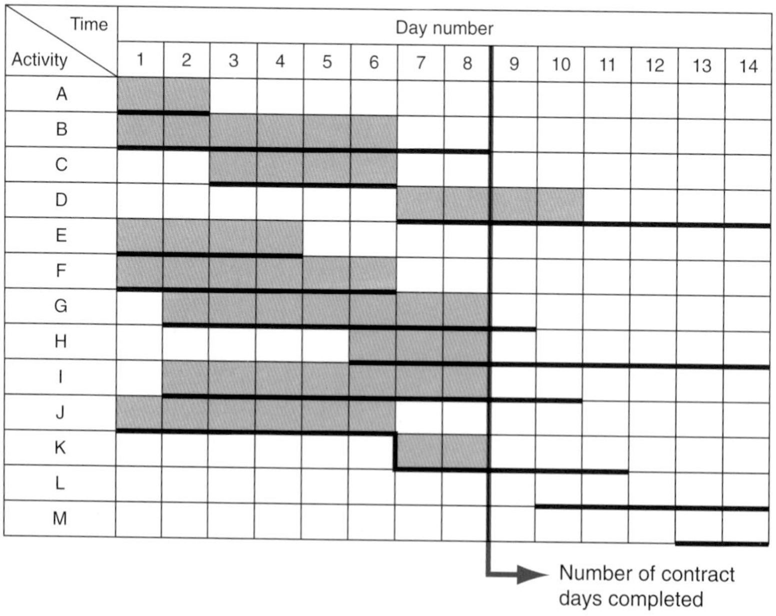

(b) A modified bar chart indicating actual work completed

FIGURE 7.3
Bar charts: (a) a simple bar chart or schedule of work; (b) a modified bar chart indicating actual work completed.

The simple bar chart shown in Fig. 7.3(a) shows a particular activity A which is estimated to last 2 days, while activity B lasts 8 days. Activity C lasts 4 days and should be started on day 3. The remaining activities çan be interpreted in the same way.

With the aid of colours, codes, symbols and a little imagination, much additional information can be included on this basic chart. For example, the actual work completed can be indicated by shading above the activity line as shown in Fig. 7.3(b) with a vertical line indicating the number of contract days completed; the activities which are on time, ahead of or behind time can easily be identified. Activity B in Fig. 7.3(b) is 2 days behind schedule, while activity D is 2 days ahead of schedule. All other activities are on time. Some activities must be completed before others can start. For example, all conduit work must be completely erected before the cables are drawn in. This is shown in Fig. 7.3(b) by activities J and K. The short vertical line between the two activities indicates that activity J must be completed before K can commence.

Useful and informative as the bar chart is, there is one aspect of the contract which it cannot display. It cannot indicate clearly the interdependence of the various activities upon each other, and it is unable to identify those activities which must strictly adhere to the time schedule if the overall contract is to be completed on time, and those activities in which some flexibility is acceptable. To overcome this limitation, in 1959 the Central Electricity Generating Board (CEGB) developed the critical path network diagram which we will now consider.

NETWORK ANALYSIS

Definition

A *network diagram* can be used to co-ordinate all the interrelated activities of the most complex project.

In large or complex contracts there are a large number of separate jobs or activities to be performed. Some can be completed at the same time, while others cannot be started until others are completed. A **network diagram** can be used to co-ordinate all the interrelated activities of the most complex project in such a way that all sequential relationships between the various activities, and the restraints imposed by one job on another, are allowed for. It also provides a method of calculating the time required to complete an individual activity and will identify those activities which are the key to meeting the completion date, called the critical path. Before considering the method of constructing a network diagram, let us define some of the terms and conventions we shall be using.

Definition

Critical path is the path taken from the start event to the end event which takes the longest time.

Critical path

Critical path is the path taken from the start event to the end event which takes the longest time. This path denotes the time required for completion of the whole contract.

Definition

Float time, slack time or time in hand is the time remaining to complete the contract after completion of a particular activity.

Float time

Float time, slack time or time in hand is the time remaining to complete the contract after completion of a particular activity.

$$\text{Float time} = \text{Critical path time} - \text{Activity time}$$

The total float time for any activity is the total leeway available for all activities in the particular path of activities in which it appears. If the float time is used up by one of the early activities in the path, there will be no float left for the remaining activities and they will become critical.

Definition

Activities are represented by an arrow, the tail of which indicates the commencement, and the head the completion of the activity.

ACTIVITIES

Activities are represented by an arrow, the tail of which indicates the commencement, and the head the completion of the activity. The length and direction of the arrows have no significance: they are not vectors or phasors. Activities require time, manpower and facilities. They lead up to or emerge from events.

Definition

Dummy activities are represented by an arrow with a dashed line.

DUMMY ACTIVITIES

Dummy activities are represented by an arrow with a dashed line. They signify a logical link only, require no time and denote no specific action or work.

Definition

An *event* is a point in time, a milestone or stage in the contract when the preceding activities are finished.

EVENT

An **event** is a point in time, a milestone or stage in the contract when the preceding activities are finished. Each activity begins and ends in an event. An event has no time duration and is represented by a circle which sometimes includes an identifying number or letter.

Time may be recorded to a horizontal scale or shown on the activity arrows. For example, the activity from event A to B takes 9 hours in the network diagram shown in Fig. 7.4.

Example 1

Identify the three possible paths from the start event A to the finish event F for the contract shown by the network diagram in Fig. 7.4. Identify the critical path and the float time in each path.

The three possible paths are:

1. event A–B–D–F
2. event A–C–D–F
3. event A–C–E–F.

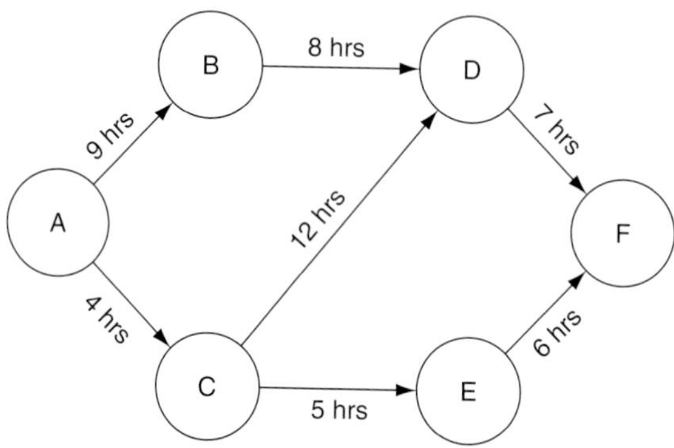

FIGURE 7.4
A network diagram for Example 1.

The times taken to complete these activities are:

1. path A–B–D–F = 9 + 8 + 7 = 24 hours
2. path A–C–D–F = 4 + 12 + 7 = 23 hours
3. path A–C–E–F = 4 + 5 + 6 = 15 hours.

The longest time from the start event to the finish event is 24 hours, and therefore the critical path is A–B–D–F.

The float time is given by:

$$\text{Float time} = \text{Critical path} - \text{Activity time}$$

For path 1, A–B–D–F,

$$\text{Float time} = 24\ \text{hours} - 24\ \text{hours} = 0\ \text{hours}$$

There can be no float time in any of the activities which form a part of the critical path since a delay on any of these activities would delay completion of the contract. On the other two paths some delay could occur without affecting the overall contract time. For path 2, A–C–D–F,

$$\text{Float time} = 24\ \text{hours} - 23\ \text{hours} = 1\ \text{hour}$$

For path 3, A–C–E–F,

$$\text{Float time} = 24\ \text{hours} - 15\ \text{hours} = 9\ \text{hours}$$

Example 2

Identify the time taken to complete each activity in the network diagram shown in Fig. 7.5. Identify the three possible paths from the start event A to the final event G and state which path is the critical path.

The time taken to complete each activity using the horizontal scale is:

activity A–B = 2 days
activity A–C = 3 days
activity A–D = 5 days
activity B–E = 5 days
activity C–F = 5 days
activity E–G = 3 days
activity D–G = 0 days
activity F–G = 0 days

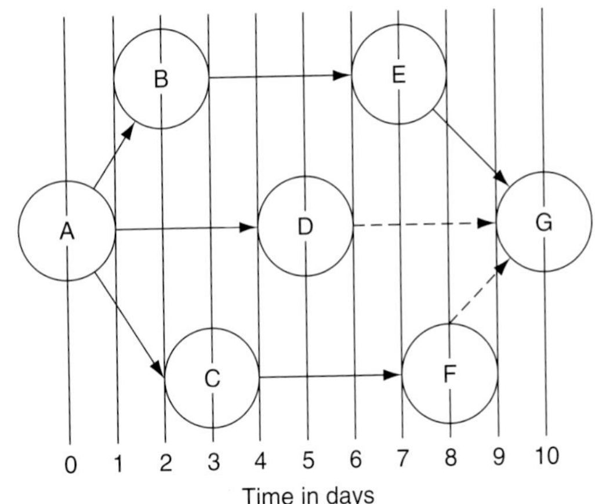

FIGURE 7.5
A network diagram for Example 2.

Activities D–G and F–G are dummy activities which take no time to complete but indicate a logical link only. This means that in this case once the activities preceding events D and F have been completed, the contract will not be held up by work associated with these particular paths and they will progress naturally to the finish event.

The three possible paths are:

1. A–B–E–G
2. A–D–G
3. A–C–F–G.

The times taken to complete the activities in each of the three paths are:

path 1, A–B–E–G = 2 + 5 + 3 = 10 days
path 2, A–D–G = 5 + 0 = 5 days
path 3, A–C–F–G = 3 + 5 + 0 = 8 days.

The critical path is path 1, A–B–E–G.

CONSTRUCTING A NETWORK

The first step in constructing a network diagram is to identify and draw up a list of all the individual jobs, or activities, which require time for their completion and which must be completed to advance the contract from start to completion.

The next step is to build up the arrow network showing schematically the precise relationship of the various activities between the start and end event. The designer of the network must ask these questions:

1. Which activities must be completed before others can commence? These activities are then drawn in a similar way to a series circuit but with event circles instead of resistor symbols.

2. Which activities can proceed at the same time? These can be drawn in a similar way to parallel circuits but with event circles instead of resistor symbols.

Commencing with the start event at the left-hand side of a sheet of paper, the arrows representing the various activities are built up step by step until the final event is reached. A number of attempts may be necessary to achieve a well-balanced and symmetrical network diagram showing the best possible flow of work and information, but this time is well spent when it produces a diagram which can be easily understood by those involved in the management of the particular contract.

Example 3

A particular electrical contract is made up of activities A–F as described below:

A = an activity taking 2 weeks commencing in week 1
B = an activity taking 3 weeks commencing in week 1
C = an activity taking 3 weeks commencing in week 4
D = an activity taking 4 weeks commencing in week 7
E = an activity taking 6 weeks commencing in week 3
F = an activity taking 4 weeks commencing in week 1.

Certain constraints are placed on some activities because of the availability of men and materials and because some work must be completed before other work can commence as follows:

Activity C can only commence when B is completed.
Activity D can only commence when C is completed.
Activity E can only commence when A is completed.
Activity F does not restrict any other activity.

(a) Produce a simple bar chart to display the activities of this particular contract.
(b) Produce a network diagram of the programme and describe each event.
(c) Identify the critical path and the total contract time.
(d) State the maximum delay which would be possible on activity E without delaying the completion of the contract.
(e) State the float time in activity F.

(a) A simple bar chart for this contract is shown in Fig. 7.6(a).
(b) The network diagram is shown in Fig. 7.6(b). The events may be described as follows:
Event 1 = the commencement of the contract.
Event 2 = the completion of activity A and the commencement of activity E.
Event 3 = the completion of activity B and the commencement of activity C.
Event 4 = the completion of activity F.
Event 5 = the completion of activity E.
Event 6 = the completion of activity C.
Event 7 = the completion of activity D and the whole contract.

(c) There are three possible paths:
1 via events 1–2–5–7
2 via events 1–4–7
3 via events 1–3–6–7.

The time taken for each path is:

path 1 = 2 weeks + 6 weeks = 8 weeks
path 2 = 4 weeks = 4 weeks
path 3 = 3 weeks + 3 weeks + 4 weeks = 10 weeks.

The critical path is therefore path 3, via events 1–3–6–7, and the total contract time is 10 weeks.

(d) We have that:

Float time = Critical path time − Activity time

Activity E is on path 1 via events 1–2–5–7 having a total activity time of 8 weeks.

Float time = 10 weeks − 8 weeks = 2 weeks.

Activity E could be delayed for a maximum of 2 weeks without delaying the completion date of the whole contract.

(e) Activity F is on path 2 via events 1–4–7 having a total activity time of 4 weeks.

Float time = 10 weeks − 4 weeks = 6 weeks.

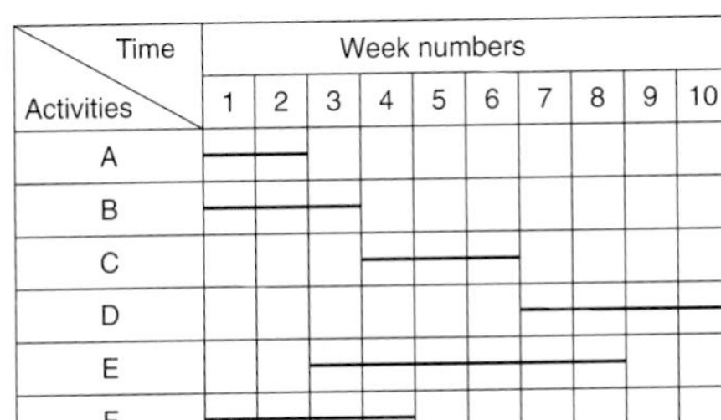

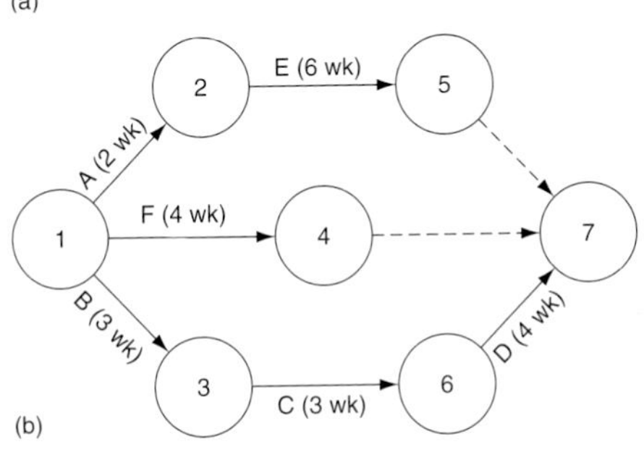

FIGURE 7.6
(a) Bar chart and (b) network diagram for Example 3.

On-site communications

Good communication is about transferring information from one person to another. Electricians and other professionals in the construction trades communicate with each other and the general public by means of drawings, sketches and symbols in addition to what we say and do.

DRAWINGS AND DIAGRAMS

Many different types of electrical drawing and diagram can be identified: layout, schematic, block, wiring and circuit diagrams. The type of diagram to be used in any particular application is the one which most clearly communicates the desired information.

Definition

These are scale drawings based upon the architect's site plan of the building and show the positions of the electrical equipment which is to be installed.

LAYOUT DRAWINGS OR SITE PLAN

These are scale drawings based upon the architect's site plan of the building and show the positions of the electrical equipment which is to be installed. The electrical equipment is identified by a graphical symbol.

The standard symbols used by the electrical contracting industry are those recommended by the British Standard BS EN 60617, *Graphical Symbols for Electrical Power, Telecommunications and Electronic Diagrams.* Some of the more common electrical installation symbols are given in Fig. 7.7.

The site plan or layout drawing will be drawn to a scale, smaller than the actual size of the building, so to find the actual measurements you must measure the distance on the drawing and then multiply by the scale.

For example, if the site plan is drawn to a scale of 1:100, then 10 mm on the site plan represents 1 m measured in the building.

A layout drawing is shown in Fig. 7.8 of a small domestic extension. It can be seen that the mains intake position, probably a consumer's unit, is situated in the store room which also contains one light controlled by a switch at the door. The bathroom contains one lighting point controlled by a one-way switch at the door. The kitchen has two doors and a switch is installed at each door to control the fluorescent luminaire. There are also three double sockets situated around the kitchen. The sitting room has a two-way switch at each door controlling the centre lighting point. Two wall lights with built in switches are to be wired, one at each side of the window. Two double sockets and one switched socket are also to be installed in the sitting room. The bedroom has two lighting points controlled independently by two one-way switches at the door.

The wiring diagrams and installation procedures for all these circuits can be found in Chapter 14 of *Basic Electrical Installation Work,* 5th Edition.

Definition

When the installation is completed a set of drawings should be produced which indicate the final positions of all the electrical equipment.

AS-FITTED DRAWINGS

When the installation is completed a set of drawings should be produced which indicate the final positions of all the electrical equipment. As the building and electrical installation progresses, it is sometimes necessary to modify the positions of equipment indicated on the layout drawing because,

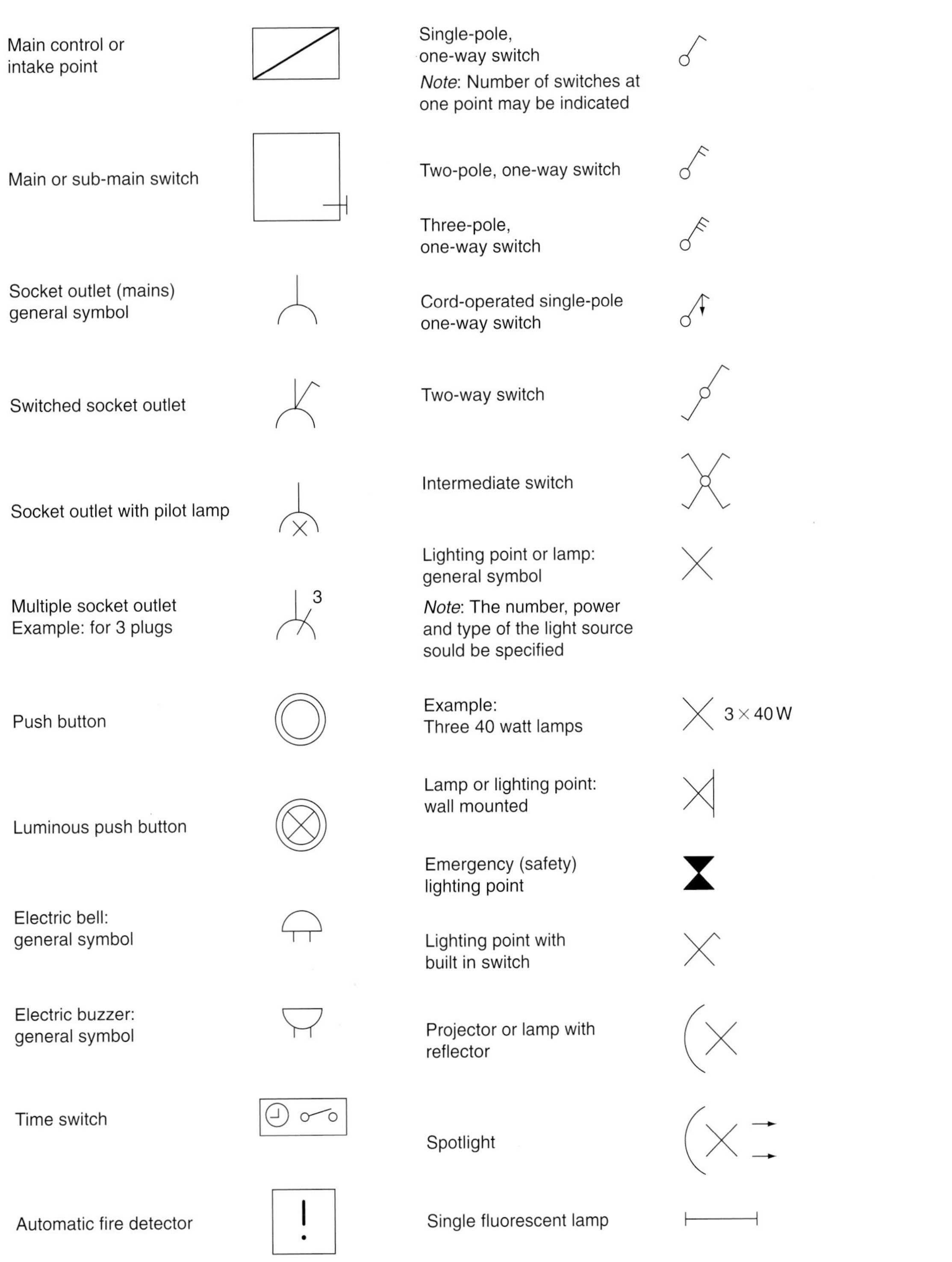

FIGURE 7.7
Some BS EN 60617 installation symbols.

for example, the position of a doorway has been changed. The layout drawings indicate the original intentions for the positions of equipment, while the 'as-fitted' drawing indicates the actual positions of equipment upon completion of the job.

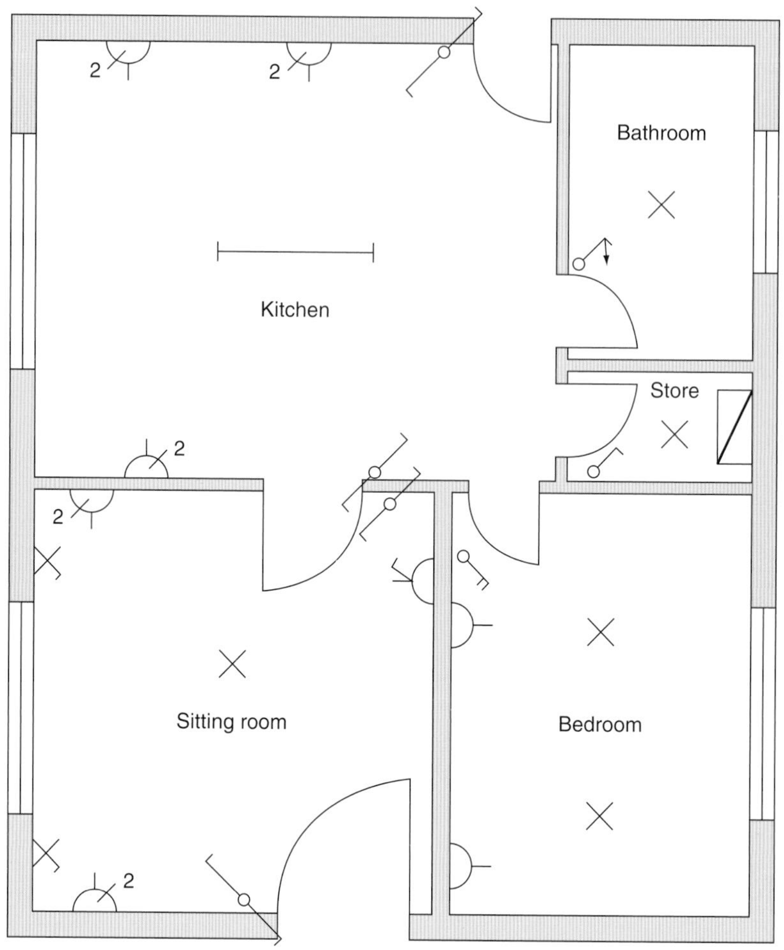

FIGURE 7.8
Layout drawing or site plan for electrical installation.

DETAIL DRAWINGS AND ASSEMBLY DRAWINGS

These are additional drawings produced by the architect to clarify some point of detail. For example, a drawing might be produced to give a fuller description of the suspended ceiling arrangements.

Definition

These are additional drawings produced by the architect to clarify some point of detail.

SCHEMATIC DIAGRAMS

A **schematic diagram** is a diagram in outline of, for example, a motor starter circuit. It uses graphical symbols to indicate the interrelationship of the electrical elements in a circuit. These help us to understand the working operation of the circuit.

Definition

A *schematic diagram* is a diagram in outline of, for example, a motor starter circuit.

An electrical schematic diagram looks very much like a circuit diagram. A mechanical schematic diagram gives a more complex description of the individual elements in the system, indicating, for example, acceleration, velocity, position, force sensing and viscous damping.

BLOCK DIAGRAMS

A **block diagram** is a very simple diagram in which the various items or pieces of equipment are represented by a square or rectangular box. The

Definition

A *block diagram* is a very simple diagram in which the various items or pieces of equipment are represented by a square or rectangular box.

purpose of the block diagram is to show how the components of the circuit relate to each other and therefore the individual circuit connections are not shown.

Definition

A *wiring diagram* or connection diagram shows the detailed connections between components or items of equipment.

WIRING DIAGRAMS

A **wiring diagram** or connection diagram shows the detailed connections between components or items of equipment. They do not indicate how a piece of equipment or circuit works. The purpose of a wiring diagram is to help someone with the actual wiring of the circuit. Figure 7.9 shows the wiring diagram for a two-way lighting circuit.

Try This

Drawing

The next time you are on site:

- ask your supervisor to show you the site plans,
- ask him to show you how the scale works.

Definition

A *circuit diagram* shows most clearly how a circuit works.

CIRCUIT DIAGRAMS

A **circuit diagram** shows most clearly how a circuit works. All the essential parts and connections are represented by their graphical symbols. The purpose of a circuit diagram is to help our understanding of the circuit. It will be laid out as clearly as possible, without regard to the physical layout of the actual components, and therefore it may not indicate the most convenient way to wire the circuit. Chapter 3 and Figs 3.1–3.6 cover this topic in more detail in *Basic Electrical Installation Work* 5th Edition.

TELEPHONE COMMUNICATIONS

Telephones today play one of the most important roles in enabling people to communicate with each other. You are never alone when you have a

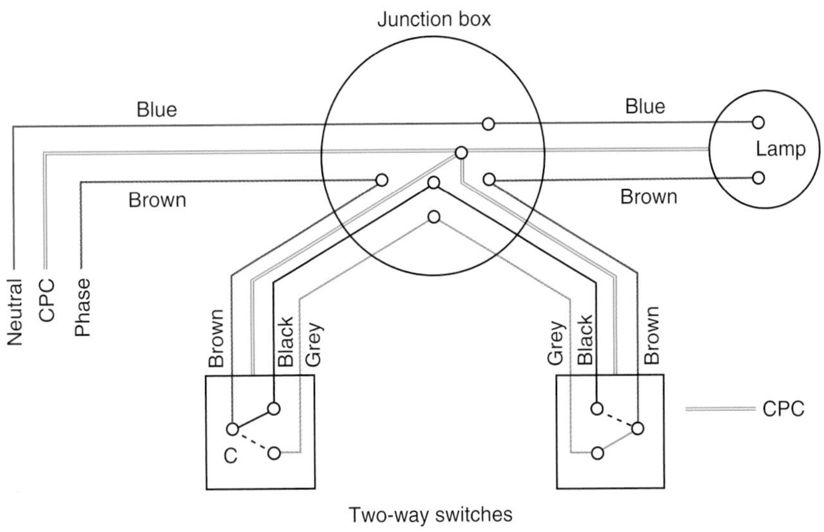

FIGURE 7.9
Wiring diagram of two-way switch control.

telephone. If there is a problem, you can ring your supervisor or foreman for help. The advantage of a telephone message over a written message is its speed; the disadvantage is that no record is kept of an agreement made over the telephone. Therefore, business agreements made on the telephone are often followed up by written confirmation.

When *taking* a telephone call, remember that you cannot be seen and, therefore, gestures and facial expressions will not help to make you understood. Always be polite and helpful when answering your company's telephone – you are your company's most important representative at that moment. Speak clearly and loud enough to be heard without shouting, sound cheerful and write down messages if asked. Always read back what you have written down to make sure that you are passing on what the caller intended.

Many companies now use standard telephone message pads such as that shown in Fig. 7.10 because they prompt people to collect all the relevant information. In this case John Gall wants Dave Twem to pick up the Megger from Jim on Saturday and take it to the Bispham site on Monday. The person taking the call and relaying the message is Dave Low.

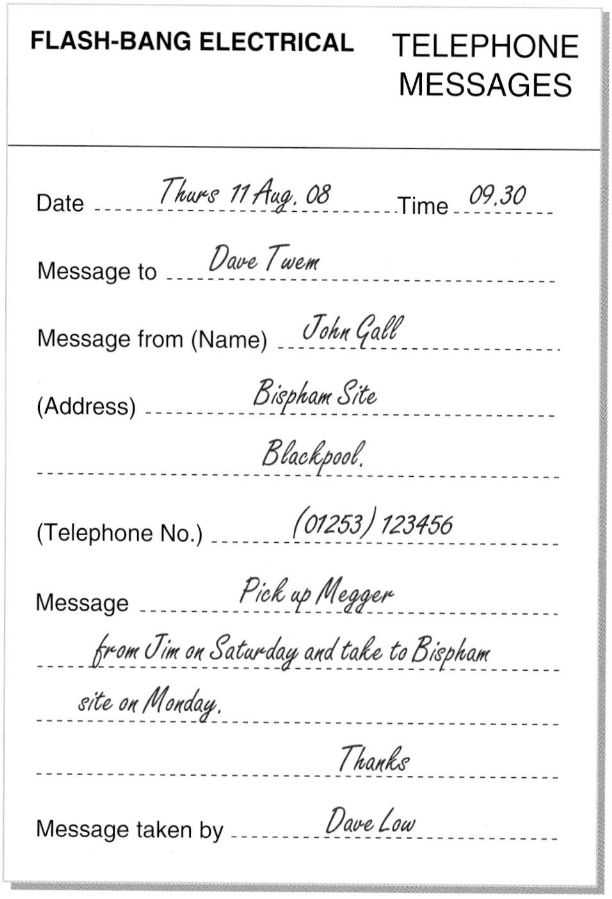

FLASH-BANG ELECTRICAL TELEPHONE MESSAGES

Date Thurs 11 Aug. 08 Time 09.30

Message to Dave Twem

Message from (Name) John Gall

(Address) Bispham Site

Blackpool.

(Telephone No.) (01253) 123456

Message Pick up Megger

from Jim on Saturday and take to Bispham

site on Monday.

Thanks

Message taken by Dave Low

FIGURE 7.10

Typical standard telephone message pad.

When *making* a telephone call, make sure you know what you want to say or ask. Make notes so that you have times, dates and any other relevant information ready before you make the call.

WRITTEN MESSAGES

Definition

A lot of communications between and within larger organizations take place by completing standard forms or sending internal memos.

A lot of communications between and within larger organizations take place by completing standard forms or sending internal memos. Written messages have the advantage of being 'auditable'. An auditor can follow the paperwork trail to see, for example, who was responsible for ordering certain materials.

When completing standard forms, follow the instructions given and ensure that your writing is legible. Do not leave blank spaces on the form, always specifying 'not applicable' or 'N/A' whenever necessary. Sign or give your name and the date as asked for on the form. Finally, read through the form again to make sure you have answered all the relevant sections correctly.

Internal memos are forms of written communication used within an organization; they are not normally used for communicating with customers or suppliers. Figure 7.11 shows the layout of a typical standard memo form used by Dave Twem to notify John Gall that he has ordered the hammer drill.

Letters provide a permanent record of communications between organizations and individuals. They may be handwritten, but formal business letters give a better impression of the organization if they are type-written. A letter should be written using simple concise language, and the tone of the letter should always be polite even if it is one of the complaints. Always include the date of the correspondence. The greeting on a formal letter should be 'Dear Sir/Madam' and concluded with 'Yours faithfully'. A less formal greeting would be 'Dear Mr Smith' and concluded 'Yours sincerely'. Your name and status should be typed below your signature.

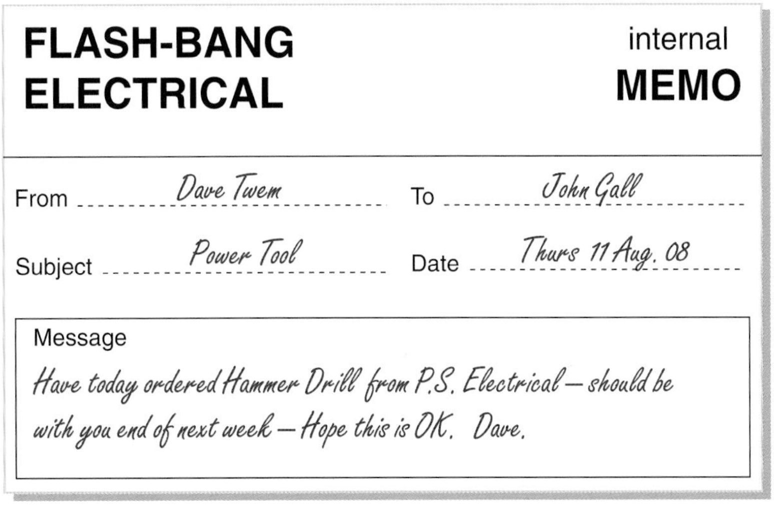

FLASH-BANG ELECTRICAL — internal **MEMO**

From: Dave Twem — To: John Gall

Subject: Power Tool — Date: Thurs 11 Aug. 08

Message

Have today ordered Hammer Drill from P.S. Electrical – should be with you end of next week – Hope this is OK. Dave.

FIGURE 7.11
Typical standard memo form.

DELIVERY NOTES

When materials are delivered to site, the person receiving the goods is required to sign the driver's 'delivery note'. This record is used to confirm that goods have been delivered by the supplier, who will then send out an invoice requesting payment, usually at the end of the month.

The person receiving the goods must carefully check that all the items stated on the delivery note have been delivered in good condition. Any missing or damaged items must be clearly indicated on the **delivery note** before signing, because, by signing the delivery note the person is saying 'yes, these items were delivered to me as my company's representative on that date and in good condition and I am now responsible for these goods'. Suppliers will replace materials damaged in transit provided that they are notified within a set time period, usually 3 days. The person receiving the goods should try to quickly determine their condition. Has the packaging been damaged, does the container 'sound' like it might contain broken items? It is best to check at the time of delivery if possible, or as soon as possible after delivery and within the notifiable period. Electrical goods delivered to site should be handled carefully and stored securely until they are installed. Copies of delivery notes are sent to head office so that payment can be made for the goods received.

Definition

By signing the *delivery note* the person is saying 'yes, these items were delivered to me as my company's representative on that date and in good condition and I am now responsible for these goods'.

TIME SHEETS

A **time sheet** is a standard form completed by each employee to inform the employer of the actual time spent working on a particular contract or site. This helps the employer to bill the hours of work to an individual job. It is usually a weekly document and includes the number of hours worked, the name of the job and any travelling expenses claimed. Office personnel require time sheets so that wages can be made up.

Definition

A *time sheet* is a standard form completed by each employee to inform the employer of the actual time spent working on a particular contract or site.

JOB SHEETS

A **job sheet** or **job card** such as that shown in Fig. 7.12 carries information about a job which needs to be done, usually a small job. It gives the name and address of the customer, contact telephone numbers, often a job reference number and a brief description of the work to be carried out. A typical job sheet work description might be:

- Job 1: Upstairs lights not working.
- Job 2: Funny fishy smell from kettle socket in kitchen.

Definition

A *job sheet* or *job card* such as that shown in Fig. 7.12 carries information about a job which needs to be done, usually a small job.

An electrician might typically have a 'jobbing day' where he picks up a number of job sheets from the office and carries out the work specified.

Job 1, for example, might be the result of a blown fuse which is easily rectified, but the electrician must search a little further for the fault which caused the fuse to blow in the first place. The actual fault might, for example, be a decayed flex on a pendant drop which has become shorted out, blowing the fuse. The pendant drop would be re-flexed or replaced, along

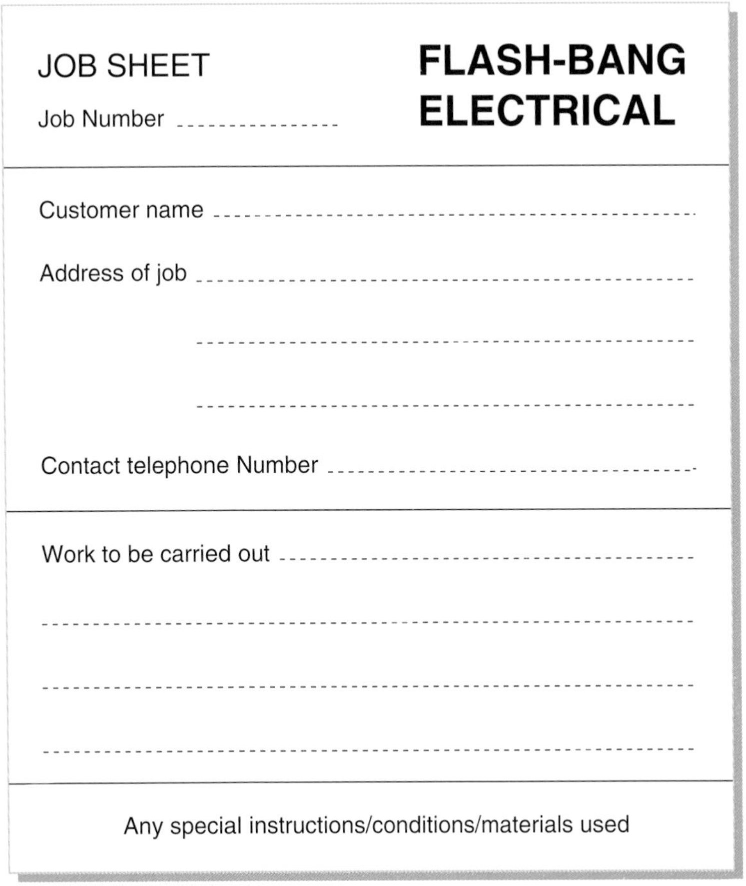

JOB SHEET Job Number	**FLASH-BANG ELECTRICAL**
Customer name	
Address of job	
Contact telephone Number	
Work to be carried out	
Any special instructions/conditions/materials used	

FIGURE 7.12
Typical time sheet.

with any others in poor condition. The installation would then be tested for correct operation and the customer given an account of what has been done to correct the fault. General information and assurances about the condition of the installation as a whole might be requested and given before setting off to job 2.

The kettle socket outlet at job 2 is probably getting warm and, therefore, giving off that 'fishy' bakelite smell because loose connections are causing the bakelite socket to burn locally. A visual inspection would confirm the diagnosis. A typical solution would be to replace the socket and repair any damage to the conductors inside the socket box. Check the kettle plug top for damage and loose connections. Make sure all connections are tight before reassuring the customer that all is well; then, off to the next job or back to the office.

The time spent on each job and the materials used are sometimes recorded on the job sheet, but alternatively a daywork sheet can be used. This will depend upon what is normal practice for the particular electrical company. This information can then be used to 'bill' the customer for work carried out.

Definition

Daywork is one way of recording *variations to a contract*, that is, work done which is outside the scope of the original contract. It is *extra* work.

DAYWORK SHEETS OR VARIATION ORDER

Daywork is one way of recording **variations to a contract**, that is, work done which is outside the scope of the original contract. It is *extra* work. If daywork is to be carried out, the site supervisor must first obtain a signature from the client's representative, for example, the architect, to **authorize the extra work**. A careful record must then be kept on the daywork sheets of all extra time and materials used so that the client can be billed for the extra work. A typical daywork sheet is shown in Fig. 7.13.

FLASH-BANG ELECTRICAL

VARIATION ORDER OR DAYWORK SHEET

Client name ....

Job number/REF. ....

Date	Labour	Start time	Finish time	Total hours	Office use

Materials quantity	Description	Office use

Site supervisor or Flash-Bang Electrical representative responsible for carrying out work ....

Signature of person approving work and status e.g.

Client ☐ Architect ☐ Q.S. ☐ Main contractor ☐ Clerk of works ☐

Signature ....

FIGURE 7.13
Typical daywork sheet or variation order.

REPORTS

On large jobs, the foreman or supervisor is often required to keep a report of the relevant events which happen on the site for example, how many people from your company are working on site each day, what goods were delivered, whether there were any breakages or accidents, and records of site meetings attended. Some firms have two separate documents, a site diary to record daily events and a weekly report which is a summary of the week's events extracted from the site diary. The site diary remains on site and the weekly report is sent to head office to keep managers informed of the work's progress.

PERSONAL COMMUNICATIONS

Remember that it is the customers who actually pay the wages of everyone employed in your company. You should always be polite and listen carefully to their wishes. They may be elderly or of a different religion or cultural background than you. In a domestic situation, the playing of loud music on a radio may not be approved of. Treat the property in which you are working with the utmost care. When working in houses, shops and offices use dust sheets to protect floor coverings and furnishings. Clean up periodically and make a special effort when the job is completed.

Dress appropriately: an unkempt or untidy appearance will encourage the customer to think that your work will be of poor quality.

The electrical installation in a building is often carried out alongside other trades. It makes good sense to help other trades where possible and to develop good working relationships with other employees. The customer will be most happy if the workers give an impression of working together as a team for the successful completion of the project.

Finally, remember that the customer will probably see more of the electrician and the electrical trainee than the managing director of your firm and, therefore, the image presented by you will be assumed to reflect the policy of the company. You are, therefore, your company's most important representative. Always give the impression of being capable and in command of the situation, because this gives customers confidence in the company's ability to meet their needs. However, if a problem does occur which is outside your previous experience and you do not feel confident to solve it successfully, then contact your supervisor for professional help and guidance. It is not unreasonable for a young member of the company's team to seek help and guidance from those employees with more experience. This approach would be preferred by most companies rather than having to meet the cost of an expensive blunder.

Construction site – safe working practice

In Chapter 1 we looked at some of the laws and regulations that affect our working environment. We looked at safety signs and personal protective equipment (PPE), and how to recognize and use different types of fire extinguishers. The structure of companies within the electrotechnical industry and the ways in which they communicate information by drawings, symbols and standard forms was discussed earlier in this chapter.

If your career in the electrotechnical industry is to be a long, happy and safe one, you must always wear appropriate PPE such as footwear, and head protection and behave responsibly and sensibly in order to maintain a safe working environment. Before starting work, make a safety assessment. What is going to be hazardous, will you require PPE, do you need any special access equipment.

Construction sites can be hazardous because of the temporary nature of the construction process. The surroundings and systems are always changing as the construction process moves to its completion date when everything is finally in place.

Safe methods of working must be demonstrated by everyone at every stage. 'Employees have a duty of care to protect their own health and safety and that of others who might be affected by their work activities'.

To make the work area safe before starting work and during work activities, it may be necessary to:

- use barriers or tapes to screen off potential hazards,
- place warning signs as appropriate,
- inform those who may be affected by any potential hazard,
- use a safe isolation procedure before working on live equipment or circuits,
- obtain any necessary 'permits to work' before work begins.

Try This

Communications

Make a list of all the different types of standard forms which your employer uses. Let me start the list for you with 'Time Sheets'.

Get into the habit of always working safely and being aware of the potential hazards around you when you are working.

Having chosen an appropriate wiring system which meets the intended use and structure of the building and satisfies the environmental conditions of the installation, you must install the system conductors, accessories and equipment in a safe and competent manner.

The structure of the building must be made good if it is damaged during the installation of the wiring system. For example, where conduits and trunking are run through walls and floors.

All connections in the wiring system must be both electrically and mechanically sound. All conductors must be chosen so that they will carry the design current under the installed conditions.

If the wiring system is damaged during installation it must be made good to prevent future corrosion. For example, where galvanized conduit trunking or tray is cut or damaged by pipe vices, it must be made good to prevent localized corrosion.

All tools must be used safely and sensibly. Cutting tools should be sharpened and screwdrivers ground to a sharp square end on a grindstone.

It is particularly important to check that the plug top and cables of hand held electrically powered tools and extension leads are in good condition. Damaged plug tops and cables must be repaired before you use them. All electrical power tools of 110 and 230V must be tested with a portable appliance tester (PAT) in accordance with the company's Health and Safety procedures, but probably at least once each year.

Tools and equipment that are left lying about in the workplace can become damaged or stolen and may also be the cause of people slipping, tripping or falling. Tidy up regularly and put power tools back in their boxes. You personally may have no control over the condition of the workplace in general, but keeping your own work area clean and tidy is the mark of a skilled and conscientious craftsman.

Finally, when the job is finished, clean up and dispose of all waste material responsibly as described in Chapter 8 of *Basic Electrical Installation Work*, 5th Edition under the heading Disposing of Waste.

Safe electrical installations

We know from earlier chapters in this book that using electricity is one of the causes of accidents in the workplace. Using electricity is a hazard because it has the potential, the possibility to cause harm. Therefore, the provision of protective devices in an electrical installation is fundamental to the whole concept of the safe use of electricity in buildings. The electrical installation as a whole must be protected against overload or short circuit and the people using the building must be protected against the risk of shock, fire or other risks arising from their own misuse of the installation or from a fault. The installation and maintenance of adequate and appropriate protective measures is a vital part of the safe use of electrical energy. I want to look at protection against an electric shock by both basic and fault protection, at protection by equipotential bonding and automatic disconnection of the supply, and protection against excess current.

Definition

Earth: The conductive mass of the earth whose electrical potential is taken as zero.

Let us first define some of the words we will be using. Chapter 54 of the IEE Regulations describes the earthing arrangements for an electrical installation. It gives the following definitions:

Definition

Earthing: The act of connecting the exposed conductive parts of an installation to the main protective earthing terminal of the installation.

Earth – the conductive mass of the earth whose electrical potential is taken as zero.

Earthing – the act of connecting the exposed conductive parts of an installation to the main protective earthing terminal of the installation.

Definition

Bonding conductor: A protective conductor providing equipotential bonding.

Definition

Bonding: The linking together of the exposed or extraneous metal parts of an electrical installation.

Definition

Circuit protective conductor: A protective conductor connecting exposed conductive parts of equipment to the main earthing terminal.

Definition

Exposed conductive parts: This is the metalwork of an electrical appliance or the trunking and conduit of an electrical system.

Definition

Extraneous conductive parts: This is the structural steelwork of a building and other service pipes such as gas, water, radiators and sinks.

Definition

Shock protection: Protection from electric shock is provided by basic protection and fault protection.

Definition

Basic protection: This is provided by the insulation of live parts in accordance with Section 416 of the IEE Regulations.

Bonding conductor – a protective conductor providing equipotential bonding.

Bonding – the linking together of the exposed or extraneous metal parts of an electrical installation.

Circuit protective conductor (CPC) – a protective conductor connecting exposed conductive parts of equipment to the main earthing terminal. This is the green and yellow insulated conductor in twin and earth cable.

Exposed conductive parts – this is the metalwork of an electrical appliance or the trunking and conduit of an electrical system which can be touched because they are not normally live, but which may become live under fault conditions.

Extraneous conductive parts – this is the structural steelwork of a building and other service pipes such as gas, water, radiators and sinks. They do not form a part of the electrical installation but may introduce a potential, generally earth potential, to the electrical installation.

Shock protection – protection from electric shock is provided by basic protection and fault protection.

Basic protection – this is provided by the insulation of live parts in accordance with Section 416 of the IEE Regulations.

Fault protection – this is provided by protective equipotential bonding and automatic disconnection of the supply (by a fuse or MCB) in accordance with IEE Regulations 411.3–6.

Protective equipotential bonding – this is equipotential bonding for the purpose of safety and shown in Figs 7.15–7.17.

Basic protection and fault protection

The human body's movements are controlled by the nervous system. Very tiny electrical signals travel between the central nervous system and the muscles, stimulating operation of the muscles, which enable us to walk, talk and run and remember that the heart is also a muscle.

If the body becomes part of a more powerful external circuit, such as the electrical mains, and current flows through it, the body's normal electrical operations are disrupted. The shock current causes unnatural operation of the muscles and the result may be that the person is unable to release the live conductor causing the shock, or the person may be thrown across the room. The current which flows through the body is determined by the resistance of the human body and the surface resistance of the skin on the hands and feet.

This leads to the consideration of exceptional precautions where people with wet skin or wet surfaces are involved, and the need for special consideration in bathroom installations.

Two types of contact will result in a person receiving an electric shock. Direct contact with live parts which involves touching a terminal or line

Definition

Fault protection: This is provided by protective equipotential bonding and automatic disconnection of the supply (by a fuse or MCB) in accordance with IEE Regulations 411.3–6.

Definition

Protective equipotential bonding: This is equipotential bonding for the purpose of safety.

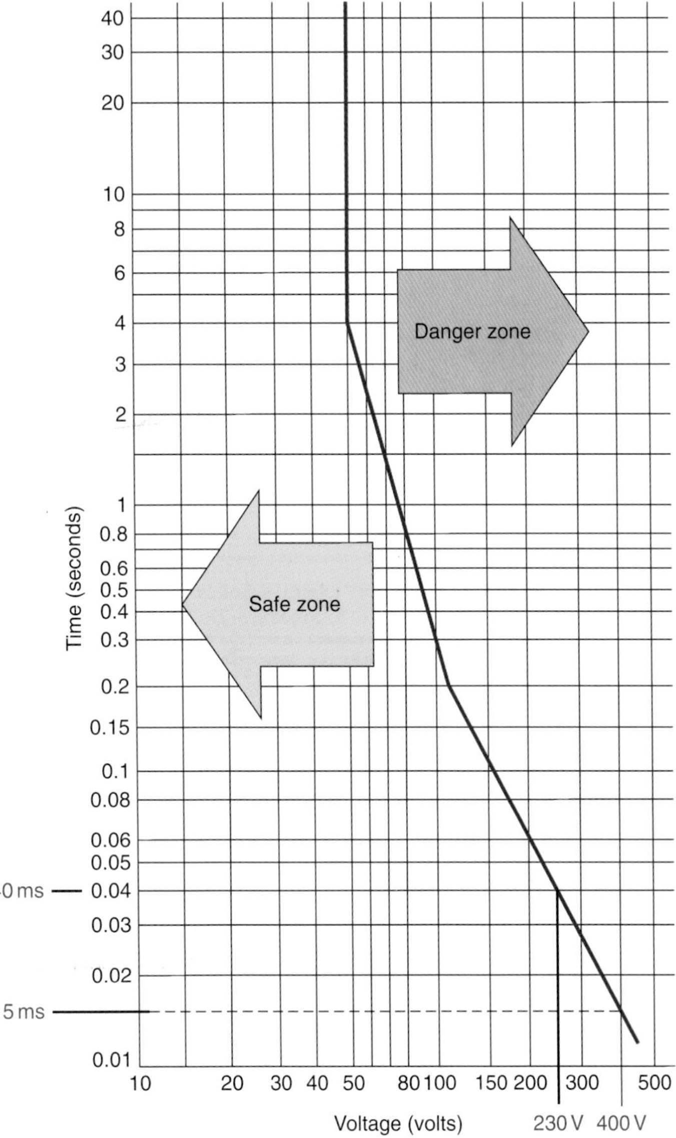

FIGURE 7.14
Touch voltage curve.

conductor that is actually live. The Regulations call this Basic Protection (131.2.1). Indirect contact results from contact with an exposed conductive part such as the metal structure of a piece of equipment that has become live as a result of a fault. The Regulations call this Fault Protection (131.2.2).

The touch voltage curve in Fig. 7.14 shows that a person in contact with 230V must be released from this danger in 40 ms if harmful effects are to be avoided. Similarly, a person in contact with 400V must be released in 15 ms to avoid being harmed.

In installations operating at normal mains voltage, the primary method of protections against direct contact is by insulation. All live parts are enclosed in insulating material such as rubber or plastic, which prevents contact with those parts. The insulating material must, of course, be suitable for the

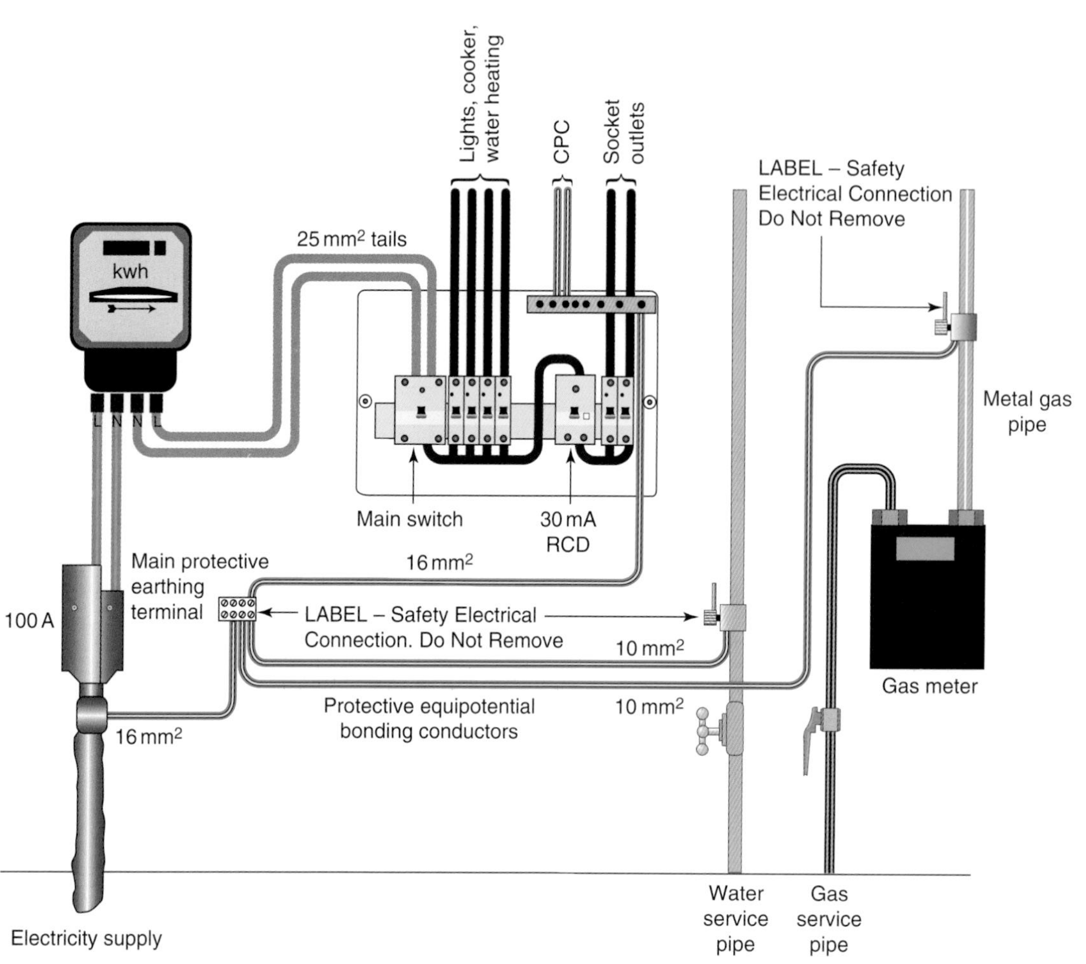

FIGURE 7.15
Cable sheath earth supply (TN-S system) showing earthing and bonding arrangements.

circumstances in which they will be used and the stresses to which they will be subjected.

Other methods of basic protection include the provision of barriers or enclosures which can only be opened by the use of a tool, or when the supply is first disconnected. Protection can also be provided by fixed obstacles such as a guardrail around an open switchboard or by placing live parts out of reach as with overhead lines.

Fault protection

Protection against indirect contact, called fault protection (IEE Regulation 131.2.2) is achieved by connecting exposed conductive parts of equipment to the protective earthing terminal as shown in Figs 7.15–7.17.

In Chapter 13 of the IEE Regulations we are told that where the metalwork of electrical equipment may become charged with electricity in such a manner as to cause danger, that metalwork will be connected with earth so as to discharge the electrical energy without danger.

In connection with fault protection, equipotential bonding is one of the important principles for safety.

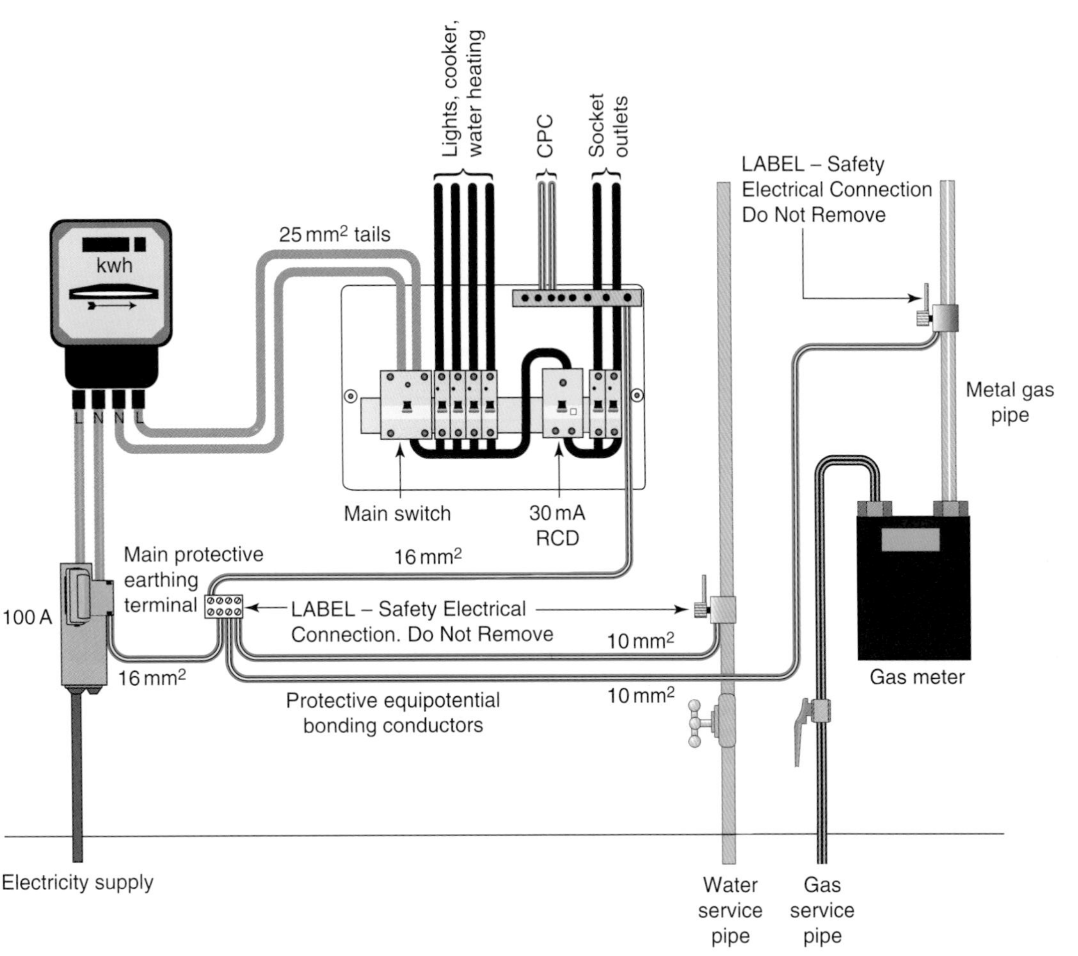

FIGURE 7.16

Protective multiple earth supply (TN-C-S system) showing earthing and bonding arrangements.

There are five methods of protection against contact with metalwork which has become unintentionally live, that is, indirect contact with exposed conductive parts recognized by the IEE Regulations. These are:

1. Protective equipotential bonding coupled with automatic disconnection of the supply.
2. The use of Class II (double insulated) equipment.
3. The provision of a non-conducting location.
4. The use of earth free equipotential bonding.
5. Electrical separation.

Methods 3 and 4 are limited to special situations under the effective supervision of trained personnel.

Method 5, electrical separation, is little used but does find an application in the domestic electric shaver supply unit which incorporates an isolating transformer.

Method 2, the use of Class II insulated equipment is limited to single pieces of equipment such as tools used on construction sites, because it

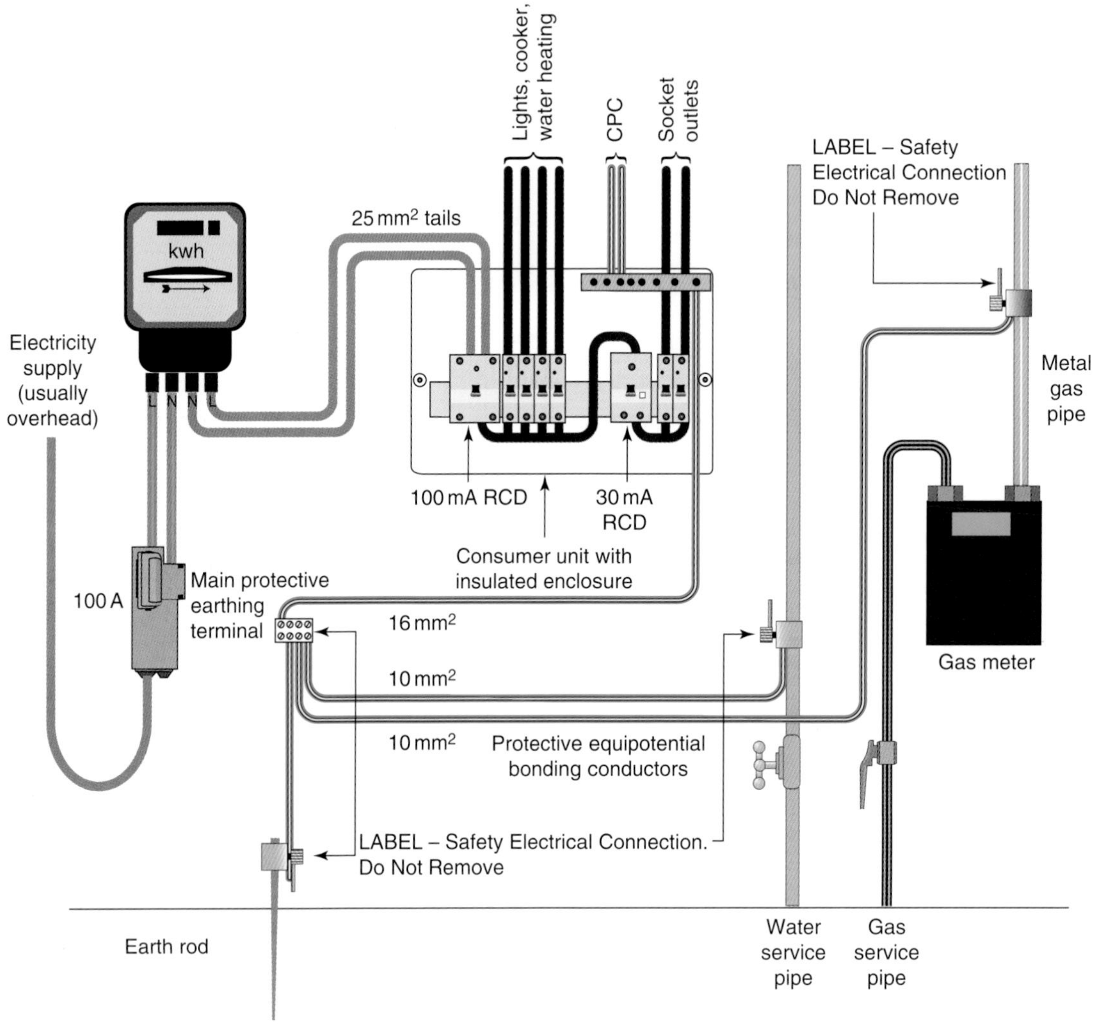

FIGURE 7.17
No earth provided supply (TT systems) showing earthing and bonding arrangements.

relies upon effective supervision to ensure that no metallic equipment or extraneous earthed metalwork enters the area of the installation.

The method which is most universally used in the United Kingdom is, therefore, Method 1 – protective equipotential bonding coupled with automatic disconnection of the supply.

This method relies upon all exposed metalwork being electrically connected together to an effective earth connection. Not only must all the metalwork associated with the electrical installation be so connected, that is conduits, trunking, metal switches and the metalwork of electrical appliances, but also Regulation 411.3.1.2 tells us to connect the extraneous metalwork of water service pipes, gas and other service pipes and ducting, central heating and air conditioning systems, exposed metallic structural parts of the building and lightning protective systems to the main protective earthing terminal. In this way the possibility of a voltage appearing between two exposed metal parts is removed. Protective equipotential bonding is shown in Figs 7.15–7.17.

Key Fact

Bonding

When carrying out earthing and bonding activities:

- use a suitable bonding clamp,
- connect to a cleaned pipe,
- make sure all connections are tight,
- fix a label 'Safety Electrical Connection' close to the connection,
- IEE Regulation 514.13.

The second element of this protection method is the provision of a means of automatic disconnection of the supply in the event of a fault occurring that causes the exposed metalwork to become live. IEE Regulation 411.3.2 tells us that for final circuits not exceeding 32 A, the maximum disconnection time shall not exceed 0.4 seconds.

The achievement of these disconnection times is dependent upon the type of protective device used, fuse or circuit breaker, the circuit conductors to the fault and the provision of adequate protective equipotential bonding. The resistance, or we call it the impedance; of the earth fault loop must be less than the values given in Tables 41.2–41.4 of the IEE Regulations. (Table 7.3 shows the maximum value of the earth fault loop impedance for circuits protected by miniature circuit breakers – MCBs to BS EN 60898.) We will look at this again later in this chapter under the heading 'Earth Fault Loop Impedance Z_S'. Chapter 54 of the IEE Regulations gives details of the earthing arrangements to be incorporated in the supply system to meet these Regulations and these are described below.

Supply system earthing arrangements

The British government agreed on 1 January 1995 that the electricity supplies in the United Kingdom would be harmonized with those of the rest of Europe. Thus the voltages used previously in low-voltage supply systems of 415 and 240V have become 400V for three-phase supplies and 230V for single-phase supplies. The Electricity Supply Regulations 1988 have also been amended to permit a range of variation from the new declared nominal voltage. From January 1995 the permitted tolerance is the nominal voltage +10% or −6%. Previously it was ±6%. This gives a voltage range of 216–253V for a nominal voltage of 230V and 376–440V for a nominal voltage of 400V (IEE Regulations Appendix 2).

It is further proposed that the tolerance levels will be adjusted to ±10% of the declared nominal voltage. All EU countries will adjust their voltages to comply with a nominal voltage of 230V single-phase and 400V three-phase.

The supply to a domestic, commercial or small industrial consumer's installation is usually protected at the incoming service cable position with a 100 A high breaking capacity (HBC) fuse. Other items of equipment at this position are the energy meter and the consumer's distribution unit, providing the protection for the final circuits and the earthing arrangements for the installation.

An efficient and effective earthing system is essential to allow protective devices to operate. The limiting values of earth fault loop impedance are given in Tables 41.2–41.4 of the IEE Regulations, and Chapter 54 and wiring systems of Part 2 of the regulations gives details of the earthing arrangements to be incorporated in the supply system to meet the requirements of the Regulations. Five systems are described in the definitions but only the TN-S, TN-C-S and TT systems are suitable for public supplies.

A system consists of an electrical installation connected to a supply. Systems are classified by a capital letter designation.

The supply earthing

The supply earthing arrangements are indicated by the first letter, where T means one or more points of the supply are directly connected to earth and I means the supply is not earthed or one point is earthed through a fault-limiting impedance.

The installation earthing

The installation earthing arrangements are indicated by the second letter, where T means the exposed conductive parts are connected directly to earth and N means the exposed conductive parts are connected directly to the earthed point of the source of the electrical supply.

The earthed supply conductor

The earthed supply conductor arrangements are indicated by the third letter, where S means a separate neutral and protective conductor and C means that the neutral and protective conductors are combined in a single conductor.

CABLE SHEATH EARTH SUPPLY (TN-S SYSTEM)

This is one of the most common types of supply system to be found in the United Kingdom where the electricity companies' supply is provided by underground cables. The neutral and protective conductors are separate throughout the system. The protective earth conductor (PE) is the metal sheath and armour of the underground cable, and this is connected to the consumer's main earthing terminal. All exposed conductive parts of the installation, gas pipes, water pipes and any lightning protective system are connected to the protective conductor via the main earthing terminal of the installation. The arrangement is shown in Fig. 7.15.

PROTECTIVE MULTIPLE EARTHING SUPPLY (TN-C-S SYSTEM)

This type of underground supply is becoming increasingly popular to supply new installations in the United Kingdom. It is more commonly referred to as protective multiple earthing (PME). The supply cable uses a combined protective earth and neutral (PEN) conductor. At the supply intake point a consumer's main earthing terminal is formed by connecting the earthing terminal to the neutral conductor. All exposed conductive parts of the installation, gas pipes, water pipes and any lightning protective system are then connected to the main earthing terminals. Thus phase to earth faults are effectively converted into phase to neutral faults. The arrangement is shown in Fig. 7.16.

NO EARTH PROVIDED SUPPLY (TT SYSTEM)

This is the type of supply more often found when the installation is fed from overhead cables. The supply authorities do not provide an earth terminal and the installation's CPCs must be connected to earth via an earth electrode provided by the consumer. An effective earth connection is sometimes difficult to obtain and in most cases a residual current device (RCD) is provided when this type of supply is used. The arrangement is shown in Fig. 7.17.

Figures 7.15–7.17 shows the layout of a typical domestic service position for these three supply systems. The TN-C and IT systems of supply do not comply with the supply regulations and therefore cannot be used for public supplies. Their use is restricted to private generating plants and for this reason I shall not include them here, but they can be seen in Part 2 of the IEE Regulations.

Residual current protection

The IEE Regulations recognize the particular problems created when electrical equipment such as lawnmowers, hedge-trimmers, drills and lights are used outside buildings. In these circumstances the availability of an adequate earth return path is a matter of chance. The Regulations, therefore, require that any socket outlet with a rated current not exceeding 20 A, for use by ordinary people and intended for general use shall have the additional protection of an RCD, which has a rated operating current of not more than 30 milliamperes (mA) (IEE Regulation 411.3.3).

Definition

An *RCD* is a type of circuit breaker that continuously compares the current in the line and neutral conductors of the circuit.

An **RCD** is a type of circuit breaker that continuously compares the current in the line and neutral conductors of the circuit. The currents in a healthy circuit will be equal, but in a circuit that develops a fault, some current will flow to earth and the line and neutral currents will no longer balance. The RCD detects the imbalance and disconnects the circuit. Figure 7.18 shows an RCD.

The Regulations recognize RCDs as 'additional protection' in the event of a failure of the provision for basic protection, fault protection or the carelessness by the users of the installation (IEE Regulation 415.1.1).

Isolation and switching

Part 4 of the IEE Regulations deals with the application of protective measures for safety and Chapter 53 with the Regulations for switching devices or switchgear required for protection, isolation and switching of a consumer's installation.

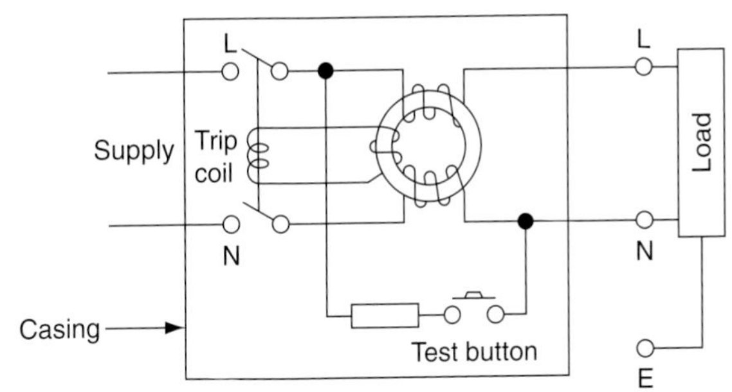

FIGURE 7.18
Construction of an RCD.

The consumer's main switchgear must be readily accessible to the consumer and be able to:

- isolate the complete installation from the supply,
- protect against overcurrent,
- cut off the current in the event of a serious fault occurring.

The Regulations identify four separate types of switching: switching for isolation; switching for mechanical maintenance; emergency switching and functional switching.

Definition

Isolation is defined as cutting off the electrical supply to a circuit or item of equipment in order to ensure the safety of those working on the equipment by making dead those parts which are live in normal service.

Isolation is defined as cutting off the electrical supply to a circuit or item of equipment in order to ensure the safety of those working on the equipment by making dead those parts which are live in normal service. *The purpose* of isolation switching is to enable electrical work to be carried out safely on an isolated circuit or piece of equipment. Isolation is intended to be used by electrically skilled or supervised persons.

An isolator is a mechanical device which is operated manually and used to open or close a circuit off load. An isolator switch must be provided close to the supply point so that all equipment can be made safe for maintenance. Isolators for motor circuits must isolate the motor and the control equipment, and isolators for discharge lighting luminaires must be an integral part of the luminaire so that it is isolated when the cover is removed or be provided with effective local isolation (Regulations 537.2.1.6). Devices which are suitable for isolation are isolation switches, fuse links, circuit breakers, plugs and socket outlets. They must isolate all live supply conductors and provision must be made to secure the isolation (IEE Regulation 537.2.2.4).

Isolation at the consumer's service position can be achieved by a double pole switch which opens or closes all conductors simultaneously. On three-phase supplies the switch needs to only break the live conductors with a solid link in the neutral, provided that the neutral link cannot be removed before opening the switch.

Definition

The *switching for mechanical maintenance* requirements is similar to those for isolation except that the control switch must be capable of switching the full load current of the circuit or piece of equipment.

The **switching for mechanical maintenance** requirements is similar to those for isolation except that the control switch must be capable of switching the full load current of the circuit or piece of equipment. *The purpose* of switching for mechanical maintenance is to enable non-electrical work to be carried out safely on the switched circuit or piece of equipment. Mechanical maintenance switching is intended for use by skilled but non-electrical persons.

Key Fact

Overcurrent

- Overcurrent is a current which exceeds the current carrying capacity of the circuit conductors.
- Overcurrent might be an overload or a short-circuit current.

Switches for mechanical maintenance must be manually operated, not have exposed live parts when the appliance is opened, must be connected in the main electrical circuit and have a reliable on/off indication or visible contact gap (Regulations 537.3.2.2). Devices which are suitable for switching off for mechanical maintenance are switches, circuit breakers, plug and socket outlets.

Key Fact

Functional switching

Functional switching provides control of electrical circuits and equipment in normal service.

Definition

Functional switching provides control of electrical circuits and equipment in normal service.

Definition

Emergency switching involves the rapid disconnection of the electrical supply by a single action to remove or prevent danger.

Definition

Functional switching involves the switching on or off or varying the supply of electrically operated equipment in normal service.

Emergency switching involves the rapid disconnection of the electrical supply by a single action to remove or prevent danger. *The purpose* of emergency switching is to cut off the electricity *rapidly* to remove an unexpected hazard. Emergency switching is for use by anyone. The device used for emergency switching must be immediately accessible and identifiable, and be capable of cutting off the full load current.

Electrical machines must be provided with a means of emergency switching, and a person operating an electrically driven machine must have access to an emergency switch so that the machine can be stopped in an emergency. The remote stop/start arrangement shown in Fig. 6.18 could meet this requirement for an electrically driven machine (Regulation 537.4.2.2). Devices which are suitable for emergency switching are switches, circuit breakers and contactors. Where contactors are operated by remote control they should *open* when the coil is de-energized, that is, fail safe. Pushbuttons used for emergency switching must be coloured red and latch in the stop or off position. They should be installed where danger may arise and be clearly identified as emergency switches. Plugs and socket outlets cannot be considered appropriate for emergency disconnection of supplies.

Functional switching involves the switching on or off or varying the supply of electrically operated equipment in normal service. *The purpose* of functional switching is to provide control of electrical circuits and equipment in normal service. Functional switching is for the user of the electrical installation or equipment. The device must be capable of interrupting the total steady current of the circuit or appliance. When the device controls a discharge lighting circuit it must have a current rating capable of switching an inductive load. The Regulations acknowledge the growth in the number of electronic dimmer switches being used for the control and functional switching of lighting circuits. The functional switch must be capable of performing the most demanding duty it may be called upon to perform (IEE Regulations 537.5.2.1 and 2).

Overcurrent protection

The consumer's mains equipment must provide protection against overcurrent; that is, a current exceeding the rated value (Regulation 430.3). Fuses provide overcurrent protection when situated in the live conductors; they must not be connected in the neutral conductor. Circuit breakers may be used in place of fuses, in which case the circuit breaker may also provide the means of isolation, although a further means of isolation is usually provided so that maintenance can be carried out on the circuit breakers themselves.

When selecting a protective device we must give consideration to the following factors:

- The prospective fault current.
- The circuit load characteristics.
- The current carrying capacity of the cable.
- The disconnection time requirements for the circuit.

Safety First

Isolation

Isolation and switching for protection is an important safety measure when working with electricity.

Definition

An *overload current* can be defined as a current which exceeds the rated value in an otherwise healthy circuit.

Definition

A *short circuit* is an overcurrent resulting from a fault of negligible impedance connected between conductors.

The essential requirements for a device designed to protect against overcurrent are:

- it must operate automatically under fault conditions,
- it must have a current rating matched to the circuit design current,
- have a disconnection time that is within the design parameters,
- have an adequate fault breaking capacity,
- be suitably located and identified.

Overcurrent can be subdivided into overload current, and short-circuit current. An overload current can be defined as a current which exceeds the rated value in an otherwise healthy circuit. **Overload currents** usually occur because the circuit is abused or because it has been badly designed or modified. A **short circuit** is an overcurrent resulting from a fault of negligible impedance connected between conductors. Short circuits usually occur as a result of an accident which could not have been predicted before the event.

An overload may result in currents of two or three times the rated current flowing in the circuit. Short-circuit currents may be hundreds of times greater than the rated current. In both cases the basic requirements for protection are that the fault currents should be interrupted quickly and the circuit isolated safely before the fault current causes a temperature rise or mechanical effects which might damage the insulation connections, joints and terminations of the circuit conductors or their surroundings (IEE Regulation 130.3).

The selected protective device should have a current rating which is not less than the full load current of the circuit but which does not exceed the cable current rating. The cable is then fully protected against both overload and short-circuit faults (Regulation 435.1). Devices which provide overcurrent protection are:

- *HBC fuses to BS 88-6*: These are for industrial applications having a maximum fault capacity of 80 kA.
- *Cartridge fuses to BS 1361*: These are used for a.c. circuits on industrial and domestic installations having a fault capacity of about 30 kA.
- *Cartridge fuses to BS 1362*: These are used in 13 A plug tops and have a maximum fault capacity of about 6 kA.
- *Semi-enclosed fuses to BS 3036*: These were previously called rewirable fuses and are used mainly on domestic installations having a maximum fault capacity of about 4 kA.
- *MCBs to BS EN 60898*: These are MCBs which may be used as an alternative to fuses for some installations. The British Standard includes ratings up to 100 A and maximum fault capacities of 9 kA. They are graded according to their instantaneous tripping currents – that is, the current at which they will trip within 100 ms. This is less than the time taken to blink an eye.

MCB Type B to BS EN 60898 will trip instantly between three and five times its rated current and is also suitable for domestic and commercial installations.

MCB Type C to BS EN 60898 will trip instantly between five and ten times its rated current. It is more suitable for highly inductive commercial and industrial loads.

MCB Type D to BS EN 69898 will trip instantly between 10 and 25 times its rated current. It is suitable for welding and X-ray machines where large inrush currents may occur.

We will now look at the construction, advantages and disadvantages of the various protective devices.

Semi-enclosed fuses (BS 3036)

Definition

The *semi-enclosed* fuse consists of a fuse wire, called the fuse element, secured between two screw terminals in a fuse carrier.

The **semi-enclosed fuse** consists of a fuse wire, called the fuse element, secured between two screw terminals in a fuse carrier. The fuse element is connected in series with the load and the thickness of the element is sufficient to carry the normal rated circuit current. When a fault occurs, an overcurrent flows and the fuse element becomes hot and melts or 'blows' breaking or opening the circuit in which it is inserted.

The designs of the fuse carrier and base are also important. They must not allow the heat generated from an overcurrent to dissipate too quickly from the element, otherwise a larger current would be required to 'blow' the fuse. Also if over-enclosed, heat will not escape and the fuse will 'blow' at a lower current. This type of fuse is illustrated in Fig. 7.19. The fuse element should consist of a single strand of plain or tinned copper wire having a diameter appropriate to the current rating as given in Table 7.1.

ADVANTAGES OF SEMI-ENCLOSED FUSES

- They are very cheap compared with other protective devices both to install and to replace.
- There are no mechanical moving parts.
- It is easy to identify a 'blown fuse'.

DISADVANTAGES OF SEMI-ENCLOSED FUSES

- The fuse element may be replaced with wire of the wrong size either deliberately or by accident.

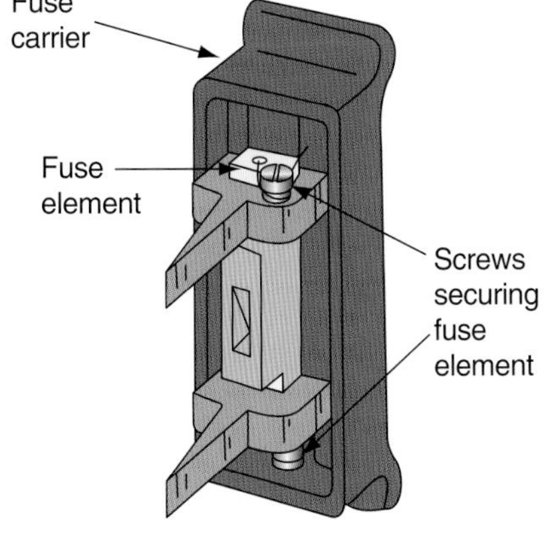

FIGURE 7.19
A semi-enclosed fuse.

Table 7.1 Size of Fuse Element

Current rating (A)	Wire diameter (mm)
5	0.20
10	0.35
15	0.50
20	0.60
30	0.85

- The fuse element weakens with age due to oxidization, which may result in a failure under normal operating conditions.
- The circuit cannot be restored quickly since the fuse element requires screw fixing.
- They have low-breaking capacity since, in the event of a severe fault, the fault current may vaporize the fuse element and continue to flow in the form of an arc across the fuse terminals.
- There is a danger from scattering hot metal if the fuse carrier is inserted into the base when the circuit is faulty.

Definition

The *cartridge fuse* breaks a faulty circuit in the same way as a semi-enclosed fuse, but its construction eliminates some of the disadvantages experienced with an open-fuse element.

Definition

The *fuse element* is encased in a glass or ceramic tube and secured to end-caps which are firmly attached to the body of the fuse so that they do not blow off when the fuse operates.

Cartridge fuses (BS 1361)

The **cartridge fuse** breaks a faulty circuit in the same way as a semi-enclosed fuse, but its construction eliminates some of the disadvantages experienced with an open-fuse element.

The **fuse element** is encased in a glass or ceramic tube and secured to end-caps which are firmly attached to the body of the fuse so that they do not blow off when the fuse operates. Cartridge fuse construction is illustrated in Fig. 7.20. With larger-size cartridge fuses, lugs or tags are sometimes brazed on to the end-caps to fix the fuse cartridge mechanically to the carrier. They may also be filled with quartz sand to absorb and extinguish the energy of the arc when the cartridge is brought into operation.

ADVANTAGES OF CARTRIDGE FUSES

- They have no mechanical moving parts.
- The declared rating is accurate.
- The element does not weaken with age.
- They have small physical size and no external arcing which permits their use in plug tops and small fuse carriers.
- Their operation is more rapid than semi-enclosed fuses. Operating time is inversely proportional to the fault current.

DISADVANTAGES OF CARTRIDGE FUSES

- They are more expensive to replace than rewirable fuse elements.
- They can be replaced with an incorrect cartridge.

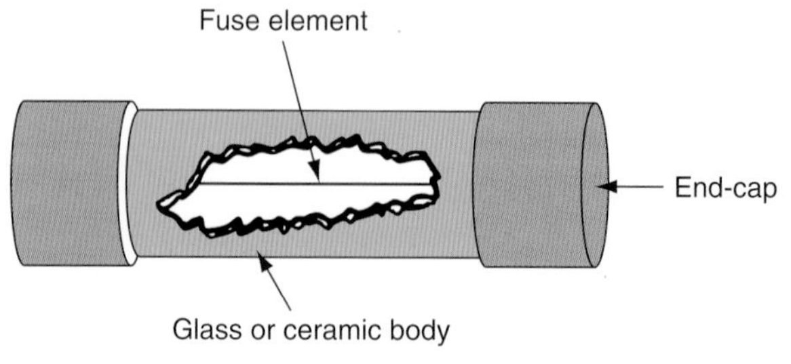

f0200

FIGURE 7.20
A cartridge fuse.

- The cartridge may be shorted out by wire or silver foil in extreme cases of bad practice.
- They are not suitable where extremely high-fault currents may develop.

HBC FUSES (BS 88-6)

As the name might imply, these **HBC (High Breaking Capacity) cartridge fuses** are for protecting circuits where extremely high-fault currents may develop such as on industrial installations or distribution systems.

Definition

As the name might imply, these *HBC cartridge fuses* are for protecting circuits where extremely high-fault currents may develop such as on industrial installations or distribution systems.

The fuse element consists of several parallel strips of pure silver encased in a substantial ceramic cylinder, the ends of which are sealed with tinned brass end-caps incorporating fixing lugs. The cartridge is filled with silica sand to ensure quick arc extraction. Incorporated on the body is an indicating device to show when the fuse has blown. HBC fuse construction is shown in Fig. 7.21.

ADVANTAGES OF HBC FUSES

- They have no mechanical moving parts.
- The declared rating is accurate.
- The element does not weaken with age.
- Their operation is very rapid under fault conditions.
- They are capable of breaking very heavy fault currents safely.
- They are capable of discriminating between a persistent fault and a transient fault such as the large starting current taken by motors.
- It is difficult to insert an incorrect size of cartridge fuse since different ratings are made to different physical sizes.

DISADVANTAGES OF HBC FUSES

- They are very expensive compared to semi-enclosed fuses.

MCBs BS EN 60898

The disadvantage of all fuses is that when they have operated they must be replaced. An MCB overcomes this problem since it is an automatic switch

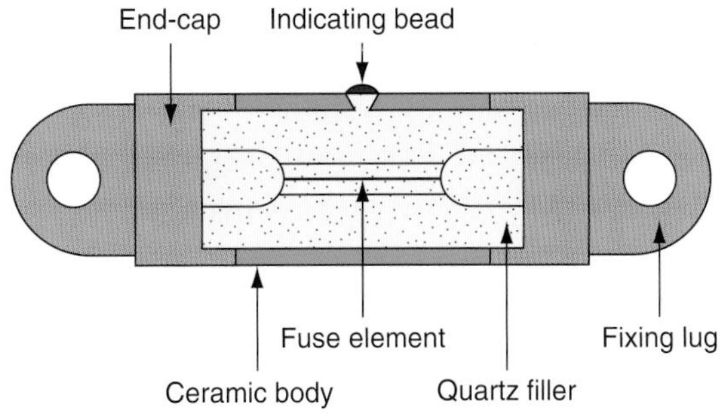

FIGURE 7.21
HBC fuse.

FIGURE 7.22
MCBs – B Breaker, fits Wylex Standard consumer unit. *Courtesy* of Wylex.

which opens in the event of an excessive current flowing in the circuit and can be closed when the circuit returns to normal.

An MCB of the type shown in Fig. 7.22 incorporates a thermal and magnetic tripping device. The load current flows through the thermal and the electromagnetic mechanisms. In normal operation the current is insufficient to operate either device, but when an overload occurs, the bimetal strip heats up, bends and trips the mechanism. The time taken for this action to occur provides an MCB with the ability to discriminate between an overload which persists for a very short time, for example the starting current of a motor, and an overload due to a fault. The device only trips when a fault current occurs. This slow operating time is ideal for overloads but when a short circuit occurs it is important to break the faulty circuit very quickly. This is achieved by the coil electromagnetic device.

When a large fault current (above about eight times the rated current) flows through the coil a strong magnetic flux is set up which trips the mechanisms almost instantly. The circuit can be restored when the fault is removed by pressing the ON toggle. This latches the various mechanisms within the MCB and 'makes' the switch contact. The toggle switch can also be used to disconnect the circuit for maintenance or isolation or to test the MCB for satisfactory operation.

ADVANTAGES OF MCBs

- Tripping characteristics and therefore circuit protection are set by installer.
- The circuit protection is difficult to interfere with.
- The circuit is provided with discrimination.
- A faulty circuit may be easily and quickly restored.
- The supply may be safely restored by an unskilled operator.

DISADVANTAGES OF MCBS

- They are relatively expensive compared to rewirable fuses but look at the advantages to see why they are so popular.
- They contain mechanical moving parts and therefore require regular testing to ensure satisfactory operation under fault conditions.

Fusing factor

The speed with which a protective device will operate under fault conditions gives an indication of the level of protection being offered by that device. This level of protection or fusing performance is given by the fusing factor of the device:

$$\text{Fusing factor} = \frac{\text{Minimum fusing current}}{\text{Current rating}}$$

The minimum fusing current of a device is the current which will cause the fuse or MCB to blow or trip in a given time (BS 88 gives this operating time as 4 hours). The current rating of a device is the current which it will carry continuously without deteriorating.

Thus, a 10A fuse which operates when 15A flows will have a fusing factor of $15 \div 10 = 1.5$.

Since the protective device must carry the rated current it follows that the fusing factor must always be greater than one. The closer the fusing factor is to one, the better is the protection offered by that device. The fusing factors of the protective devices previously considered are:

- semi-enclosed fuses: between 1.5 and 2,
- cartridge fuses: between 1.25 and 1.75,
- HBC fuses: less than 1.25,
- MCBs: less than 1.5.

In order to give protection to the conductors of an installation:

- The current rating of the protective device must be equal to or less than the current carrying capacity of the conductor;
- The current causing the protective device to operate must not be greater than 1.45 times the current carrying capacity of the conductor to be protected.

The current carrying capacities of cables given in the tables of Appendix 4 of the IEE Regulations assume that the circuit will comply with these requirements and that the circuit protective device will have a fusing factor of 1.45 or less. Cartridge fuses, HBC fuses and MCBs do have a fusing factor less than 1.45 and therefore when this type of protection is afforded the current carrying capacities of cables may be read directly from the tables.

However, semi-enclosed fuses can have a fusing factor of 2. The wiring regulations require that the rated current of a rewirable fuse must not exceed

0.725 times the current carrying capacity of the conductor it is to protect. This factor is derived as follows:

The maximum fusing factor of a rewirable fuse is 2.

Now, if:

I_n = current rating of the protective device
I_z = current carrying capacity of conductor
I_2 = current causing the protective device to operate.

Then:

$$I_2 = 2\, I_n \leq 1.45\, I_z$$

Therefore:

$$I_n \leq \frac{1.45\, I_z}{2}$$

Or:

$$I_n \leq 0.725\, I_z$$

When rewirable fuses are used, the current carrying capacity of the cables given in the tables is reduced by a factor of 0.725, as detailed in Appendix 4 item 4 of the Regulations.

Installing overcurrent protective devices

Isolation, switching and protective devices can be found at the consumers' mains equipment position such as that shown in Figs 7.15–7.17. The general principle to be followed is that a protective device must be placed at a point where a reduction occurs in the current carrying capacity of the circuit conductors (IEE Regulations 433.2 and 434.2). A reduction may occur because of a change in the size or type of conductor or because of a change in the method of installation or a change in the environmental conditions. The only exceptions to this rule are where an overload protective device opening a circuit might cause a greater danger than the overload itself – for example, a circuit feeding an overhead electromagnet in a scrapyard.

Fault protection

The overcurrent protection device protecting circuits not exceeding 32 A shall have a disconnection time not exceeding 0.4 seconds (IEE Regulation 411.3.2.2).

The IEE Regulations permit us to assume that where an overload protective device is also intended to provide short-circuit protection, and has a rated breaking capacity greater than the prospective short-circuit current at the point of its installation, the conductors on the load side of the protective device are considered to be adequately protected against short-circuit currents without further proof. This is because the cable rating and the overload rating of the device are compatible. However, if this condition is not met or if there is some doubt, it must be verified that fault currents will be interrupted quickly before they can cause a dangerously high-temperature rise in the circuit conductors. Regulation 434.5.2 provides an equation for calculating the maximum operating time of the protective device to prevent the permitted conductor temperature rise being exceeded as follows:

$$t = \frac{k^2 S^2}{I^2} \text{ (seconds)}$$

where

t = duration time in seconds

S = cross-sectional area of conductor in square millimetres

I = short-circuit rms current in amperes

k = a constant dependent upon the conductor metal and type of insulation (see Table 43A of the IEE Regulations).

Example

A 10 mm PVC sheathed MI copper cable is short-circuited when connected to a 400 V supply. The impedance of the short-circuit path is 0.1 Ω. Calculate the maximum permissible disconnection time and show that a 50 A Type B MCB to BS EN 60898 will meet this requirement:

$$I = \frac{V}{Z} \text{ (A)}$$

$$\therefore I = \frac{400\,\text{V}}{0.1\,\Omega} = 4000\,\text{A}$$

$$\therefore \text{ Fault current} = 4000\,\text{A}$$

For PVC sheathed MI copper cables, Table 43.1 gives a value for k of 115. So,

$$t = \frac{k^2 S^2}{I^2} \text{ (s)}$$

$$\therefore t = \frac{115^2 \times 10^2\,\text{mm}^2}{4000\,\text{A}} = 82.66 \times 10^{-3}\,\text{s}$$

The maximum time that a 4000 A fault current can be applied to this 10 mm^2 cable without dangerously raising the conductor temperature is 82.66 ms. Therefore, the protective device must disconnect the supply to the cable in less than 82.66 ms under short-circuit conditions. Manufacturers' information and Appendix 3 of the IEE Regulations give the operating times of protective devices at various short-circuit currents in the form of graphs. Let us come back to this problem in a moment.

Time/current characteristics of protective devices

Disconnection times for various overcurrent devices are given in the form of a logarithmic graph. This means that each successive graduation of the axis represents a ten times change over the previous graduation.

These logarithmic scales are shown in the graphs of Figs 7.23 and 7.24. From Fig. 7.23 it can be seen that the particular protective device represented by

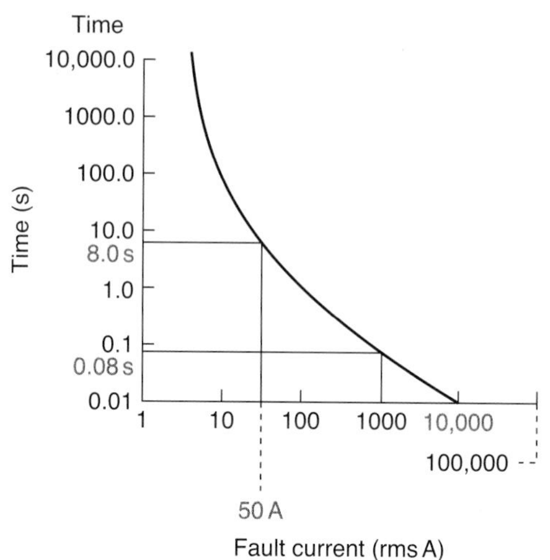

FIGURE 7.23
Time/current characteristic of an overcurrent protective device.

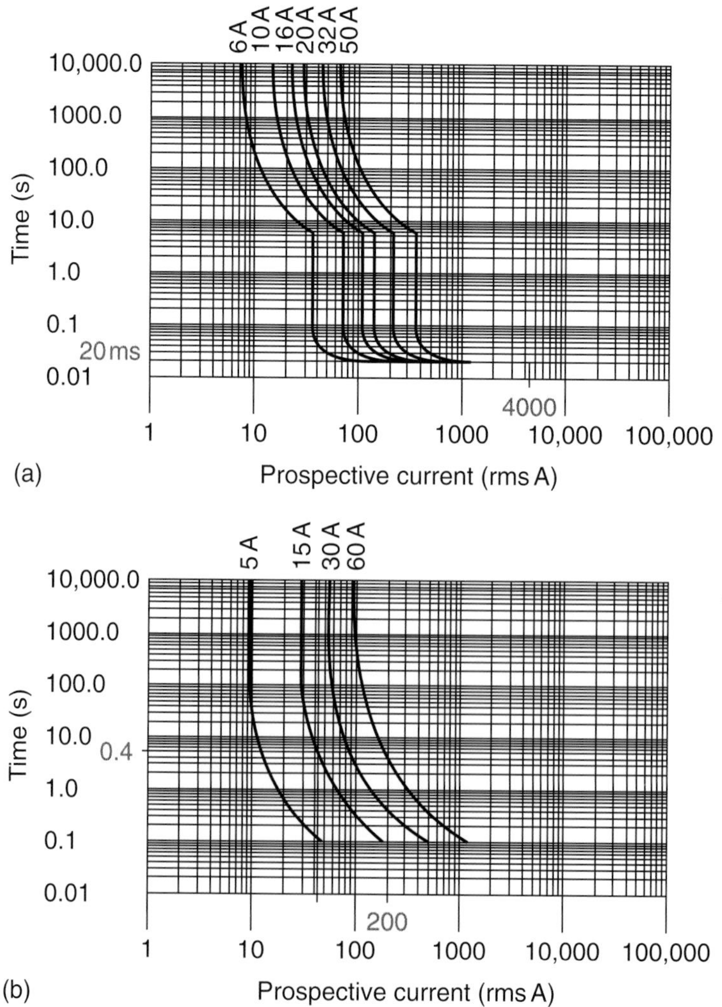

FIGURE 7.24
Time/current characteristics of (a) a Type B MCB to BS EN 60898; (b) semi-enclosed fuse to BS 3036.

this characteristic will take 8 seconds to disconnect a fault current of 50 A and 0.08 seconds to clear a fault current of 1000 A. Let us now go back to the problem and see if the Type B MCB will disconnect the supply in less than 82.66 ms.

Figure 7.24(a) shows the time/current characteristics for a Type B MCB to BS EN 60898. This graph shows that a fault current of 4000 A will trip the protective device in 20 ms. Since this is quicker than 82.66 ms, the 50 A Type B MCB is suitable and will clear the fault current before the temperature of the cable is raised to a dangerous level.

Appendix 3 of the IEE Regulations gives the time/current characteristics and specific values of prospective short-circuit current for a number of protective devices.

These indicate the value of fault current which will cause the protective device to operate in the times indicated by IEE Regulations 411.

Figures 3.1–3.3 in Appendix 3 of the IEE Regulations deal with fuses and Figs 3.4–3.6 with MCBs.

It can be seen that the prospective fault current required to trip an MCB in the required time is a multiple of the current rating of the device. The multiple depends upon the characteristics of the particular devices. Thus:

- Type B MCB to BS EN 60898 has a multiple of 5.
- Type C MCB to BS EN 60898 has a multiple of 10.
- Type D MCB to BS EN 60898 has a multiple of 20.

Example

A 6 A Type B MCB to BS EN 60898 which is used to protect a domestic lighting circuit will trip within 0.4 seconds when 6 A times a multiple of 5, that is 30 A, flows under fault conditions.

Therefore if the earth fault loop impedance is low enough to allow at least 30 A to flow in the circuit under fault conditions, the protective device will operate within the time required by Regulation 411.

The characteristics shown in Appendix 3 of the IEE Regulations give the specific values of prospective short-circuit current for all standard sizes of protective device.

Effective discrimination of protective devices

In the event of a fault occurring on an electrical installation only the protective device nearest to the fault should operate, leaving other healthy circuits unaffected. A circuit designed in this way would be considered to have effective discrimination. Effective discrimination can be achieved by graded protection since the speed of operation of the protective device increases as the rating decreases. This can be seen in Fig. 7.24(b). A fault current of 200 A will cause a 15 A semi-enclosed fuse to operate in about 0.1 seconds, a 30 A semi-enclosed fuse in about 0.4 seconds and a 60 A semi-enclosed fuse in about 5.0 seconds. If a circuit is arranged as shown

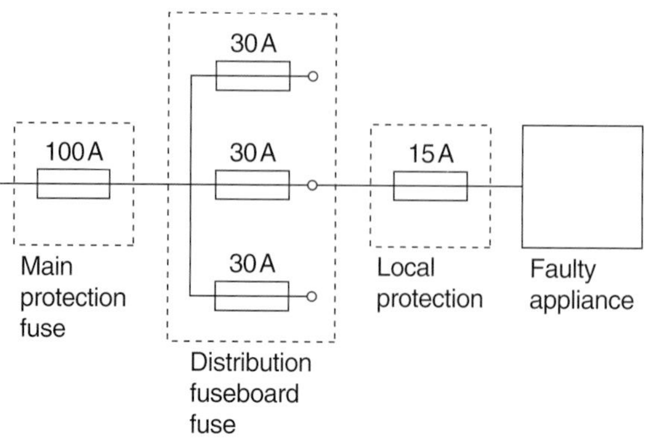

FIGURE 7.25
Effective discrimination achieved by graded protection.

in Fig. 7.25 and a fault occurs on the appliance, effective discrimination will be achieved because the 15A fuse will operate more quickly than the other protective devices if they were all semi-enclosed type fuses with the characteristics shown in Fig. 7.24(b).

Security of supply, and therefore effective discrimination, is an important consideration for an electrical designer and is also a requirement of the IEE Regulations.

Safety First

Protective devices

Effective discrimination of a protective device occurs when only the protective device closest to the fault operates under fault conditions.

Earth fault loop impedance Z_S

In order that an overcurrent protective device can operate successfully, meeting the required disconnection times, of Regulations 411.3.2.2, that is final circuits not exceeding 32A shall have a disconnection time not exceeding 0.4 seconds. To achieve this, the earth fault loop impedance value measured in ohms must be less than those values given in Tables 41.2 and 41.3 of the IEE Regulations. The value of the earth fault loop impedance may be verified by means of an earth fault loop impedance test as described in Chapter 8. The formula is:

$$Z_S = Z_E + (R_1 + R_2)\ (\Omega)$$

Here Z_E is the impedance of the supply side of the earth fault loop. The actual value will depend upon many factors: the type of supply, the ground conditions, the distance from the transformer, etc. The value can be obtained from the electricity companies in that area, but typical values are 0.35 Ω for TN-C-S (PME) supplies and 0.8 Ω for TN-S (cable sheath earth) supplies. Also in the above formula, R_1 is the resistance of the phase conductor and R_2 is the resistance of the earth conductor. The complete earth fault loop path is shown in Fig. 7.26.

Values of $R_1 + R_2$ have been calculated for copper and aluminium conductors and are shown in Table 7.2.

Example

A 20 A radial socket outlet circuit is wired in 2.5 mm^2 PVC cable incorporating a 1.5 mm^2 CPC. The cable length is 30 m installed in an ambient temperature of 20°C and the consumer's

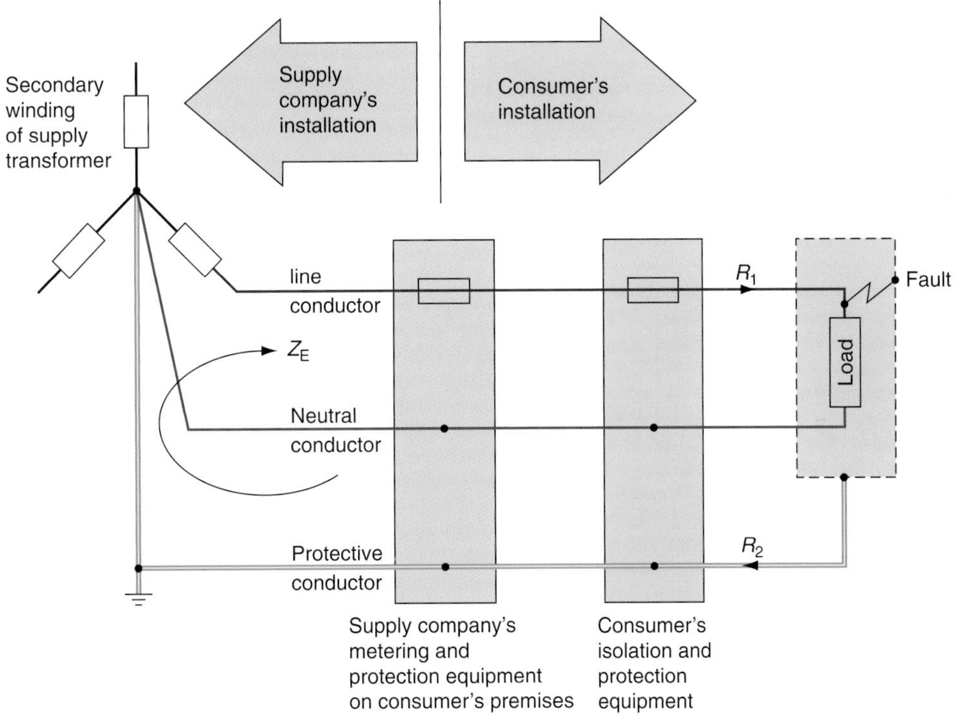

FIGURE 7.26
Earth fault loop path for a TN-S system.

protection is by 20 A MCB Type B to BS EN 60898. The earth fault loop impedance of the supply is 0.5 Ω. Calculate the total earth fault loop impedance Z_S, and establish that the value is less than the maximum value permissible for this type of circuit.

We have:

$$Z_S = Z_E + (R_1 + R_2)\ (\Omega)$$
$$Z_E = 0.5\Omega \text{ (value given in the question)}$$

From the value given in Table 7.2 a 2.5 mm phase conductor with a 1.5 mm protective conductor has an $(R_1 + R_2)$ value of $19.51 \times 10^{-3}\ \Omega/\text{m}$

$$(R_1 + R_2) = 19.51 \times 10^{-3}\ \Omega/\text{m} \times 30\text{m} = 0.585\Omega$$

However, under fault conditions, the temperature and therefore the cable resistance will increase. To take account of this, we must multiply the value of cable resistance by the factor given in Table 9C of the *On Site Guide*. In this case the factor is 1.20 and therefore the cable resistance under fault conditions will be:

$$0.585\Omega \times 1.20 = 0.702\ \Omega$$

The total earth fault loop impedance is therefore:

$$Z_S = 0.5\ \Omega + 0.702\ \Omega = 1.202\ \Omega$$

The maximum permitted value given for a 20 A MCB protecting a socket outlet is 2.3 Ω as shown in Table 7.3. The circuit earth fault loop impedance is less than this value and therefore the protective device will operate within the required disconnection time of 0.4 seconds.

Table 7.2 This shows resistance values per metre. Adapted from the IEE *On Site Guide* by Kind Permission of the Institution of Electrical Engineers

Cross-sectional area (mm^2)		Resistance/metre or ($R_1 + R_2$)/metre (mΩ/m)	
Phase conductor	Protective conductor	Copper	Aluminium
1	—	18.10	
1	1	36.20	
1.5	—	12.10	
1.5	1	30.20	
1.5	1.5	24.20	
2.5	—	7.41	
2.5	1	25.51	
2.5	1.5	19.51	
2.5	2.5	14.82	
4	—	4.61	
4	1.5	16.71	
4	2.5	12.02	
4	4	9.22	
6	—	3.08	
6	2.5	10.49	
6	4	7.69	
6	6	6.16	
10	—	1.83	
10	4	6.44	
10	6	4.91	
10	10	3.66	
16	—	1.15	1.91
16	6	4.23	—
16	10	2.98	—
16	16	2.30	3.82
25	—	0.727	1.20
25	10	2.557	—
25	16	1.877	—
25	25	1.454	2.40
35	—	0.524	0.87
35	16	1.674	2.78
35	25	1.251	2.07
35	35	1.048	1.74
50	—	0.387	0.64
50	25	1.114	1.84
50	35	0.911	1.51
50	50	0.774	1.28

Size of protective conductor

The CPC forms an integral part of the total earth fault loop impedance, so it is necessary to check that the cross-section of this conductor is adequate. If the cross-section of the CPC complies with Table 54.7 of the IEE Regulations, there is no need to carry out further checks. Where line and protective conductors are made from the same material, Table 54.7 tells us that:

- for line conductors equal to or less than 16 mm^2, the protective conductor should equal the line conductor;

Table 7.3 Maximum Earth Fault Loop Impedance Z. Adapted from the IEE *On Site Guide* by Kind Permission of the Institution of Electrical Engineers

Type B MCB
Maximum measured earth fault loop impedance (in ohms) overcurrent protective device is Type B MCB to BSEN 60898

	MCB rating (amperes)							
	6	10	16	20	25	32	40	50
For 0.4 second disconnection Z_S (ohms)	7.67	4.6	2.87	2.3	1.84	1.44	1.15	0.92

- for line conductors greater than 16 mm^2 but less than 35 mm^2, the protective conductor should have a cross-sectional area of 16 mm^2;
- for line conductors greater than 35 mm^2, the protective conductor should be half the size of the line conductor.

However, where the conductor cross-section does not comply with this table, then the formula given in Regulation 543.1.3 must be used:

$$S = \frac{\sqrt{I^2 t}}{k} (\mathrm{mm}^2)$$

where

S = cross-sectional area in mm^2

I = value of maximum fault current in amperes

t = operating time of the protective device

k = a factor for the particular protective conductor (see Tables 54.2–54.4 of the IEE Regulations).

Example 1

A 230V ring main circuit of socket outlets is wired in 2.5 mm single PVC copper cables in a plastic conduit with a separate 1.5 mm CPC. An earth fault loop impedance test identifies Z_S as 1.15 Ω. Verify that the 1.5 mm CPC meets the requirements of Regulation 543.1.3 when the protective device is a 30 A semi-enclosed fuse.

$$I = \text{Maximum fault current} = \frac{V}{Z_S} \text{ (A)}$$

$$\therefore I = \frac{230}{1.15} = 200\text{ A}$$

t = Maximum operating time of the protective device for a circuit not exceeding 32A is 0.4 seconds from Regulation 411.3.2.2. From Fig. 7.24(b) you can see that the time taken to clear a fault of 200 A is about 0.4 seconds.

$$k = 115 \text{ (from Table 54.3).}$$

$$S = \frac{\sqrt{I^2 t}}{k} (\text{mm}^2)$$

$$S = \frac{\sqrt{(200\text{ A})^2 \times 0.4\text{ s}}}{115} = 1.10\text{ mm}^2$$

A 1.5 mm^2 CPC is acceptable since this is the nearest standard-size conductor above the minimum cross-sectional area of 1.10 mm^2 found by calculation.

Example 2

A TN supply feeds a domestic immersion heater wired in 2.5 mm^2 PVC insulated copper cable and incorporates a 1.5 mm^2 CPC. The circuit is correctly protected with a 15 A semi-enclosed fuse to BS 3036. Establish by calculation that the CPC is of an adequate size to meet the requirements of Regulation 543.1.3. The characteristics of the protective device are given in IEE Regulations Table 3.2A.

For final circuits less than 32 A the maximum operating time of the protective device is 0.4 seconds. From Table 3.2A it can be seen that a current of about 90 A will trip the 15 A fuse in 0.4 seconds.

The small insert table on the top right of Table 3.2A of the IEE Regulations gives the value of the prospective fault current required to operate the device within the various disconnection times given.

So, in this case the table states that 90 A will trip a 15 A fuse in 0.4 seconds.

$$\therefore I = 90\text{ A}$$

$$t = 0.4\text{ s}$$

$$k = 115 \text{ (from Table 54.3)}$$

$$S = \frac{\sqrt{I^2 t}}{k} (\text{mm}^2) \text{ (from Regulation 543.1.3)}$$

$$S = \frac{\sqrt{(90\text{ A})^2 \times 0.4\text{ s}}}{115} = 0.49\text{mm}^2$$

The CPC of the cable is greater than 0.49 mm^2 and is therefore suitable. If the protective conductor is a separate conductor, that is, it does not form part of a cable as in this example and is not enclosed in a wiring system as in Example 1, the cross-section of the protective conductor must be not less than 2.5 mm^2 where mechanical protection is provided or 4.0 mm^2 where mechanical protection is *not* provided in order to comply with Regulation 544.2.3.

Cable selection/calculation

The size of a cable to be used for an installation depends upon:

- the current rating of the cable under defined installation conditions,
- the maximum permitted drop in voltage as defined by Regulation 525.

The factors which influence the current rating are:

1. the design current – the cable must carry the full load current;
2. the type of cable – PVC, MICC, copper conductors or aluminium conductors;
3. the installed conditions – clipped to a surface or installed with other cables in a trunking;
4. the surrounding temperature – cable resistance increases as temperature increases and insulation may melt if the temperature is too high;
5. the type of protection – for how long will the cable have to carry a fault current?

Regulation 525 states that the drop in voltage from the supply terminals to the fixed current-using equipment must not exceed 3% for lighting circuits and 5% for other uses of the mains voltage. That is, a maximum of 6.9 V for lighting circuits and 11.5 V for other uses on a 230 V installation. The volt drop for a particular cable may be found from:

$$VD = \text{Factor} \times \text{Design current} \times \text{Length of run}$$

The factor is given in the tables of Appendix 4 of the IEE Regulations and Appendix 6 of the *On Site Guide* (they are also given in Table 7.4).

The cable rating, denoted I_t, may be determined as follows:

$$I_\text{t} = \frac{\text{Current rating of protective device}}{\text{Any applicable correction factors}}$$

The cable rating must be chosen to comply with Regulation 433.1. The correction factors which may need applying are given below as:

Ca = the ambient or surrounding temperature correction factor, which is given in Tables 4B1 and 4B2 of Appendix 4 of the IEE Regulations which is shown in Table 7.5.

Cg = the grouping correction factor given in Tables 4C1–4C5 of the IEE Regulations and 6C of the *On Site Guide*.

Cc = the 0.725 correction factor to be applied when semi-enclosed fuses protect the circuit as described in item 5.1.1 of the preface to Appendix 4 of the IEE Regulations.

Ci = the correction factor to be used when cables are enclosed in thermal insulation. Regulation 523.6.6 gives us three possible correction values:

Table 7.4 Voltage Drop in Cables Factor. Adapted from the IEE *On site Guide* by Kind Permission of the Institution of Electrical Engineers

Table 4D1B IEE Regulations

Voltage drop: (per ampere per metre)		Conductor operating temperature: 70°C	
Conductor cross-sectional area 1	Two-core cable, d.c. 2	Two-core cable, single-phase a.c. 3	Three- or four-core cable, three-phase 4
mm^2	mV/A/m	mV/A/m	mV/A/m
1	44	44	38
1.5	29	29	25
2.5	18	18	15
4	11	11	9.5
6	7.3	7.3	6.4
10	4.4	4.4	3.8
16	2.8	2.8	2.4
		r	r
25	1.75	1.80	1.50
35	1.25	1.30	1.10
50	0.93	0.95	0.81
70	0.63	0.65	0.56
95	0.46	0.49	0.42

Note: For a fuller treatment see Appendix 4 of BS 7671

Table 7.5 Ambient Temperature Correction Factors. Adapted from the IEE *On Site Guide* by Kind Permission of the Institution of Electrical Engineers

Table 4B1 of the IEE Regulations ambient temperature factors
Correction factors for ambient temperature where protection is against short-circuit and overload

Type of insulation	Operating temperature	Ambient temperature (°C)								
		25	30	35	40	45	50	55	60	65
Thermoplastic (general purpose PVC)	70°C	1.03	1.0	0.94	0.87	0.79	0.71	0.61	0.50	NA

- Where one side of the cable is in contact with thermal insulation we must read the current rating from the column in the table which relates to reference method A (see Table 7.6).
- Where the cable is *totally* surrounded over a length greater than 0.5 m we must apply a factor of 0.5.
- Where the cable is *totally* surrounded over a short length, the appropriate factor given in Table 52.2 of the IEE Regulations or Table 6B of the *On Site Guide* should be applied.

Note: A cable should preferably NOT be installed in thermal insulation.

Table 7.6 Current Carrying Capacity of Cables. Adapted from the IEE *On Site Guide* by Kind Permission of the Institution of Electrical Engineers

IEE Table 4D1A Multicore cables having thermoplastic (pvc) or thermosetting insulation (note 1), non-armoured, (Copper conductors)

Ambient temperature: 30°C. Conductor operating temperature: 70°C

Current-carrying capacity (amperes): BS 6004, BS 7629

Conductor cross-sectional area	Reference Method A (enclosed in conduit in an insulated wall, etc.)		Reference Method B (enclosed in conduit on a wall or ceiling, or in trunking)		Reference Method C (clipped direct)		Reference Method F (on a perforated cable tray) or (free air)	
	2 cables single-phase a.c. or d.c.	3 or 4 cables three-phase a.c.	1 two-core cable*, single-phase a.c. or d.c.	1 three-core cable* or 1 four-core cable, three-phase a.c.	1 two-core cable*, single-phase a.c. or d.c.	1 three-core cable* or 1 four-core cable, three-phase a.c.	1 two-core cable*, single-phase a.c. or d.c.	1 three-core cable* or 1 four-core cable, three-phase a.c.
1	2	3	4	5	6	7	8	9
mm^2	A	A	A	A	A	A	A	A
1	11	10	13	11.5	15.5	13.5	17	14.5
1.5	14.5	13	16.5	15	20	17.5	22	18.5
2.5	20	17.5	23	20	27	24	30	25
4	26	23	30	27	37	32	40	34
6	34	29	38	34	47	41	51	43
10	46	39	52	46	65	57	70	60
16	61	52	69	62	87	76	94	80
25	80	68	90	80	114	96	119	101
35	99	83	111	99	141	119	148	126
50	119	99	133	118	182	144	180	153
70	151	125	168	149	234	184	232	196
95	182	150	201	179	284	223	282	238

* With or without protective conductor.

Having calculated the cable rating, (I_t) the smallest cable should be chosen from the appropriate table which will carry that current. This cable must also meet the voltage drop Regulation 525 and this should be calculated as described earlier. When the calculated value is less than 3% for lighting and 5% for other uses of the mains voltage the cable may be considered suitable. If the calculated value is greater than this value, the next larger cable size must be tested **until a cable is found which meets both the current rating and voltage drop criteria**.

Example

A house extension has a total load of 6 kW installed some 18 m away from the mains consumer unit for lighting. A PVC insulated and sheathed twin and earth cable will provide a sub-main to this load and be clipped to the side of the ceiling joists over much of its length in a roof space which is anticipated to reach 35°C in the summer and where insulation is installed up to the top of the joists. Calculate the minimum cable size if the circuit is to be protected by a Type B MCB to BS EN 60878. Assume a TN-S supply, that is, a supply having a separate neutral and protective conductor throughout.

Let us solve this question using only the tables given in the *On Site Guide*. The tables in the Regulations will give the same values, but this will simplify the problem because we can refer to Tables 7.4–7.6 in this book.

$$\text{Design current } I_b = \frac{\text{Power}}{\text{Volts}} = \frac{6000\text{ W}}{230\text{ V}} = 26.09\text{ A}$$

Nominal current setting of the protection for this load I_n = 32 A.

The cable rating, I_t is given by:

$$I_t = \frac{\text{Current rating of protective device } (I_n)}{\text{The product of the correction factors}}$$

The correction factors to be included in this calculation are:

Ca = ambient temperature; from the Table 4B1 shown in Table 7.5 the correction factor for 35°C is 0.94.

Cg = grouping factors need not be applied.

Cc = since protection is by MCB no factor need be applied.

Ci = thermal insulation demands that we assume installed method A (see Table 7.6).

The design current is 26.09 A and we will therefore choose a 32 A MCB for the nominal current setting of the protective device, I_n.

$$\text{Cable rating} = I_t = \frac{32}{0.94} = 34.04\text{ A}$$

From column 2 of the Table 4D1A shown in Table 7.6, a 10 mm cable, having a rating of 46 A, is required to carry this current.

Key Fact

Volt drop

Maximum permissible volt drop on 230V supplies

- 3% for lighting = 6.9V
- 5% for other uses = 11.5V
- IEE Regulations 525 and Appendix 12.

Now test for volt drop: from the Table 4D1B which is shown in Table 7.4 the volt drop per ampere per metre for a 10mm cable is 4.4mV. So the volt drop for this cable length and load is equal to

$$4.4 \times 10^{-3}\ \text{V/(A m)} \times 26.09\ \text{A} \times 18\ \text{m} = 2.06\ \text{V}$$

Since this is less than the maximum permissible value for a lighting circuit of 6.9V a 10mm cable satisfies the current and drop in voltage requirements when the circuit is protected by an MCB. This cable is run in a loft that gets hot in summer and has thermal insulation touching one side of the cable. We must, therefore, use installed method A of Table 7.6. If we were able to route the cable under the floor, or clipped direct, or in a conduit or trunking on a wall, we may be able to use a 6mm cable for this load. You can see how the current carrying capacity of a cable varies with the installed method by looking at Table 7.6. Compare the values in column 2 with those in column 6. When the cable is clipped directly on to a wall or surface the current rating is higher because the cable is cooler than run in a loft with insulation on one side. If the alternative route was longer, you would need to test for volt drop before choosing the cable. These are some of the decisions which the electrical contractor must make when designing an installation which meets the requirements of the customer and the IEE Regulations.

If you are unsure of the standard fuse and MCB rating of protective devices, you can refer to Fig. 3.4 in Appendix 3 of the IEE Regulations.

Check your Understanding

When you have completed these questions, check out the answers at the back of the book.
Note: more than one multiple choice answer may be correct.

1. From the list below, identify those parties who would usually be considered the main contractor' for a construction project:
   a. the architect
   b. the building contractor
   c. the electrical contractor
   d. the mechanical services contractor.
2. From the list below, identify those parties who would usually be considered to be a 'sub-contractor' for a construction project:
   a. the architect
   b. the building contractor
   c. the electrical contractor
   d. the mechanical services contractor.
3. Identify those parties who would usually act as the client's agent for a construction project:
   a. the architect
   b. the building contractor
   c. the electrical contractor
   d. the mechanical services contractor.
4. A formal document that sets out the terms of agreement between a client or main contractor and a sub-contractor is called a:
   a. bar chart
   b. formal contract
   c. network analysis
   d. variation order.
5. The document that establishes the sequence and timing of the various activities involved in a contract is called a:
   a. bar chart
   b. formal contract
   c. network analysis
   d. variation order.

6. A set of drawings that identify the final position of the electrical equipment is called:
   a. assembly drawing
   b. as-fitted drawing
   c. site plan
   d. layout drawing.
7. The protection provided by the insulation of live parts is called:
   a. basic protection
   b. fault protection
   c. overcurrent protection
   d. overload protection.
8. The protection provided by equipotential bonding and automatic disconnection of the supply is called:
   a. basic protection
   b. fault protection
   c. overcurrent protection
   d. overload.
9. State four electrical installation work activities.
10. Using electricity is one of the causes of accidents at work. State five activities that you could carry out that would demonstrate 'a safe method of working' on a construction site and that would reduce the risk of an electrical accident occurring.
11. State the four factors that you must consider when selecting a protective device.
12. State the five essential requirements for a device designed to protect against overcurrent.
13. State three factors affecting the size of a conductor chosen to supply a particular load in a building.
14. Use a sketch and briefly describe the operation of an RCD under normal operating conditions and under fault conditions.

15. Very briefly describe isolation switching and switching for mechanical maintenance.

16. Use a sketch to show the earth fault loop path for a TN-S system.

17. State the maximum earth fault loop impedance for a ring main of socket outlets protected by a 32 A Type B MCB (Table 7.3 will be useful here).

18. State the maximum earth fault loop impedance for a lighting circuit protected by a 6 A Type B MCB.

19. Use a sketch to describe what we mean by effective discrimination of protective devices.

20. The Regulations inform us that a consumer's equipment must be capable of

    a. isolating the installation from the supply
    b. protecting against overcurrent
    c. cutting off the current in the event of a serious fault occurring.

    Sketch the consumer's equipment at a typical domestic service position and identify the parts that provide:

    a. isolation facilities
    b. protection against overcurrent
    c. protection against fault current.

Finally, identify on your sketch the:

a protective equipotential bonding conductors
b. extraneous conductive parts
c. main protective earthing terminal.

Inspection, testing and commissioning

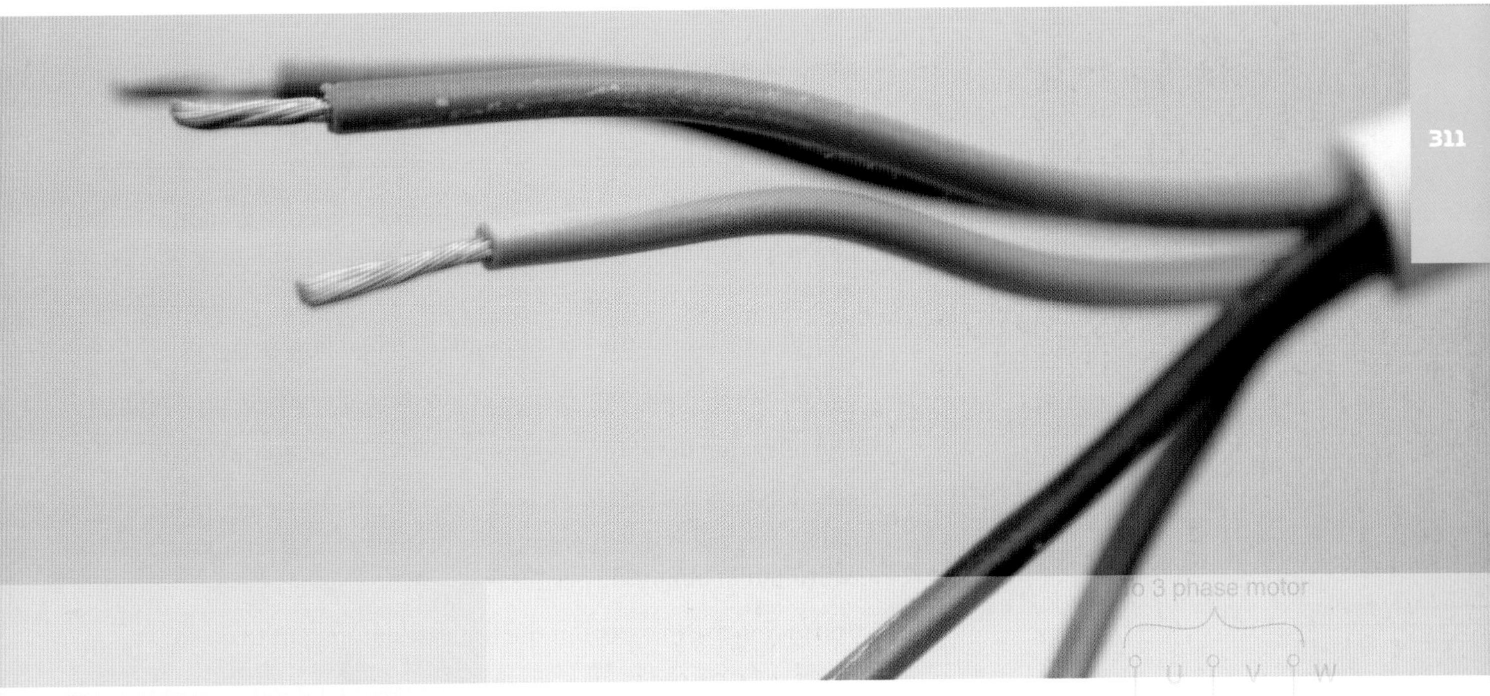

Unit 2 – Installation (buildings and structures): inspection, testing and commissioning – Outcome 2

Underpinning knowledge: when you have completed this chapter you should be able to:

- state the reasons for testing an electrical installation
- state the factors to be considered when carrying out electrical tests safely
- list some of the IEE checklist items associated with visual inspections (Reg. 611.3)
- state the need for regular calibration of electrical test instruments
- describe how to carry out basic electrical installation testing
- describe the certification process for an electrical installation
- state the meaning of commissioning
- describe portable appliance testing (PAT)

Electrical testing

The electrical contractor is charged with a responsibility to carry out a number of tests on an electrical installation and electrical equipment. The individual tests are dealt with in Part 6 of the IEE Regulations and described later in this chapter.

The reasons for testing the installation are:

- to ensure that the installation complies with the Regulations,
- to ensure that the installation meets the specification,
- to ensure that the installation is safe to use.

Those who are to carry out the electrical tests must first consider the following safety factors:

- An assessment of safe working practice must be made before testing begins.
- All safety precautions must be put in place before testing begins.
- Everyone must be notified that the test process is about to take place, for example the client and other workers who may be affected by the tests.
- 'Permits-to-Work' must be obtained where relevant.
- All sources of information relevant to the tests have been obtained.
- The relevant circuits and equipment have been identified.
- Safe isolation procedures have been carried out – care must be exercised here, in occupied premises, not to switch off computer systems without first obtaining permission.
- Those who are to carry out the tests are competent to do so.

The electrical contractor is charged by the IEE Regulations for Electrical Installations to test all new installations and major extensions during erection and upon completion before being put into service. The contractor may also be called upon to test installations and equipment in order to identify and remove faults. These requirements imply the use of appropriate test instruments, and in order to take accurate readings consideration should be given to the following points:

- Is the instrument suitable for this test?
- Have the correct scales been selected?
- Is the test instrument correctly connected to the circuit?

Many commercial instruments are capable of making more than one test or have a range of scales to choose from. A range selector switch is usually used to choose the appropriate scale. A scale range should be chosen which suits the range of the current, voltage or resistance being measured. For example, when taking a reading in the 8 or 9V range the obvious scale

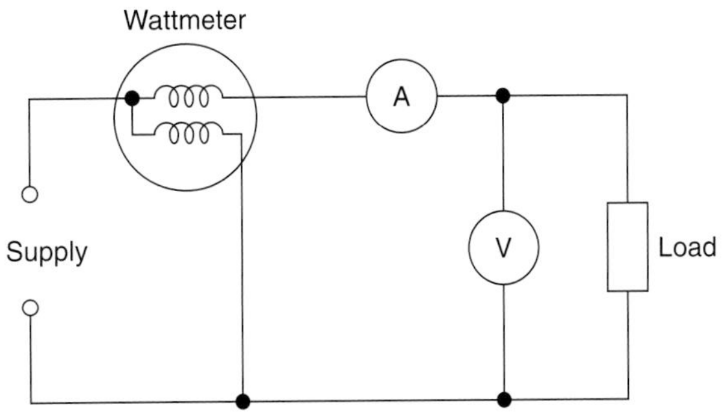

FIGURE 8.1
Wattmeter, ammeter and voltmeter correctly connected to a load.

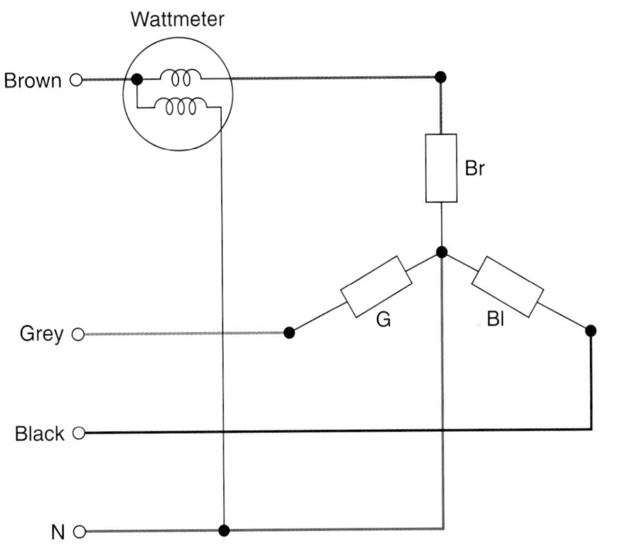

FIGURE 8.2
One-wattmeter measurement of power.

choice would be one giving 10V full scale deflection. To make this reading on an instrument with 100V full scale deflection would lead to errors, because the deflection is too small.

Ammeters must be connected in series with the load, and voltmeters in parallel across the load as shown in Fig. 8.1. The power in a resistive load may be calculated from the readings of voltage and current since $P = VI$. This will give accurate calculations on both a.c. and d.c. supplies, but when measuring the power of an a.c. circuit which contains inductance or capacitance a wattmeter must be used because the voltage and current will be out of phase.

Measurement of power in a three-phase circuit

ONE-WATTMETER METHOD

When three-phase loads are balanced, for example in motor circuits, one wattmeter may be connected into any phase, as shown in Fig. 8.2. This

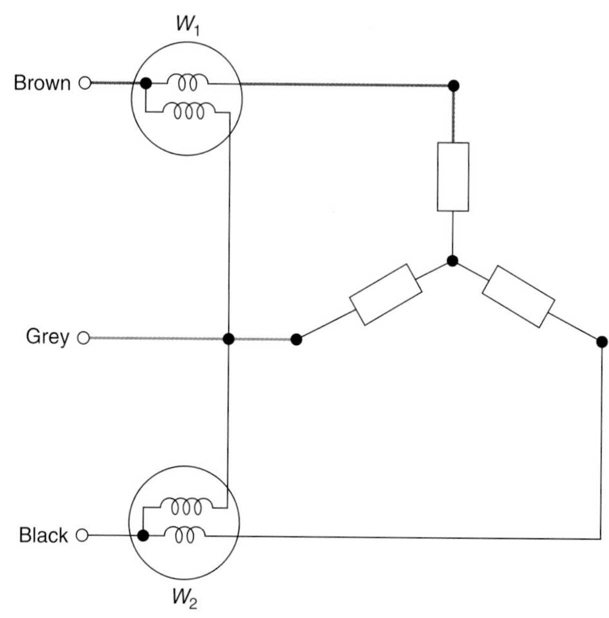

FIGURE 8.3
Two-wattmeter measurement of power.

wattmeter will indicate the power in that phase and since the load is balanced the total power in the three-phase circuit will be given by:

$$\text{Total power} = 3 \times \text{Wattmeter reading}$$

TWO-WATTMETER METHOD

This is the most commonly used method for measuring power in a three-phase, three-wire system since it can be used for both balanced and unbalanced loads connected in either star or delta. The current coils are connected to any two of the lines, and the voltage coils are connected to the other line, the one without a current coil connection, as shown in Fig. 8.3. Then,

$$\text{Total power} = W_1 + W_2$$

This equation is true for any three-phase load, balanced or unbalanced, star or delta connection, provided there is no fourth wire in the system.

THREE-WATTMETER METHOD

If the installation is four-wire, and the load on each phase is unbalanced, then three-wattmeter readings are necessary, connected as shown in Fig. 8.4. Each wattmeter measures the power in one phase and the total power will be given by:

$$\text{Total power} = W_1 + W_2 + W_3$$

Tong tester

The tong tester or clip-on ammeter works on the same principle as the transformer. The laminated core of the transformer can be opened and passed over the busbar or single-core cable. In this way a measurement of

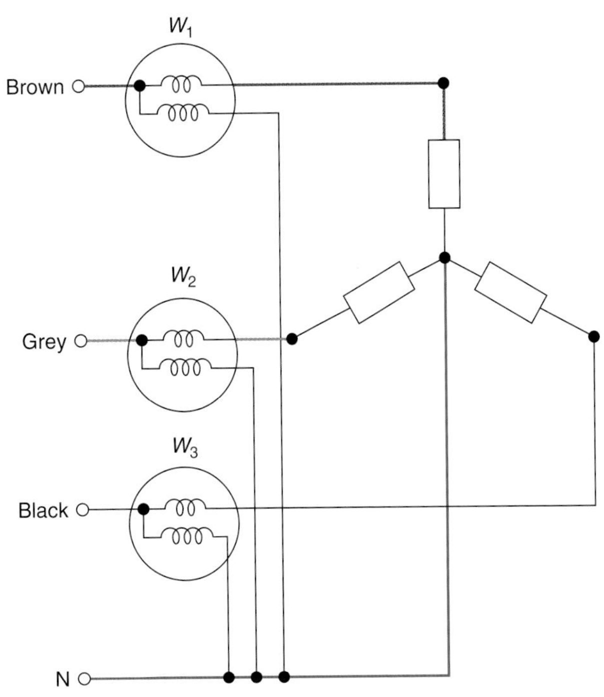

FIGURE 8.4
Three-wattmeter measurement of power.

FIGURE 8.5
Tong tester or clip-on ammeter.

the current being carried can be made without disconnection of the supply. The construction is shown in Fig. 8.5.

Phase sequence testers

Phase sequence is the order in which each phase of a three-phase supply reaches its maximum value. The normal phase sequence for a three-phase supply is brown–black–grey, which means that first brown, then black and finally the grey phase reaches its maximum value.

Phase sequence has an important application in the connection of three-phase transformers. The secondary terminals of a three-phase transformer must not be connected in parallel until the phase sequence is the same.

A phase sequence tester can be an indicator which is, in effect, a miniature induction motor, with three clearly colour-coded connection leads. A rotating disc with a pointed arrow shows the normal rotation for phase sequence brown–black–grey. If the sequence is reversed the disc rotates in the opposite direction to the arrow. However, an on-site phase sequence tester can be made by connecting four 230V by 100 W lamps and a p.f. correction capacitor from a fluorescent luminaire as shown in Fig. 8.6.

The capacitor takes a leading current which results in a phase displacement in the other two phases. The phasor addition of the voltage in the circuit results in one pair of lamps illuminating brightly while the other pairs are illuminated dimly. Two lamps must be connected in series as shown in Fig. 8.6 because most of the line voltage will be across them during the test.

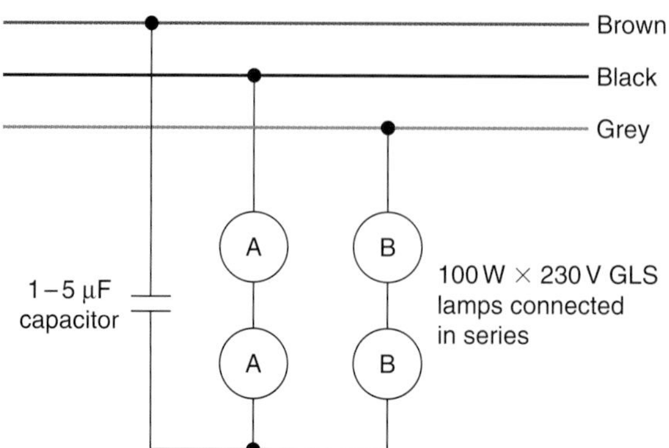

FIGURE 8.6
Phase sequence test by the lamps bright, lamps dim method.

Test equipment used by electricians

The Health and Safety Executive (HSE) has published Guidance Notes (GS 38) which advise electricians and other electrically competent people on the selection of suitable test probes, voltage indicating devices and measuring instruments. This is because they consider suitably constructed test equipment to be as vital for personal safety as the training and practical skills of the electrician. In the past, unsatisfactory test probes and voltage indicators have frequently been the cause of accidents, and therefore all test probes must now incorporate the following features:

1. The probes must have finger barriers or be shaped so that the hand or fingers cannot make contact with the live conductors under test.
2. The probe tip must not protrude more than 2 mm, and preferably only 1 mm, be spring-loaded and screened.
3. The lead must be adequately insulated and coloured so that one lead is readily distinguished from the other.
4. The lead must be flexible and sufficiently robust.
5. The lead must be long enough to serve its purpose but not too long.
6. The lead must not have accessible exposed conductors even if it becomes detached from the probe or from the instrument.
7. Where the leads are to be used in conjunction with a voltage detector they must be protected by a fuse.

A suitable probe and lead is shown in Fig. 8.7.

GS 38 also tells us that where the test is being made simply to establish the presence or absence of a voltage, the preferred method is to use a proprietary test lamp or voltage indicator which is suitable for the working voltage, rather than a multimeter. Accident history has shown that incorrectly set multimeters or makeshift devices for voltage detection have frequently

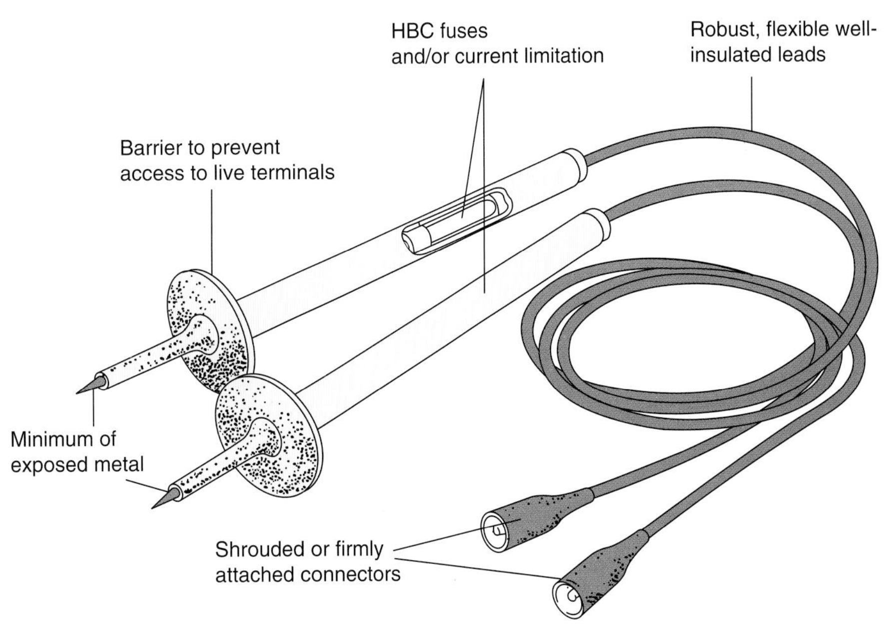

FIGURE 8.7
Recommended type of test probe and leads.

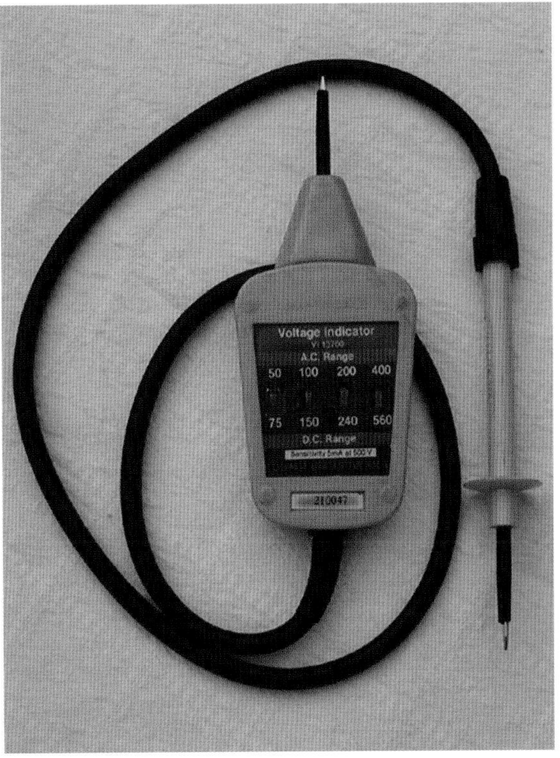

FIGURE 8.8
Typical voltage indicator.

caused accidents. Fig. 8.8 shows a suitable voltage indicator. Test lamps and voltage indicators are not fail-safe, and therefore GS 38 recommends that they should be regularly proved, preferably before and after use, as described in the flowchart for a safe isolation procedure shown in Fig 8.11.

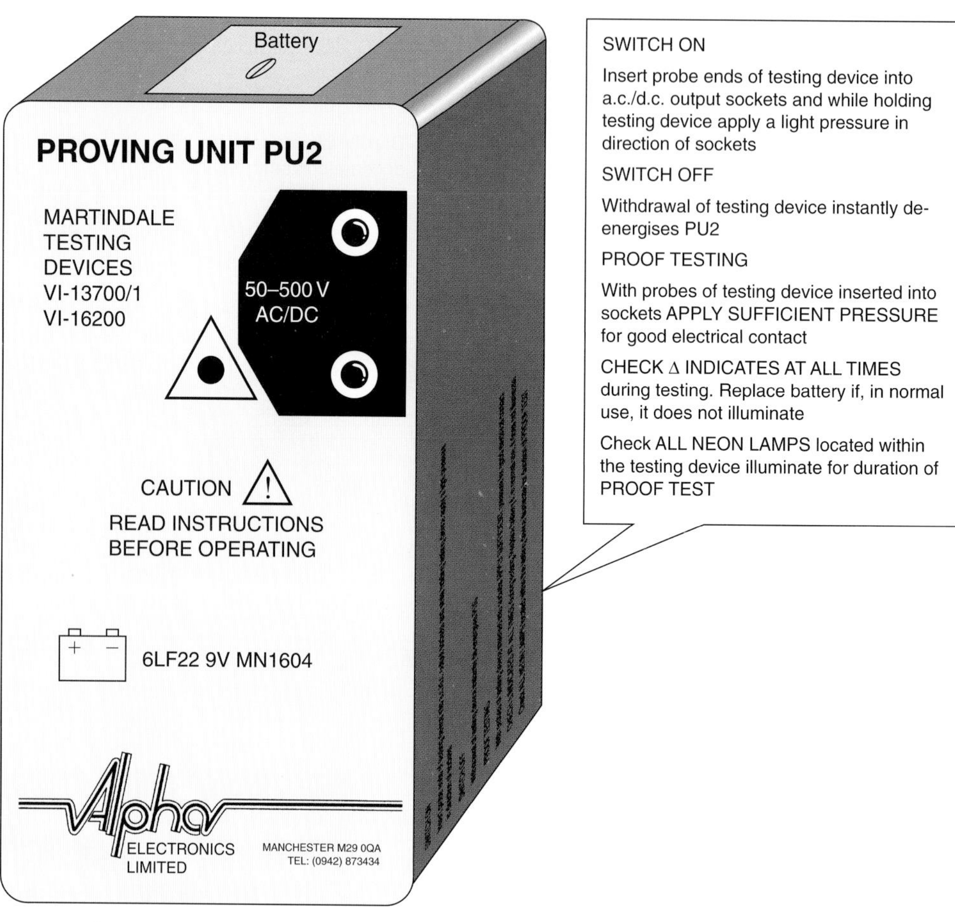

FIGURE 8.9
Voltage proving unit.

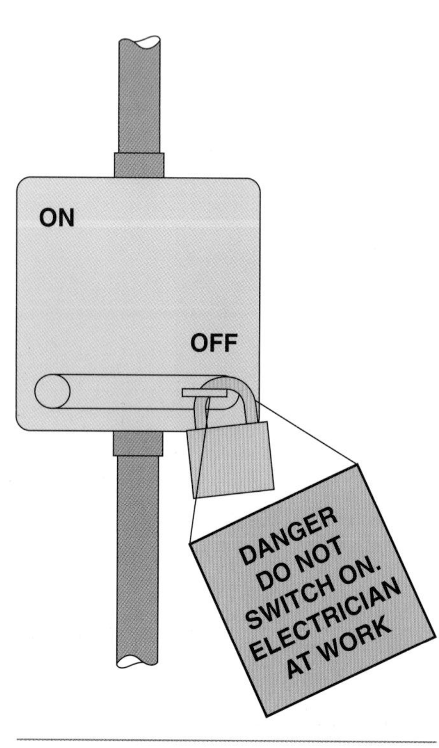

FIGURE 8.10
Secure isolation of a supply.

Test procedures

1. The circuits must be isolated using a 'safe isolation procedure', such as that described below, before beginning to test.
2. All test equipment must be 'approved' and connected to the test circuits by recommended test probes as described by the HSE Guidance Notes GS 38. The test equipment used must also be 'proved' on a known supply or by means of a proving unit such as that shown in Fig. 8.9.
3. Isolation devices must be 'secured' in the 'off' position as shown in Fig. 8.10.
4. Warning notices must be posted.
5. All relevant safety and functional tests must be completed before restoring the supply.

Live testing

The **Electricity at Work Act** tells us that it is 'preferable' that supplies be made dead before work commences (Regulation 4(3)). However, it does acknowledge that some work, such as fault-finding and testing, may require the electrical equipment to remain energized. Therefore, if the fault finding

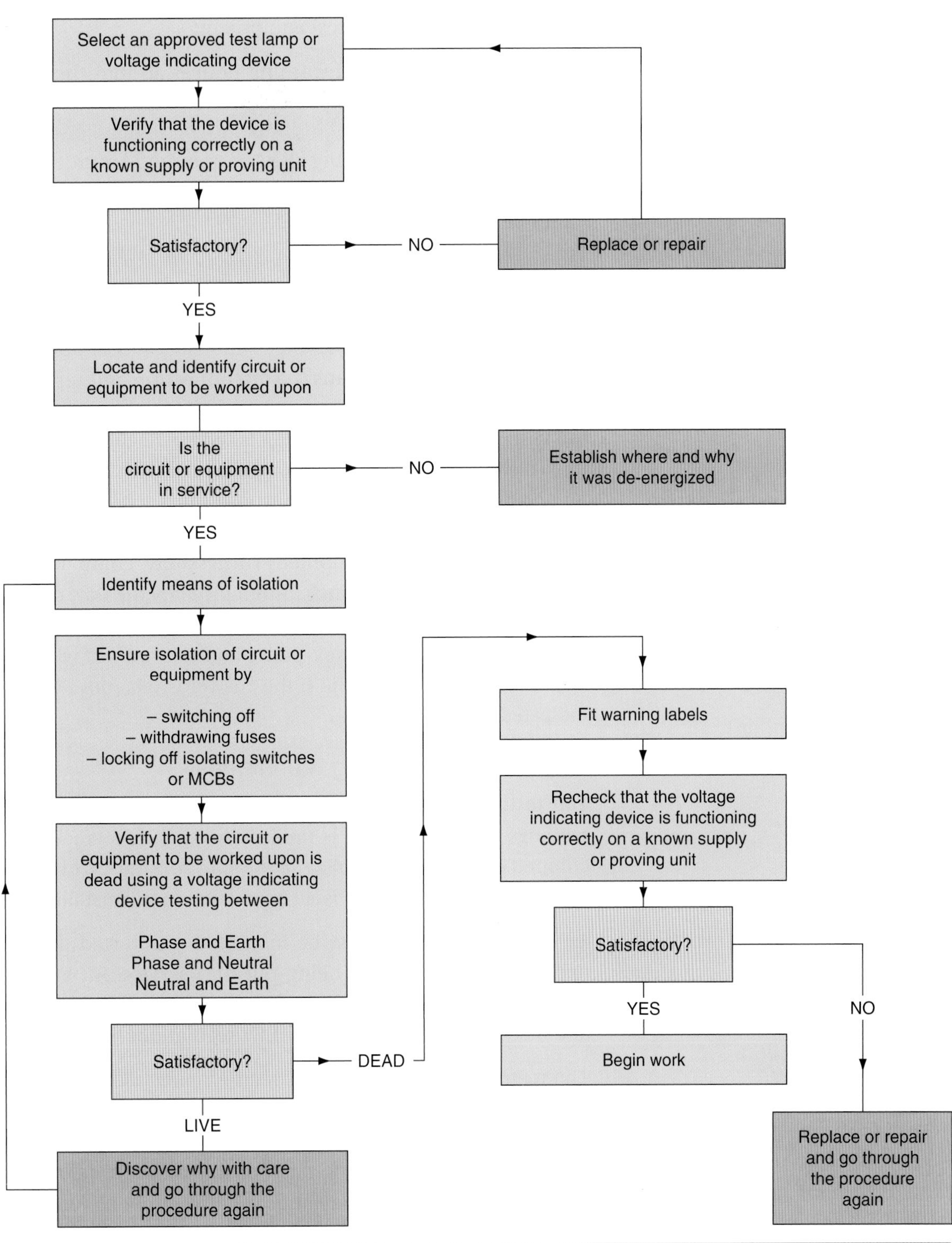

FIGURE 8.11
Flowchart for a secure isolation procedure.

Definition

The *Electricity at Work Act* tells us that it is 'preferable' that supplies be made dead before work commences (Regulation 4(3)).

and testing can only be successfully carried out 'live', then the person carrying out the fault diagnosis must:

- be trained so that he understands the equipment and the potential hazards of working live and can, therefore, be deemed to be 'competent' to carry out the activity;

- only use approved test equipment;
- set up barriers and warning notices so that the work activity does not create a situation dangerous to others.

Definition

Note that while live testing may be required in order to find the fault, live repair work must not be carried out. The individual circuit or item of equipment must first be isolated.

Definition

Isolation means the disconnection and separation of the electrical equipment from every source of electrical energy in such a way that this disconnection and separation is secure.

Note that while live testing may be required in order to find the fault, live repair work must not be carried out. The individual circuit or item of equipment must first be isolated.

Isolation of supply

The Electricity at Work Regulations are very specific in describing the procedure to be used for isolation of the electrical supply. Regulation 12(1) tells us that **isolation** means the disconnection and separation of the electrical equipment from every source of electrical energy in such a way that this disconnection and separation is secure. Regulation 4(3) tells us that we must also prove the conductors dead before work commences and that the test instrument used for this purpose must itself be proved immediately before and immediately after testing the conductors. To isolate an individual circuit or item of equipment successfully, competently and safely we must follow a procedure such as that given by the flow diagram in Fig. 8.11. Start at the top and work your way down the flowchart. When you get to the heavy-outlined amber boxes, pause and ask yourself whether everything is satisfactory up to this point. If the answer is yes, move on. If no, go back as indicated by the diagram.

Inspection and testing techniques

The testing of an installation implies the use of instruments to obtain readings. However, a test is unlikely to identify a cracked socket outlet, a chipped or loose switch plate, a missing conduit-box lid or saddle, so it is also necessary to make a visual inspection of the installation.

All new installations must be inspected and tested during erection and upon completion before being put into service. All existing installations should be periodically inspected and tested to ensure that they are safe and meet the regulations of the IEE (Regulations 610–634).

The method used to test an installation may inject a current into the system. This current must not cause danger to any person or equipment in contact with the installation, even if the circuit being tested is faulty. The test results must be compared with any relevant data, including the IEE Regulation tables, and the test procedures must be followed carefully and in the correct sequence, as indicated by Regulation 612.1. This ensures that the protective conductors are correctly connected and secure before the circuit is energized.

Definition

The aim of the *visual inspection* is to confirm that all equipment and accessories are undamaged and comply with the relevant British and European Standards, and also that the installation has been securely and correctly erected.

VISUAL INSPECTION

The installation must be visually inspected before testing begins. The aim of the **visual inspection** is to confirm that all equipment and accessories are undamaged and comply with the relevant British and European Standards, and also that the installation has been securely and correctly erected. Regulation 611.3 gives a checklist for the initial visual inspection of an installation, including:

- connection of conductors;
- identification of conductors;
- routing of cables in safe zones;
- selection of conductors for current carrying capacity and volt drop;
- connection of single-pole devices for protection or switching in phase conductors only;
- correct connection of socket outlets, lampholders, accessories and equipment;
- presence of fire barriers, suitable seals and protection against thermal effects;
- methods of 'Basic protection' against electric shock, including the insulation of live parts and placement of live parts out of reach by fitting appropriate barriers and enclosures;
- methods of 'Fault Protection' against electric shock including the presence of earthing conductors for both protective bonding and supplementary bonding.
- prevention of detrimental influences (e.g. corrosion);
- presence of appropriate devices for isolation and switching;
- presence of undervoltage protection devices;
- choice and setting of protective devices;
- labelling of circuits, fuses, switches and terminals;
- selection of equipment and protective measures appropriate to external influences;
- adequate access to switchgear and equipment;
- presence of danger notices and other warning notices;
- presence of diagrams, instructions and similar information;
- appropriate erection method.

The checklist is a guide, it is not exhaustive or detailed, and should be used to identify relevant items for inspection, which can then be expanded upon. For example, the first item on the checklist, connection of conductors, might be further expanded to include the following:

- Are connections secure?
- Are connections correct? (conductor identification)
- Is the cable adequately supported so that no strain is placed on the connections?
- Does the outer sheath enter the accessory?
- Is the insulation undamaged?
- Does the insulation proceed up to but not *into* the connection?

This is repeated for each appropriate item on the checklist.

Those tests which are relevant to the installation must then be carried out in the sequence given in Regulation 612.1 for reasons of safety and accuracy. These tests are as follows:

Before the supply is connected

1. Test for continuity of protective conductors, including protective equipotential and supplementary bonding.
2. Test the continuity of all ring final circuit conductors.
3. Test for insulation resistance.
4. Test for polarity using the continuity method.
5. Test the earth electrode resistance.

With the supply connected

6. Recheck polarity using a voltmeter or approved test lamp.
7. Test the earth fault loop impedance.
8. Carry out additional protection testing (e.g. operation of residual current devices, RCDs).

If any test fails to comply with the Regulations, then *all* the preceding tests must be repeated after the fault has been rectified. This is because the earlier test results may have been influenced by the fault (Regulation 612.1).

There is an increased use of electronic devices in electrical installation work, for example, in dimmer switches and ignitor circuits of discharge lamps. These devices should temporarily be disconnected so that they are not damaged by the test voltage of, for example, the insulation resistance test (Regulation 612.3).

APPROVED TEST INSTRUMENTS

The **test instruments and test leads** used by the electrician for testing an electrical installation must meet all the requirements of the relevant regulations. The HSE has published Guidance Notes GS 38 for test equipment used by electricians. The IEE Regulations (BS 7671) also specify the test voltage or current required to carry out particular tests satisfactorily. All test equipment must be chosen to comply with the relevant parts of BS EN 61557. *All testing must, therefore, be carried out using an 'approved' test instrument if the test results are to be valid. The test instrument must also carry a calibration certificate, otherwise the recorded results may be void.* **Calibration certificates** usually last for a year. Test instruments must, therefore, be tested and recalibrated each year by an approved supplier. This will maintain the accuracy of the instrument to an acceptable level, usually within 2% of the true value.

Definition

The *test instruments and test leads* used by the electrician for testing an electrical installation must meet all the requirements of the relevant regulations.

Definition

Calibration certificates usually last for a year. Test instruments must, therefore, be tested and recalibrated each year by an approved supplier.

Modern digital test instruments are reasonably robust, but to maintain them in good working order they must be treated with care. An approved test instrument costs equally as much as a good-quality camera; it should, therefore, receive the same care and consideration.

Let us now look at the requirements of four often used test meters.

Continuity tester

To measure accurately the resistance of the conductors in an electrical installation we must use an instrument which is capable of producing an open circuit voltage of between 4 and 24V a.c. or d.c., and deliver a short-circuit current of not less than 200 mA (Regulation 612.2.1). The functions of continuity testing and insulation resistance testing are usually combined in one test instrument.

Insulation resistance tester

The test instrument must be capable of detecting insulation leakage between live conductors and between live conductors and earth. To do this and comply with Regulation 612.3 the test instrument must be capable of producing a test voltage of 250, 500 or 1000 V and deliver an output current of not less than 1 mA at its normal voltage.

Earth fault loop impedance tester

The test instrument must be capable of delivering fault currents as high as 25 A for up to 40 ms using the supply voltage. During the test, the instrument does an Ohm's law calculation and displays the test result as a resistance reading.

RCD tester

Where circuits are protected by an RCD we must carry out a test to ensure that the device will operate very quickly under fault conditions and within the time limits set by the IEE Regulations. The instrument must, therefore, simulate a fault and measure the time taken for the RCD to operate. The instrument is, therefore, calibrated to give a reading measured in milliseconds to an in-service accuracy of 10%.

If you purchase good-quality 'approved' test instruments and leads from specialist manufacturers they will meet all the Regulations and Standards and therefore give valid test results. However, to carry out all the tests required by the IEE Regulations will require a number of test instruments and this will represent a major capital investment in the region of £1000.

Let us now consider the individual tests.

1 TESTING FOR CONTINUITY OF PROTECTIVE CONDUCTORS (612.2.1)

The object of the test is to ensure that the circuit protective conductor (CPC) is correctly connected, is electrically sound and has a total resistance which is low enough to permit the overcurrent protective device to operate within the disconnection time requirements of Regulation 411.4.6, should an earth fault occur. Every protective conductor must be separately tested from the consumer's main protective earthing terminal to verify that it is electrically sound and correctly connected, including the protective equipotential and supplementary bonding conductors.

Key Fact

Testing

- A new installations must be inspected and tested during erection and upon completion.
- All existing installations must be inspected and tested periodically.
- IEE Regulations 610–634.

Safety First

Live working

- **NEVER** work **LIVE**
- Some 'live testing' is allowed by 'competent persons'
- Otherwise, isolate and secure the isolation
- Prove the supply dead before starting work.

Definition

A 'competent person' is someone who has the necessary technical skills, training and expertise to safely carry out a particular activity.

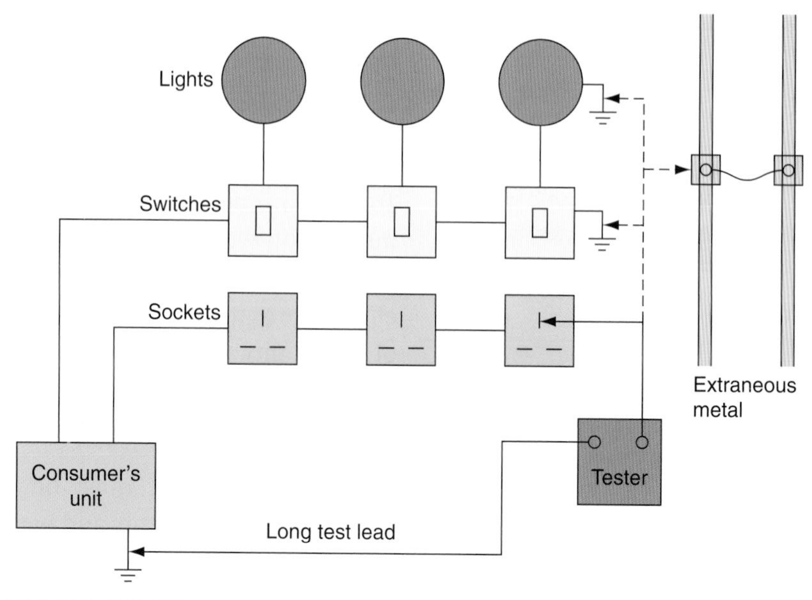

FIGURE 8.12
Testing continuity of protective conductors.

A d.c. test using an ohmmeter continuity tester is suitable where the protective conductors are of copper or aluminium up to 35 mm^2. The test is made with the supply disconnected, measuring from the consumer's main protective earthing terminal to the far end of each CPC, as shown in Fig. 8.12. The resistance of the long test lead is subtracted from these readings to give the resistance value of the CPC. The result is recorded on an installation schedule such as that given in Appendix 6 of the IEE Regulations.

Where steel conduit or trunking forms the protective conductor, the standard test described above may be used, but additionally the enclosure must be visually checked along its length to verify the integrity of all the joints.

If the inspecting engineer has grounds to question the soundness and quality of these joints then the phase earth loop impedance test described later in this chapter should be carried out.

If, after carrying out this further test, the inspecting engineer still questions the quality and soundness of the protective conductor formed by the metallic conduit or trunking then a further test can be done using an a.c. voltage not greater than 50 V at the frequency of the installation and a current approaching 1.5 times the design current of the circuit, but not greater than 25 A.

This test can be done using a low-voltage transformer and suitably connected ammeters and voltmeters, but a number of commercial instruments are available such as the Clare tester, which give a direct reading in ohms.

Because fault currents will flow around the earth fault loop path, the measured resistance values must be low enough to allow the overcurrent protective device to operate quickly. For a satisfactory test result, the resistance of the protective conductor should be consistent with those values calculated for a line conductor of similar length and cross-sectional area.

Table 8.1 Resistance Values of Some Metallic Containers

Metallic sheath	Size (mm)	Resistance at 20°C (mΩ/m)
Conduit	20 25 32	1.25 1.14 0.85
Trunking	50 × 50 75 × 75 100 × 100	0.949 0.526 0.337

Values of resistance per metre for copper and aluminium conductors are given in Table 7.2 of Chapter 7 in this book. The resistances of some other metallic containers are given in Table 8.1.

Example

The CPC for a ring final circuit is formed by a 1.5 mm² copper conductor of 50 m approximate length. Determine a satisfactory continuity test value for the CPC using the value given in Table 7.2 of Chapter 7.

$$\text{Resistance/metre for a } 1.5\,\text{mm}^2 \text{ copper conductor} = 12.10\,\text{m}\Omega/\text{m}$$

Therefore,

$$\text{the resistance of } 50\,\text{m} = 50 \times 12.10 \times 10^{-3}$$
$$= 0.605\,\Omega$$

The protective conductor resistance values calculated by this method can only be an approximation since the length of the CPC can only be estimated. Therefore, in this case, a satisfactory test result would be obtained if the resistance of the protective conductor was about 0.6 Ω. A more precise result is indicated by the earth fault loop impedance test which is carried out later in the sequence of tests.

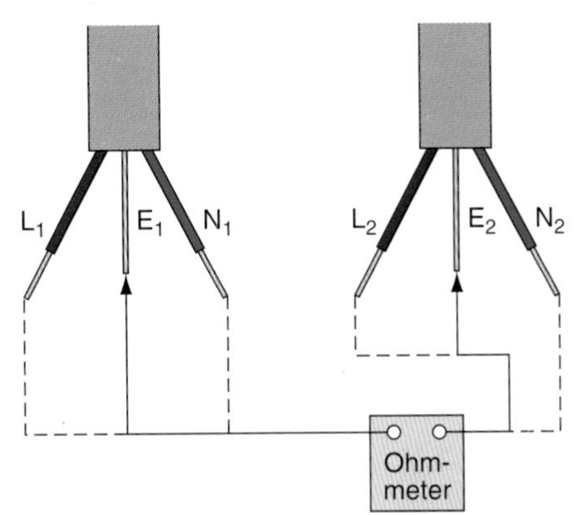

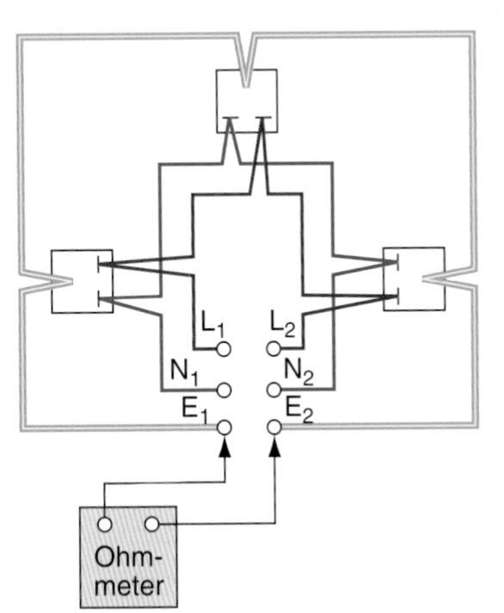

FIGURE 8.13
Step 1 test: measuring the resistance of phase, neutral and protective conductors.

2 TESTING FOR CONTINUITY OF RING FINAL CIRCUIT CONDUCTORS (612.2.2)

The object of the test is to ensure that all ring circuit cables are continuous around the ring, that is, that there are no breaks and no interconnections in the ring, and that all connections are electrically and mechanically sound. This test also verifies the polarity of each socket outlet.

The test is made with the supply disconnected, using an ohmmeter as follows:

Disconnect and separate the conductors of both legs of the ring at the main fuse. There are three steps to this test:

Step 1
Measure the resistance of the line conductors (L_1 and L_2), the neutral conductors (N_1 and N_2) and the protective conductors (E_1 and E_2) at the mains

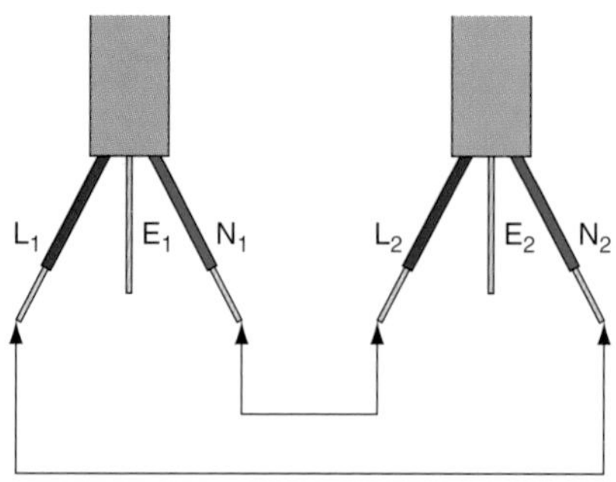

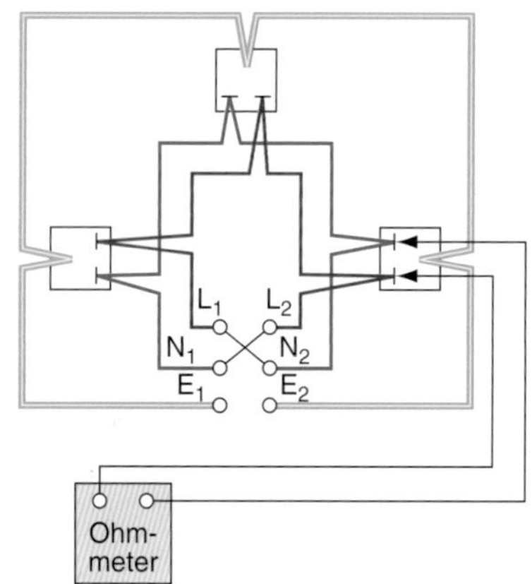

FIGURE 8.14
Step 2 test: connection of mains conductors and test circuit conditions.

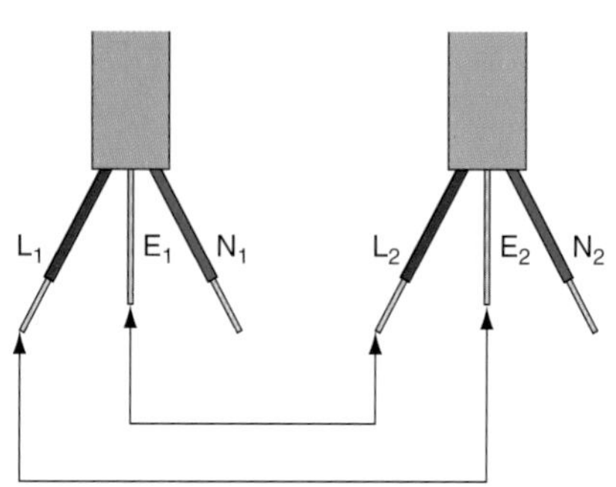

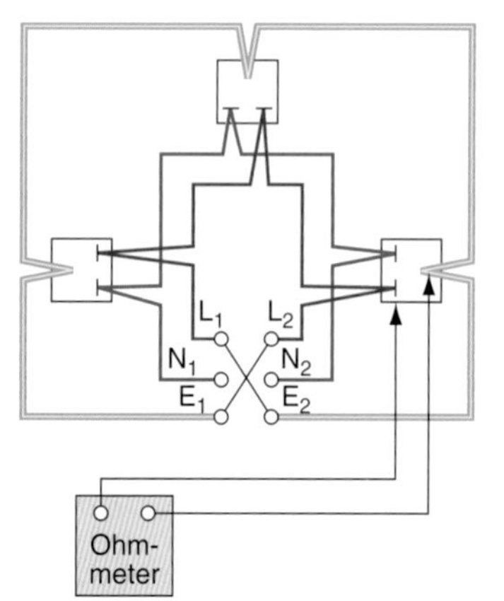

FIGURE 8.15
Step 3 test: connection of mains conductors and test circuit conditions.

Table 8.2 Table Which May Be Used to Record the Readings Taken When Carrying Out the Continuity of Ring Final Circuit Conductors Tests According to IEE Regulation 612.2.2

Test	Ohmmeter connected to	Ohmmeter readings	This gives a value for
Step 1	L_1 and L_2 N_1 and N_2 E_1 and E_2		r_1 r_2
Step 2	Live and neutral at each socket		
Step 3	Live and earth at each socket		$R_1 + R_2$
As a check $(R_1 + R_2)$ value should equal $(r_1 + r_2)/4$.			

position as shown in Fig. 8.13. End-to-end live and neutral conductor readings should be approximately the same (i.e. within $0.05\,\Omega$) if the ring is continuous. The protective conductor reading will be 1.67 times as great as these readings if 2.5/1.5 mm cable is used. Record the results on a table such as that shown in Table 8.2.

Step 2

The live and neutral conductors should now be temporarily joined together as shown in Fig. 8.14. An ohmmeter reading should then be taken between live and neutral at *every* socket outlet on the ring circuit. The readings obtained should be substantially the same, provided that there are no breaks or multiple loops in the ring. Each reading should have a value of approximately half the live and neutral ohmmeter readings measured in Step 1 of this test. Sockets connected as a spur will have a slightly higher value of resistance because they are fed by only one cable, while each socket on the ring is fed by two cables. Record the results on a table such as that shown in Table 8.2.

Step 3

Where the CPC is wired as a ring, for example where twin and earth cables or plastic conduit is used to wire the ring, temporarily join the live and CPCs together as shown in Fig. 8.15. An ohmmeter reading should then be taken between live and earth at *every* socket outlet on the ring. The readings obtained should be substantially the same provided that there are no breaks or multiple loops in the ring. This value is equal to $R_1 + R_2$ for the circuit. Record the results on an installation schedule such as that given in Appendix 6 of the IEE Regulations or a table such as that shown in Table 8.2. The Step 3 value of $R_1 + R_2$ should be equal to $(r_1 + r_2)/4$, where r_1 and r_2 are the ohmmeter readings from Step 1 of this test (see Table 8.2).

3 TESTING INSULATION RESISTANCE (612.3)

The object of the test is to verify that the quality of the insulation is satisfactory and has not deteriorated or short-circuited. The test should be made at the consumer's unit with the mains switch off, all fuses in place and all switches closed. Neon lamps, capacitors and electronic circuits should be disconnected, since they will respectively glow, charge up or be damaged by the test.

There are two tests to be carried out using an insulation resistance tester which must have a test voltage of 500 V d.c. for 230 V and 400 V installations. These are line and neutral conductors to earth and between line conductors. The procedures are:

Line and neutral conductors to earth

1. Remove all lamps.
2. Close all switches and circuit breakers.
3. Disconnect appliances.
4. Test separately between the line conductor and earth, *and* between the neutral conductor and earth, for *every* distribution circuit at the consumer's unit as shown in Fig. 8.16a. Record the results on a schedule of test results such as that given in Appendix 6 of the IEE Regulations.

Between line conductors

1. Remove all lamps.
2. Close all switches and circuit breakers.
3. Disconnect appliances.

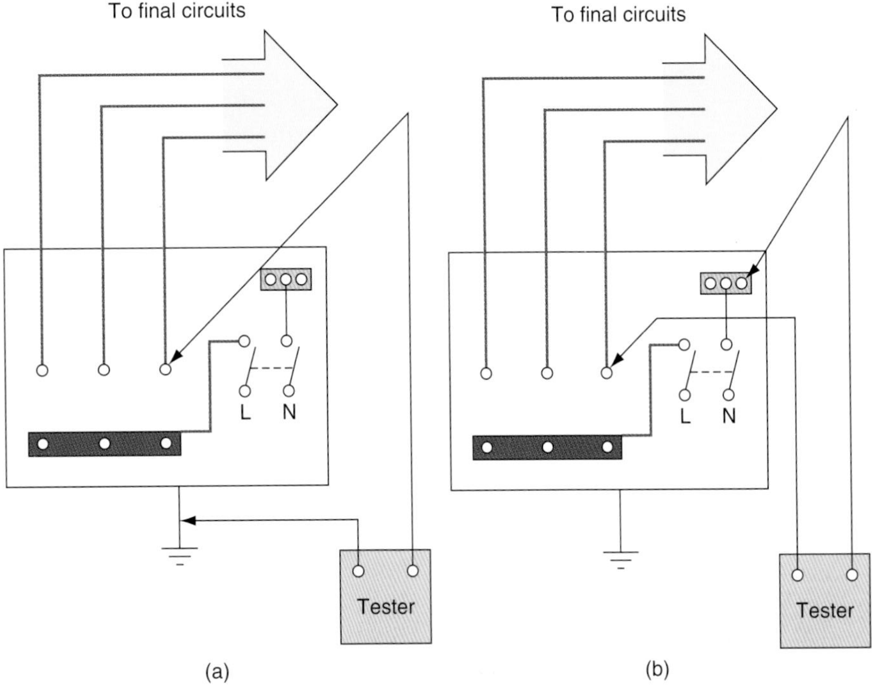

FIGURE 8.16
Insulation resistance test.

4. Test between line and neutral conductors of *every* distribution circuit at the consumer's unit as shown in Fig. 8.16b and record the result.

The insulation resistance readings for each test must be not less than 1.0 MΩ for a satisfactory result (IEE Regulation 612.3.2).

Where the circuit includes electronic equipment which might be damaged by the insulation resistance test, a measurement between all live conductors (i.e. live and neutral conductors connected together) and the earthing arrangements may be made. The insulation resistance of these tests should be not less than 1.0 MΩ (IEE Regulation 612.3.3).

Although an insulation resistance reading of 1.0 MΩ complies with the Regulations, the IEE Guidance Notes tell us that much higher values than this can be expected and that a reading of less than 2 MΩ might indicate a latent but not yet visible fault in the installation. In these cases each circuit should be separately tested to obtain a reading greater than 2 MΩ.

4 TESTING POLARITY (612.6)

The object of this test is to verify that all fuses, circuit breakers and switches are connected in the line or live conductor only, and that all socket outlets are correctly wired and Edison screw-type lampholders have the centre contact connected to the live conductor. It is important to make a polarity test on the installation since a visual inspection will only indicate conductor identification.

The test is done with the supply disconnected using an ohmmeter or continuity tester as follows:

1. Switch off the supply at the main switch.
2. Remove all lamps and appliances.
3. *Fix a temporary link* between the line and earth connections on the consumer's side of the main switch.
4. Test between the 'common' terminal and earth at each switch position.
5. Test between the centre pin of any Edison screw lampholders and any convenient earth connection.
6. Test between the live pin (i.e. the pin to the right of earth) and earth at each socket outlet as shown in Fig. 8.17.

For a satisfactory test result the ohmmeter or continuity meter should read very close to zero for each test.

Remove the test link and record the results on a schedule of test results such as that given in Appendix 6 of the IEE Regulations.

5 TESTING EARTH ELECTRODE RESISTANCE (612.7)

When an earth electrode has been sunk into the general mass of earth, it is necessary to verify the resistance of the electrode. The general mass of

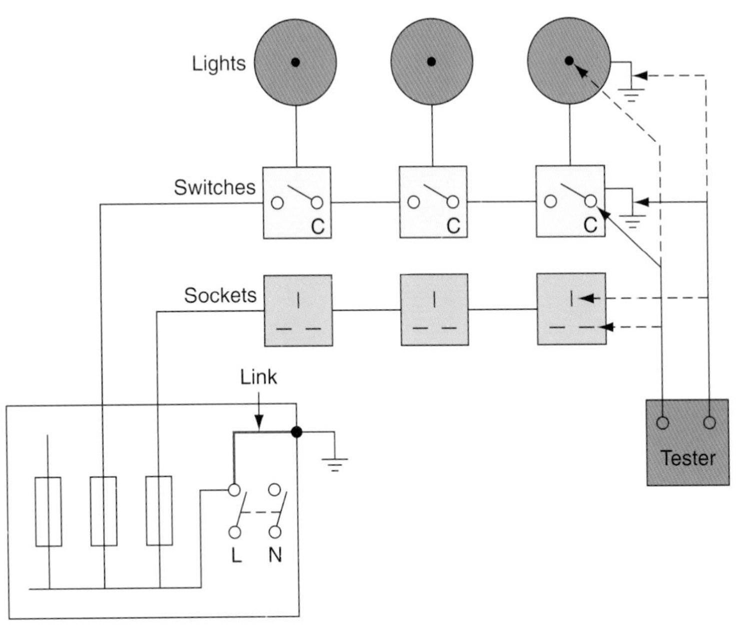

FIGURE 8.17
Polarity test.

earth can be considered as a large conductor which is at zero potential. Connection to this mass through earth electrodes provides a reference point from which all other voltage levels can be measured. This is a technique which has been used for a long time in power distribution systems.

The resistance to earth of an electrode will depend upon its shape, size and the resistance of the soil. Earth rods form the most efficient electrodes. A rod of about 1 m will have an earth electrode resistance of between 10 and 200 Ω. Even in bad earthing conditions a rod of about 2 m will normally have an earth electrode resistance which is less than 500 Ω in the United Kingdom. In countries which experience long dry periods of weather the earth electrode resistance may be thousands of ohms.

In the past, electrical engineers used the metal pipes of water mains as an earth electrode, but the recent increase in the use of PVC pipe for water mains now prevents the use of water pipes as the means of earthing in the United Kingdom, although this practice is still permitted in some countries. The IEE Regulation 542.2.1 recognizes the use of the following types of earth electrodes:

- earth rods or pipes
- earth tapes or wires
- earth plates
- earth electrodes embedded in foundations
- welded metallic reinforcement of concrete structures
- other suitable underground metalwork
- lead sheaths or other metallic coverings of cables.

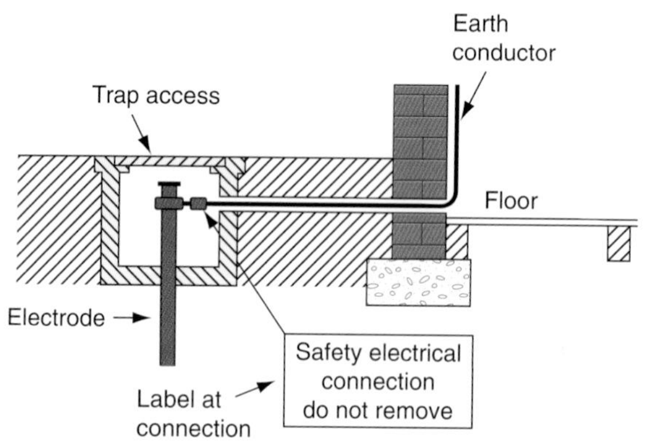

FIGURE 8.18
Termination of an earth electrode.

The earth electrode is sunk into the ground, but the point of connection should remain accessible (Regulation 542.4.2). The connection of the earthing conductor to the earth electrode must be securely made with a copper conductor complying with Table 54.1 and Regulation 542.3.2 as shown in Fig. 8.18.

The installation site must be chosen so that the resistance of the earth electrode does not increase above the required value due to climatic conditions such as the soil drying out or freezing, or from the effects of corrosion (542.2.2 and 3).

Under fault conditions the voltage appearing at the earth electrode will radiate away from the electrode like the ripples radiating away from a pebble thrown into a pond. The voltage will fall to a safe level in the first 2 or 3 m away from the point of the earth electrode.

The basic method of measuring earth electrode resistance is to pass a current into the soil through the electrode and to measure the voltage required to produce this current.

Regulation 612.9 demands that where earth electrodes are used they should be tested.

If the electrode under test forms part of the earth return for a TT installation in conjunction with an RCD, Guidance Note 3 of the IEE Regulations describes the following method:

1. Disconnect the installation protective equipotential bonding from the earth electrode to ensure that the test current passes only through the earth electrode.
2. Switch off the consumer's unit to isolate the installation.
3. Using a line earth loop impedance tester, test between the incoming line conductor and the earth electrode.
4. Reconnect the protective bonding conductors when the test is completed.

Record the result on a schedule of test results such as that given in Appendix 6 of the IEE Regulations.

The IEE Guidance Note 3 tells us that an acceptable value for the measurement of the earth electrode resistance would be less than 200 Ω.

Providing the first five tests were satisfactory, the supply may now be switched on and the final tests completed with the supply connected.

6 TESTING POLARITY – SUPPLY CONNECTED

Using an approved voltage indicator such as that shown at Fig. 8.8 or test lamp and probes which comply with the HSE Guidance Note GS 38, again carry out a polarity test to verify that all fuses, circuit breakers and switches are connected in the live conductor. Test from the common terminal of switches to earth, the live pin of each socket outlet to earth and the centre pin of any Edison screw lampholders to earth. In each case the voltmeter or test lamp should indicate the supply voltage for a satisfactory result.

7 TESTING EARTH FAULT LOOP IMPEDANCE (SUPPLY CONNECTED) (612.9)

The object of this test is to verify that the impedance of the whole earth fault current loop line to earth is low enough to allow the overcurrent protective device to operate within the disconnection time requirements of Regulations 411.3.2.2 and 411.4.6 and 411.4.7, should a fault occur.

The whole earth fault current loop examined by this test is comprised of all the installation protective conductors, the main protective earthing terminal and protective earth conductors, the earthed neutral point and the secondary winding of the supply transformer and the line conductor from the transformer to the point of the fault in the installation.

The test will, in most cases, be done with a purpose-made line earth loop impedance tester which circulates a current in excess of 10 A around the loop for a very short time, so reducing the danger of a faulty circuit. The test is made with the supply switched on, and carried out from the furthest point of *every* final circuit, including lighting, socket outlets and any fixed appliances. Record the results on a schedule of test results.

Purpose-built testers give a readout in ohms and a satisfactory result is obtained when the loop impedance does not exceed the appropriate values given in Tables 41.2 and 41.3 of the IEE Regulations.

(*Note* Table 7.3 of Chapter 7 shows the earth fault loop impedance values for a Type B MCB.)

8 ADDITIONAL PROTECTION: TESTING OF RCD – SUPPLY CONNECTED (612.13)

The object of the test is to verify the effectiveness of the RCD, that it is operating with the correct sensitivity and proving the integrity of the electrical and mechanical elements. The test must simulate an appropriate fault condition and be independent of any test facility incorporated in the device.

When carrying out the test, all loads normally supplied through the device are disconnected.

The testing of a ring circuit protected by a general-purpose RCD to BS EN 61008 in a split-board consumer unit is carried out as follows:

1. Using the standard lead supplied with the test instrument, disconnect all other loads and plug in the test lead to the socket at the centre of the ring (i.e. the socket at the furthest point from the source of supply).
2. Set the test instrument to the tripping current of the device and at a phase angle of 0°.
3. Press the test button – the RCD should trip and disconnect the supply within 200 ms.
4. Change the phase angle from 0° to 180° and press the test button once again. The RCD should again trip within 200 ms. Record the highest value of these two results on a schedule of test results such as that given in Appendix 6 of the IEE Regulations.
5. Now set the test instrument to 50% of the rated tripping current of the RCD and press the test button. The RCD should *not trip* within 2 seconds. This test is testing the RCD for inconvenience *or* nuisance tripping.
6. Finally, the effective operation of the test button incorporated within the RCD should be tested to prove the integrity of the mechanical elements in the tripping device. This test should be repeated every 3 months.

If the RCD fails any of the above tests it should be changed for a new one.

Where the RCD has a rated tripping current not exceeding 30 mA and has been installed to reduce the risk associated with 'basic' and or 'fault' protection, as indicated in Regulation 411.1, a residual current of 150 mA should cause the circuit breaker to open within 40 ms.

Certification and reporting

Following the completion of all new electrical work or additional work to an existing installation, the installation must be inspected and tested and an installation certificate issued and signed by a competent person. The 'competent person' must have a sound knowledge of the type of work undertaken, be fully versed in the inspection and testing procedures contained in the IEE Regulations (BS 7671) and employ adequate testing equipment.

A certificate and test results shall be issued to those ordering the work in the format given in Appendix 6 of the IEE Regulations.

All installations must be periodically tested and inspected, and for this purpose a periodic inspection report should be issued (IEE Regulation 631.2). The standard format is again shown in Appendix 6 of the IEE Regulations.

In both cases the certificate must include the test values which verify that the installation complies with the IEE Regulations at the time of testing.

Suggested frequency of periodic inspection intervals are given below:

- Domestic installations – 10 years
- Commercial installations – 5 years
- Industrial installations – 3 years
- Agricultural installations – 3 years
- Caravan site installations – 1 year
- Caravans – 3 years
- Temporary installations on construction sites – 3 months.

Safe working procedures when testing

Whether you are carrying out the test procedure (i) as a part of a new installation (ii) upon the completion of an extension to an existing installation (iii) because you are trying to discover the cause of a fault on an installation or (iv) because you are carrying out a periodic test and inspection of a building, you must always be aware of your safety, the safety of others using the building and the possible damage which your testing might cause to other systems in the building.

For your own safety:

- Always use 'approved' test instruments and probes.
- Ensure that the test instrument carries a valid calibration certificate otherwise the results may be invalid.
- Secure all isolation devices in the 'off' position.
- Put up warning notices so that other workers will know what is happening.
- Notify everyone in the building that testing is about to start and for approximately how long it will continue.
- Obtain a 'permit-to-work' if this is relevant.
- Obtain approval to have systems shut down which might be damaged by your testing activities. For example, computer systems may 'crash' when supplies are switched off. Ventilation and fume extraction systems will stop working when you disconnect the supplies.

For the safety of other people:

- Fix warning notices around your work area.
- Use cones and highly visible warning tape to screen off your work area.
- Make an effort to let everyone in the building know that testing is about to begin. You might be able to do this while you carry out the initial inspection of the installation.
- Obtain verbal or written authorization to shut down information technology, emergency operation or stand-by circuits.

To safeguard other systems:

- Computer systems can be severely damaged by a loss of supply or the injection of a high test voltage from, for example, an insulation resistance test. Computer systems would normally be disconnected during the test period but this will generally require some organization before the testing begins. Commercial organizations may be unable to continue to work without their computer systems and, in these circumstances it may be necessary to test outside the normal working day.
- Any resistance measurements made on electronic equipment or electronic circuits must be achieved with a battery operated ohmmeter in order to avoid damaging the electronic circuits.
- Farm animals are creatures of habit and may become very grumpy to find you testing their milking parlour equipment at milking time.
- Hospitals and factories may have emergency stand-by generators which re-energize essential circuits in the event of a mains failure. Your isolation of the circuit for testing may cause the emergency systems to operate. Discuss any special systems with the person authorizing the work before testing begins.

Portable appliance testing

A quarter of all serious electrical accidents involve portable electrical appliances, that is, equipment which has a cable lead and plug and which is normally moved around or can easily be moved from place to place. This includes, for example, floor cleaners, kettles, heaters, portable power tools, fans, televisions, desk lamps, photocopiers, fax machines and desktop computers. There is a requirement for employers under the Health and Safety at Work Act to take adequate steps to protect users of portable appliances from the hazards of electric shock and fire. The responsibility for safety applies equally to small as well as large companies. The Electricity at Work Regulations 1989 also place a duty of care upon employers to ensure that the risks associated with the use of electrical equipment are controlled.

Against this background the HSE have produced guidance notes HS(G) 107 *Maintaining Portable and Transportable Electrical Equipment* and leaflets

Maintaining Portable Electrical Equipment in Offices and Maintaining Portable Electrical Equipment in Hotels and Tourist Accommodation. In these publications the HSE recommend that a three level system of inspection can give cost effective maintenance of portable appliances. These are:

- user checking;
- visual inspection by an appointed person;
- combined inspection and testing by a competent person or contractor.

A user visually checking the equipment is probably the most important maintenance procedure. About 95% of faults or damage can be identified by just looking. The user should check for obvious damage using common sense. The use of potentially dangerous equipment can then be avoided. Possible dangers to look for are as follows:

- Damage to the power cable or lead which exposes the colours of the internal conductors, which are brown, blue and green with a yellow stripe.
- Damage to the plug top itself. The plug top pushes into the wall socket, usually a square pin 13 A socket in the United Kingdom, to make an electrical connection. With the plug top removed from the socket the equipment is usually electrically 'dead'. If the bakelite plastic casing of the plug top is cracked, broken or burned, or the contact pins are bent, do not use it.
- Non-standard joints in the power cable, such as taped joints.
- *Poor cable retention*: The outer sheath of the power cable must be secured and enter the plug top at one end and the equipment at the other. The coloured internal conductors must not be visible at either end.
- Damage to the casing of the equipment such as cracks, pieces missing, loose or missing screws or signs of melted plastic, burning, scorching or discolouration.
- Equipment which has previously been used in unsuitable conditions such as a wet or dusty environment.

If any of the above dangers are present, the equipment should not be used until the person appointed by the company to make a 'visual inspection' has had an opportunity to do so.

A **visual inspection** will be carried out by an appointed person within a company, such person having been trained to carry out this task. In addition to the user checks described above, an inspection could include the removal of the plug top cover to check that:

- a fuse of the correct rating is being used and also that a proper cartridge fuse is being used and not a piece of wire, a nail or silver paper;
- the cord grip is holding the sheath of the cable and not the coloured conductors;

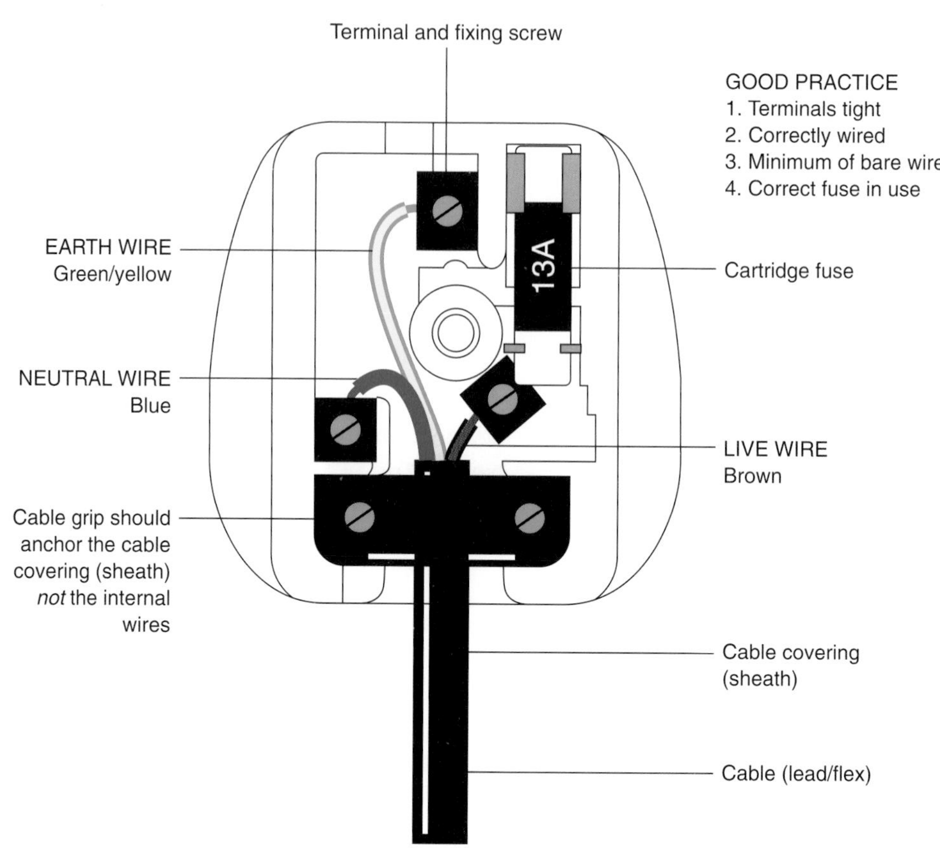

FIGURE 8.19
Correct connection of plug top.

- the wires (conductors) are connected to the correct terminals of the plug top as shown in Fig. 8.19;
- the coloured insulation of each conductor wire goes right up to the terminal so that no bare wire is visible;
- the terminal fixing screws hold the conductor wires securely and the screws are tight;
- all the conductor wires are secured within the terminal;
- there are no internal signs of damage such as overheating, excessive 'blowing' of the cartridge fuse or the intrusion of foreign bodies such as dust, dirt or liquids.

The above inspection cannot apply to 'moulded plugs', which are moulded on to the flexible cable by the manufacturer in order to prevent some of the bad practice described above. In the case of a moulded plug top, only the fuse can be checked. The visual inspection checks described above should also be applied to extension leads and their plugs. The HSE recommends that a simple procedure be written to give guidance to the 'appointed person' carrying out the visual inspection.

Combined inspection and testing is also necessary on some equipment because some faults cannot be seen by just looking – for example, the continuity and effectiveness of earth paths. For some portable appliances the earth is essential to the safe use of the equipment and, therefore, all earthed equipment and most extension leads should be periodically tested and inspected for these faults. All portable appliance test instruments (PAT Testers) will carry out two important tests, earth bonding and insulation resistance.

Earth bonding tests apply a substantial test current, typically about 25 A, down the earth pin of the plug top to an earth probe, which should be connected to any exposed metalwork on the portable appliance being tested. The PAT Tester will then calculate the resistance of the earth bond and either give an actual reading or indicate pass or fail. A satisfactory result for this test would typically be a reading of less than 0.1 Ω. The earth bond test is, of course, not required for double insulated portable appliances because there will be no earthed metalwork.

Insulation resistance tests apply a substantial test voltage, typically 500V, between the live and neutral bonded together and the earth. The PAT Tester then calculates the insulation resistance and either gives an actual reading or indicates pass or fail. A satisfactory result for this test would typically be a reading greater than 2 MΩ.

Some PAT Testers offer other tests in addition to the two described above. These are described below.

A flash test tests the insulation resistance at a higher voltage than the 500V test described above. The flash test uses 1.5kV for Class 1 portable appliances, that is earthed appliances, and 3kV for Class 2 appliances which are double insulated. The test establishes that the insulation will remain satisfactory under more stringent conditions but must be used with caution, since it may overstress the insulation and will damage electronic equipment. A satisfactory result for this test would typically be less than 3mA.

A fuse test tests that a fuse is in place and that the portable appliance is switched on prior to carrying out other tests. A visual inspection will be required to establish that the *size* of the fuse is appropriate for that particular portable appliance.

An earth leakage test measures the leakage current to earth through the insulation. It is a useful test to ensure that the portable appliance is not deteriorating and liable to become unsafe. It also ensures that the tested appliances are not responsible for nuisance 'tripping' of RCDs (RCDs – see Chapter 7). A satisfactory reading is typically less than 3mA.

An operation test proves that the preceding tests were valid (i.e. that the unit was switched on for the tests), that the appliances will work when

Table 8.3 HSE Suggested Intervals for Checking, Inspecting and Testing of Portable Appliances in Offices and Other Low-Risk Environments

Equipment/environment	User checks	Formal visual inspection	Combined visual inspection and electrical testing
Battery-operated: (less than 20 V)	No	No	No
Extra low voltage: (less than 50 V a.c.) e.g. telephone equipment, low voltage desk lights	No	No	No
Information technology: e.g. desktop computers, VDU screens	No	Yes, 2–4 years	No if double insulated – otherwise up to 5 years
Photocopiers, fax machines: *not* hand-held, rarely moved	No	Yes, 2–4 years	No if double insulated – otherwise up to 5 years
Double insulated equipment: *not* hand-held, moved occasionally, e.g. fans, table lamps, slide projectors	No	Yes, 2–4 years	No
Double insulated equipment: *hand-held*, e.g. power tools	Yes	Yes, 6 months to 1 year	No
Earthed equipment (Class 1): e.g. electric kettles, some floor cleaners, power tools	Yes	Yes, 6 months to 1 year	Yes, 1–2 years
Cables (leads) and plugs connected to the above	Yes	Yes, 6 months to 4 years depending on the type of equipment it is connected to	Yes, 1–5 years depending on the type of equipment it is connected to
Extension leads (mains voltage)			

connected to the appropriate voltage supply and not draw a dangerously high current from that supply. A satisfactory result for this test would typically be less than 3.2 kW for 230 V equipment and less than 1.8 kW for 110 V equipment.

All PAT Testers are supplied with an operating manual, giving step-by-step instructions for their use and pass and fail scale readings. The HSE suggested intervals for the three levels of checking and inspection of portable appliances in offices and other low-risk environments is given in Table 8.3.

Safety First

Power tools

- Look at the power tools that you use at work.
- Do they have a PAT Test label?
- Is it 'in date'?

WHO DOES WHAT?

When actual checking, inspecting and testing of portable appliances takes place, will depend upon the company's safety policy and risk assessments. In low-risk environments such as offices and schools, the three-level system of checking, inspection and testing recommended by the HSE should be carried out. Everyone can use common sense and carry out the user checks described earlier. Visual inspections must be carried out by a 'competent person' but that person does not need to be an electrician or electronics service engineer. Any sensible member of staff who has received training can carry out this duty. They will need to know what to look for and what to do, but more importantly, they will need to be able to avoid danger to themselves and to others. The HSE recommend that the appointed person follows a simple written procedure for each visual inspection. A simple tick sheet would meet this requirement. For example:

1. Is the correct fuse fitted? Yes/No
2. Is the cord grip holding the cable sheath? Yes/No

The tick sheet should incorporate all the appropriate visual checks and inspections described earlier.

Testing and inspection require a much greater knowledge than is required for simple checks and visual inspections. This more complex task need not necessarily be carried out by a qualified electrician or electronics service engineer. However, the person carrying out the test must be trained to use the equipment and to interpret the results. Also, greater knowledge will be required for the inspection of the range of portable appliances which might be tested.

KEEPING RECORDS

Records of the inspecting and testing of portable appliances are not required by law but within the Electricity at Work Regulations 1989, it is generally accepted that some form of recording of results is required to implement a quality control system. The control system should:

- ensure that someone is nominated to have responsibility for portable appliance inspection and testing;
- maintain a log or register of all portable appliance test results to ensure that equipment is inspected and tested when it is due;
- label tested equipment with the due date for its next inspection and test as shown in Fig. 8.20.

Any piece of equipment which fails a PAT Test should be disabled and taken out of service (usually by cutting off the plug top), labelled as faulty and sent for repair.

The register of PAT Test results will help managers to review their maintenance procedures and the frequency of future visual inspections and testing. Combined inspection and testing should be carried out where there is a reason to suspect that the equipment may be faulty, damaged or contaminated but cannot be verified by visual inspection alone. Inspection and testing should also be carried out after any repair or modification to establish the integrity of the equipment or at the start of a maintenance

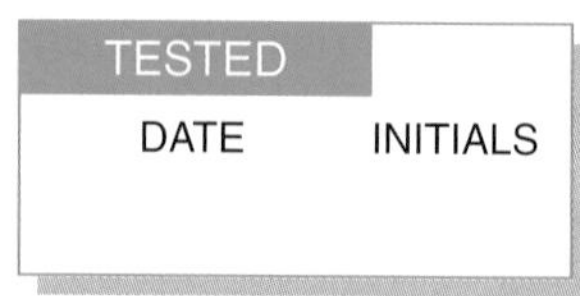

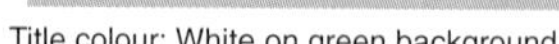

Title colour: White on green background

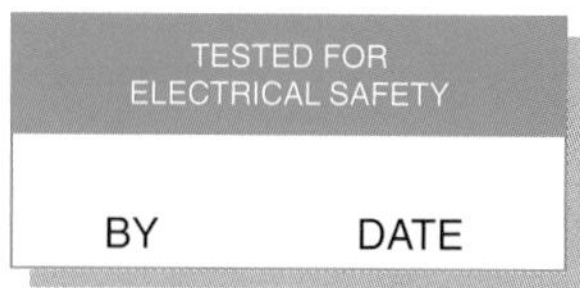

Title colour: White on green background

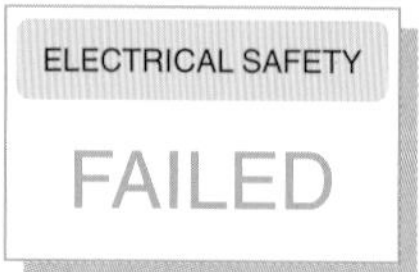

Title colour: White on red background

FIGURE 8.20
Typical PAT Test labels.

system, to establish the initial condition of the portable equipment being used by the company.

Commissioning electrical systems

The commissioning of the electrical and mechanical systems within a building is a part of the 'handing-over' process of the new building by the architect and main contractor to the client or customer in readiness for its occupation and intended use. To 'commission' means to give authority to someone to check that everything is in working order. If it is out of commission, it is not in working order.

Following the completion, inspection and testing of the new electrical installation, the functional operation of all the electrical systems must be tested before they are handed over to the customer. It is during the commissioning period that any design or equipment failures become apparent, and this testing is one of the few quality controls possible on a building services installation.

This is the role of the commissioning engineer, who must assure himself that all the systems are in working order and that they work as they were designed to work. He must also instruct the client's representative, or the staff who will use the equipment, in the correct operation of the systems, as part of the handover arrangements.

The commissioning engineer must test the operation of all the electrical systems, including the motor controls, the fan and air conditioning systems, the fire alarm and emergency lighting systems. However, before testing the emergency systems, he must first notify everyone in the building of his intentions so that alarms may be ignored during the period of testing.

Commissioning has become one of the most important functions within the building projects completion sequence. The commissioning engineer will therefore have access to all relevant contract documents, including the building specifications and the electrical installation certificates as required by the IEE Regulations (BS 7671), and have a knowledge of the requirements of the Electricity at Work Act and the Health and Safety at Work Act.

The building will only be handed over to the client if the commissioning engineer is satisfied that all the building services meet the design specification in the contract documents.

Check your Understanding

When you have completed these questions, check out the answers at the back of the book.
Note: more than one multiple choice answer may be correct.

1. A tong test instrument can also correctly be called:
   a. a continuity tester
   b. a clip-on ammeter
   c. an insulation resistance tester
   d. a voltage indicator throuout question.

2. All electrical test probes and leads must comply with the standards set by the:
   a. BS EN 60898
   b. BS 7671
   c. HSE Guidance Note GS 38
   d. IEE Regulations Part 2.

3. When making a test to determine the presence or absence of a voltage, the HSE recommends that for our own safety we should use:
   a. any old tester bought at a car-boot sale
   b. a multimeter set to the correct voltage
   c. a proprietary test lamp
   d. a voltage indicator.

4. For electrical test results to be valid the test instruments used:
   a. must be new
   b. must be of an approved type
   c. must have a calibration certificate
   d. must have a digital readout.

5. The test required by the Regulations to ascertain that the CPC is correctly connected is called:
   a. a basic protection
   b. continuity of ring final circuit conductors
   c. continuity of protective conductors
   d. earth electrode resistance.

6. One objective of the polarity test is to verify that:
   a. lampholders are correctly earthed
   b. final circuits are correctly fused
   c. the CPC is continuous throughout the installation
   d. the protective devices are connected in the live conductor.
7. When testing a 230V installation an insulation resistance tester must supply a voltage of:
   a. less than 50V
   b. 500V
   c. less than 500V
   d. greater than twice the supply voltage but less than 1000V.
8. The value of a satisfactory insulation resistance test on each final circuit of a 230V installation must be:
   a. less than 1 Ω
   b. less than 0.5 MΩ
   c. not less than 0.5 MΩ
   d. not less than 1 MΩ.
9. Instrument calibration certificates are usually valid for a period of:
   a. 3 months
   b. 1 year
   c. 3 years
   d. 5 years.
10. The maximum inspection and retest period for a domestic electrical installation is:
   a. 3 months
   b. 3 years
   c. 5 years
   d. 10 years.

11. A visual inspection of a new installation must be carried out:
    a. during the erection period
    b. during testing upon completion
    c. after testing upon completion
    d. before testing upon completion.
12. 'To ensure that all the systems within a building work as they were intended to work' is one definition of the purpose of:
    a. testing electrical equipment
    b. inspecting electrical systems
    c. commissioning electrical systems
    d. isolating electrical systems.
13. Use bullet points to state three reasons for testing a new electrical installation.
14. State five of the most important safety factors to be considered before electrical testing begins.
15. State the seven requirements of GS 38 when selecting probes, voltage indicators and measuring instruments.
16. Use bullet points to describe a safe isolation procedure of a final circuit fed from an MCB in a distribution board.
17. IEE Regulation 611.3 gives a checklist of about twenty items to be considered in the initial visual inspection of an electrical installation. Make a list of ten of the most important items to be considered in the visual inspection process. (Perhaps by joining together similar items.)
18. State three reasons why electricians must only use 'approved' test instruments.
19. State the first five tests to be carried out on a new electrical installation following a satisfactory 'inspection'. For each test:
    i. state the object (reason for) the test
    ii. a satisfactory test result.

20. State the certification process for a
    i. new electrical installation and
    ii. an electrical installation that is being re-tested as a part of the periodic inspection process
    iii. what will be indicated on the test certificates
    iv. who will receive the test certificates and
    v. who will issue the certificates
    vi. Finally, who will carry out the actual testing (A..................person).

21. State three safe working procedures relevant to your own safety when carrying out electrical testing.

22. State three safe working procedures relevant to the safety of other people when carrying out electrical testing.

23. State three safe working procedures relevant to the safety of other electrical systems when carrying out electrical testing.

24. State the two important tests that a PAT tester carries out on a portable appliance.

25. Use bullet points to state the reasons for commissioning a new building upon its completion.

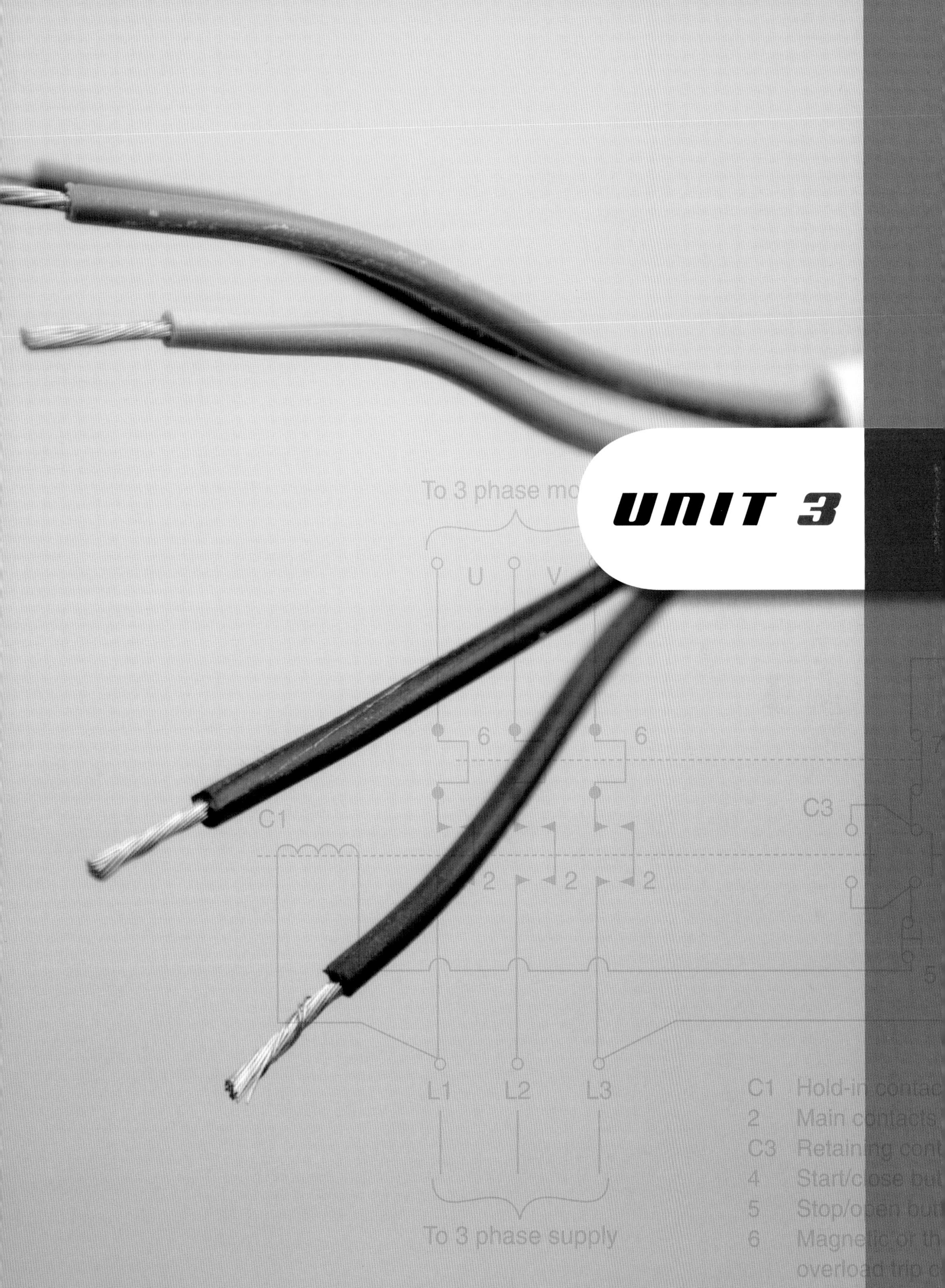
UNIT 3
U
V
6
6
C1
C3
2
2
2
L1
L2
L3
To 3 phase supply
C1
2
C3
4
5
6

CH 9

Fault diagnosis and repair

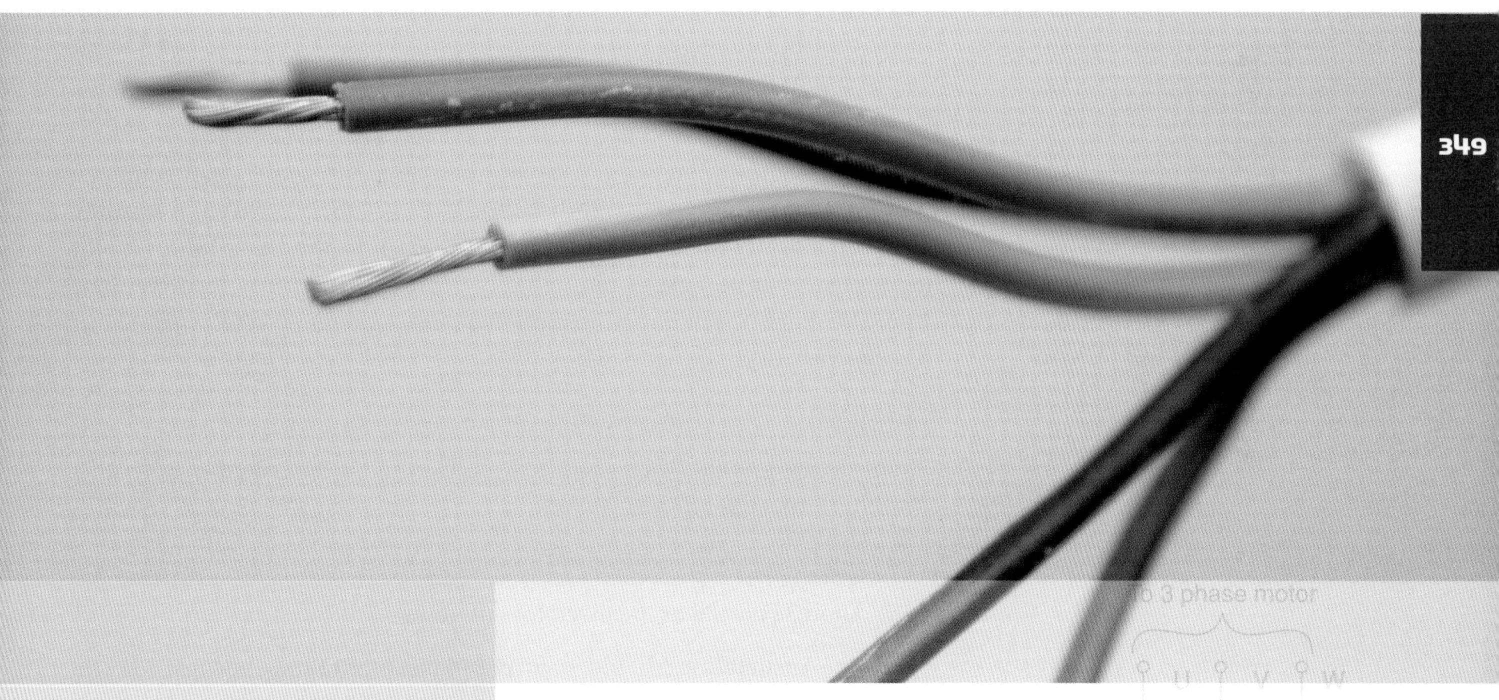

Unit 3 – Installation (buildings and structures): fault diagnosis and rectification – Outcome 1

Underpinning knowledge: when you have completed this chapter you should be able to:

- identify electrotechnical systems for power and lighting
- state the symptoms and causes of electrical faults
- identify *where* electrical faults occur in electrotechnical systems
- state four requirements for successful electrical fault finding
- state five requirements for safe working procedures when fault finding
- state four factors which influence the decision to either repair or replace faulty equipment
- state six situations where special precautions should be applied

To diagnose and find faults in electrical installations and equipment is probably one of the most difficult tasks undertaken by an electrician. The knowledge of fault finding and the diagnosis of faults can never be completely 'learned' because no two fault situations are exactly the same. As the systems we install become more complex, then the faults developed on these systems become more complicated to solve. To be successful the individual must have a thorough knowledge of the installation or piece of equipment and have a broad range of the skills and competences associated with the electrotechnical industries.

The ideal person will tackle the problem using a reasoned and logical approach, recognize his own limitations and seek help and guidance where necessary.

The tests recommended by the IEE Regulations can be used as a diagnostic tool but the safe working practices described by the Electricity at Work Act and elsewhere must always be observed during the fault-finding procedures.

If possible, fault finding should be planned ahead to avoid inconvenience to other workers and to avoid disruption of the normal working routine. **However, a faulty piece of equipment or a fault in the installation is not normally a planned event and usually occurs at the most inconvenient time.** The diagnosis and rectification of a fault is therefore often carried out in very stressful circumstances.

Symptoms of an electrical fault

The basic symptoms of an electrical fault may be described in one or a combination of the following ways:

1. There is a complete loss of power.
2. There is partial or localized loss of power.
3. The installation or piece of equipment is failing because of the following:
   - an individual component is failing;
   - the whole plant or piece of equipment is failing;
   - the insulation resistance is low;
   - the overload or protective devices operate frequently;
   - electromagnetic relays will not latch, giving an indication of undervoltage.

Definition

A *fault* is not a natural occurrence; it is an unplanned event which occurs unexpectedly.

Causes of electrical faults

A **fault** *is not a natural occurrence; it is an unplanned event which occurs unexpectedly.* The fault in an electrical installation or piece of equipment may be caused by:

- negligence – that is, lack of proper care and attention;
- misuse – that is, not using the equipment properly or correctly;
- abuse – that is, deliberate ill-treatment of the equipment.

If the installation was properly designed in the first instance to perform the tasks required of it by the user, then the *negligence, misuse or abuse* must be the fault of the user. However, if the installation does not perform the tasks required of it by the user then the negligence is due to the electrical contractor in not designing the installation to meet the needs of the user.

Negligence on the part of the user may be due to insufficient maintenance or lack of general care and attention, such as not repairing broken equipment or removing covers or enclosures which were designed to prevent the ingress of dust or moisture.

Misuse of an installation or pieces of equipment may occur because the installation is being asked to do more than it was originally designed to do, because of expansion of a company, for example. Circuits are sometimes overloaded because a company grows and a greater demand is placed on the existing installation by the introduction of new or additional machinery and equipment.

WHERE DO ELECTRICAL FAULTS OCCUR?

1. Faults occur in wiring systems, but not usually along the length of the cable, unless it has been damaged by a recent event such as an object being driven through it or a JCB digger pulling up an underground cable. Cable faults usually occur at each end, where the human hand has been at work at the point of cable inter-connections. This might result in broken conductors, trapped conductors or loose connections in joint boxes, accessories or luminaires.

   All cable connections must be made mechanically and electrically secure. They must also remain accessible for future inspection, testing and maintenance (IEE Regulation 526.3). The only exceptions to this rule are when:
   - underground cables are connected in a compound filled or encapsulated joint;
   - floor warming or ceiling warming heating systems are connected to a cold tail;
   - a joint is made by welding, brazing, soldering or compression tool.

   Since they are accessible, cable inter-connections are an obvious point of investigation when searching out the cause of a fault.

2. Faults also occur at cable terminations. The IEE Regulations require that a cable termination of any kind must securely anchor all conductors to reduce mechanical stresses on the terminal connections. All conductors of flexible cords must be terminated within the terminal connection otherwise the current carrying capacity of the conductor is reduced, which may cause local heating. Flexible cords are delicate – has the terminal screw been over-tightened, thus breaking the connection as the conductors flex or vibrate? Cables and flexible cords must be suitable for the temperature to be encountered at the point of termination or must be provided with

additional insulation sleeves to make them suitable for the surrounding temperatures (IEE Regulation 522.2).

3. Faults also occur at accessories such as switches, sockets, control gear, motor contactors or at the point of connection with electronic equipment. The source of a possible fault is again at the point of human contact with the electrical system and again the connections must be checked as described in the first two points above. Contacts that make and break a circuit are another source of wear and possible failure, so switches and motor contactors may fail after extensive use. Socket outlets that have been used extensively and loaded to capacity in say kitchens, are another source of fault due to overheating or loose connections. Electronic equipment can be damaged by the standard tests described in the IEE Regulations and must, therefore, be disconnected before testing begins.

4. Faults occur on instrumentation panels either as a result of a faulty instrument or as a result of a faulty monitoring probe connected to the instrument. Many panel instruments are standard sizes connected to CTs or VTs and this is another source of possible faults of the types described in points 1–3.

5. Faults occur in protective devices for the reasons given in points 1–3 above but also because they may have been badly selected for the job in hand and do not offer adequate protection or discrimination as described in Chapter 7 of this book.

6. Faults often occur in luminaires (light fittings) because the lamp has expired. Discharge lighting (fluorescent fittings) also require a 'starter' to be in good condition, although many fluorescent luminaires these days use starter-less electronic control gear. The points made in 1–3 about cable and flexible cord connections are also relevant to luminaire faults.

7. Faults occur when terminating flexible cords as a result of the flexible cable being of a smaller cross-section than the load demands, because it is not adequately anchored to reduce mechanical stresses on the connection or because the flexible cord is not suitable for the ambient temperature to be encountered at the point of connection. When terminating flexible cords, the insulation should be carefully removed without cutting out any flexible cord strands of wire because this effectively reduces the cross-section of the conductor. The conductor strands should be twisted together and then doubled over, if possible, and terminated in the appropriate connection. The connection screws should be opened fully so that they will not snag the flexible cord as it is eased into the connection. The insulation should go up to, but not into, the termination. The terminal screws should then be tightened.

8. Faults occur in electrical components, equipment and accessories such as motors, starters, switch gear, control gear, distribution panels,

Safety First

Cable fault

- faults do occur in wiring systems,
- but not usually along the cable length,
- faults usually occur at each end,
- where the human hand has been at work making connections.

switches, sockets and luminaires because these all have points at which electrical connections are made. It is unusual for an electrical component to become faulty when it is relatively new because it will have been manufactured and tested to comply with the appropriate British Standard. Through overuse or misuse components and equipment do become faulty but most faults are caused by poor installation techniques.

Modern electrical installations using new materials can now last longer than fifty years. Therefore, they must be properly installed. Good design, good workmanship and the use of proper materials are essential if the installation is to comply with the relevant Regulations (IEE Regulations 133.1.1 and 134.1.1).

Fault finding

Before an electrician can begin to diagnose the cause of a fault he must:

- have a thorough knowledge and understanding of the electrical installation or electrical equipment;
- collect information about the fault and the events occurring at or about the time of the fault from the people who were in the area at the time;
- begin to predict the probable cause of the fault using his own and other people's skills and expertise;
- test some of the predictions using a logical approach to identify the cause of the fault.

Most importantly, electricians must use their detailed knowledge of electrical circuits and equipment learned through training and experience and then apply this knowledge to look for a solution to the fault.

Let us, therefore, now briefly consider some of the basic wiring circuits that we first considered while studying Chapter 14 of the Level 2 Certificate of Electrotechnical Technology discussed in Chapter 14 of *Basic Electrical Installation Work* 5th Edition.

LIGHTING CIRCUITS

Table 1A in Appendix 1 of the *On Site Guide* deals with the assumed current demand of points, and states that for lighting outlets we should assume a current equivalent to a minimum of 100W per lampholder. This means that for a domestic lighting circuit rated at 5A, a maximum of 11 lighting outlets could be connected to each circuit. In practice, it is usual to divide the fixed lighting outlets into a convenient number of circuits of seven or eight outlets each. In this way the whole installation is not plunged into darkness if one lighting circuit fuses (IEE Regulation 314.1).

Lighting circuits are usually wired in 1.0 or 1.5mm cable using either a loop-in or joint-box method of installation. The loop-in method is universally employed with conduit installations or when access from above or below is prohibited after installation, as is the case with some industrial installations or blocks of flats. In this method the only joints are at the switches or lighting points, the live conductors being looped from switch to switch and the neutrals from one lighting point to another.

The use of junction boxes with fixed brass terminals is the method often adopted in domestic installations, since the joint boxes can be made accessible but are out of site in the loft area and under floorboards.

The live conductors must be broken at the switch position in order to comply with the Polarity Regulations (612.7). A ceiling rose may only be connected to installations operating at 250V maximum and must only accommodate one flexible cord unless it is specially designed to take more than one IEE Regulation 559.6.1.2 and 3. Lampholders suspended from flexible cords must be capable of suspending the mass of the luminaire fixed to the lampholder (559.6.1.5).

The type of circuit used will depend upon the installation conditions and the customer's requirements. One light controlled by one switch is called one-way switch control. A room with two access doors might benefit from a two-way switch control so that the lights may be switched on or off at either position. A long staircase with more than two switches controlling the same lights would require intermediate switching.

One-way, two-way or intermediate switches can be obtained as plate switches for wall mounting or ceiling mounted cord switches. Cord switches can provide a convenient method of control in bedrooms or bathrooms and for independently controlling an office luminaire.

SOCKET OUTLET CIRCUITS

Where portable equipment is to be used, it should be connected by a plug top to a conveniently accessible socket outlet (Regulation 553.1.7). Pressing the plug top into a socket outlet connects the appliance to the source of supply. **Socket outlets** therefore provide an easy and convenient method of connecting portable electrical appliances to a source of supply.

Definition

Socket outlets therefore provide an easy and convenient method of connecting portable electrical appliances to a source of supply.

Socket outlets can be obtained in 15, 13, 5 and 2A ratings, but the 13A flat pin type complying with BS 1363 is the most popular for domestic installations in the United Kingdom. Each 13A plug top contains a cartridge fuse to give maximum potential protection to the flexible cord and the appliances which it serves.

Socket outlets may be wired on a ring or radial circuit and in order that every appliance can be fed from an adjacent and conveniently accessible socket outlet, the number of sockets is unlimited provided that the floor area covered by the circuit does not exceed that given in Appendix 15 of the *IEE Regulations.*

In a radial circuit each socket outlet is fed from the previous one. Live is connected to live, neutral to neutral and earth to earth at each socket outlet. The fuse and cable sizes are given in Appendix 15 but circuits may also be expressed with a block diagram, as shown in Fig. 9.1. The number

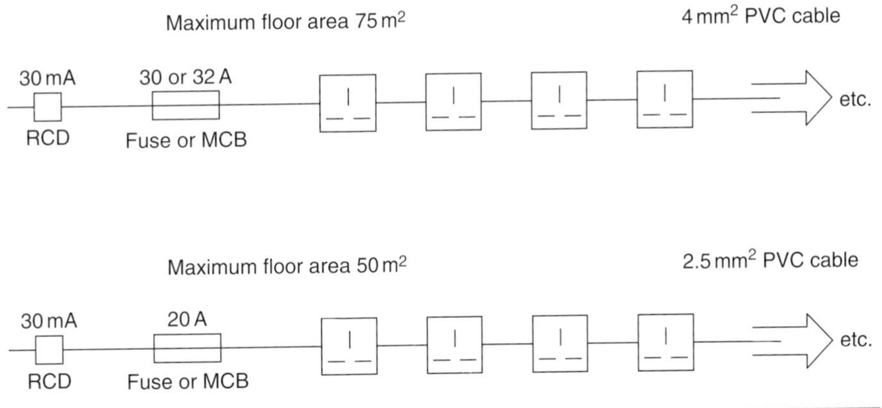

FIGURE 9.1
Block diagram of radial circuits.

of permitted socket outlets is unlimited but each radial circuit must not exceed the floor area stated and the known or estimated load.

Where two or more circuits are installed in the same premises, the socket outlets and permanently connected equipment should be reasonably shared out among the circuits, so that the total load is balanced.

When designing ring or radial circuits special consideration should be given to the loading in kitchens which may require separate circuits. This is because the maximum demand of current-using equipment in kitchens may exceed the rating of the circuit cable and protection devices.

Ring and radial circuits may be used for domestic or other premises where the maximum demand of the current using equipment is estimated not to exceed the rating of the protective devices for the chosen circuit.

Ring circuits are very similar to radial circuits in that each socket outlet is fed from the previous one, but in ring circuits the last socket is wired back to the source of supply. Each ring final circuit conductor must be looped into every socket outlet or joint box which forms the ring and must be electrically continuous throughout its length. The number of permitted socket outlets is unlimited but each ring circuit must not cover more than 100 m^2 of floor area.

The circuit details are given in Appendix 15 of the *IEE Regulations* but may also be expressed by the block diagram given in Fig. 9.2.

Additional protection by 30 mA residual current device (RCD) is now required in addition to overcurrent protection for all socket outlet circuits that are to be used by ordinary persons and intended for general use.

This additional protection is provided in case basic protection or fault protection fails or if the user of the installation is careless (IEE Regulations 411.3.3 and 415.1.1).

Note: An ordinary person is one who is neither an electrically skilled nor an instructed person.

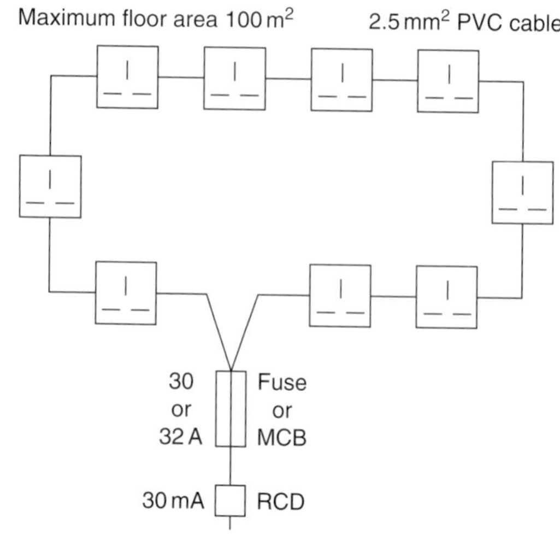

FIGURE 9.2
Block diagram of ring circuits.

Safety First

Isolation

- NEVER work 'LIVE'
- Isolate
- Secure the isolation
- Prove the supply 'dead' before starting work.

Safety First

Socket outlets

All socket outlets used by ordinary persons intended for general use must have:

- 30 mA RCD protection
- Regulations 411.3.3 and 415.1.1.

These circuits are considered in detail in Chapter 14 of *Basic Electrical Installation Work* 5th Edition and shown in Figs 14.12–14.16.

Designing out faults

The designer of the installation cannot entirely design out the possibility of a fault occurring but he can design in 'damage limitation' should a fault occur.

For example designing in two, three or four lighting and power circuits will reduce the damaging effect of any one circuit failing because not all lighting and power will be lost as a result of a fault. Limiting faults to only one of many circuits is good practice because it limits the disruption caused by a fault. Regulation 314 tells us to divide an installation into circuits as necessary so as to:

1. avoid danger and minimize inconvenience in the event of a fault occurring,
2. facilitate safe operation, inspection testing and maintenance.

Requirements for successful electrical fault finding

The steps involved in successfully finding a fault can be summarized as follows:

1. Gather *information* by talking to people and looking at relevant sources of information such as manufacturer's data, circuit diagrams, charts and schedules.
2. *Analyse* the evidence and use standard tests and a visual inspection to predict the cause of the fault.
3. *Interpret* test results and diagnose the cause of the fault.

4. *Rectify* the fault.
5. *Carry out* functional tests to verify that the installation or piece of equipment is working correctly and that the fault has been rectified.

Requirements for safe working procedures

The following five safe working procedures must be applied before undertaking the fault diagnosis.

1. The circuits must be isolated using a 'safe isolation procedure', such as that described in Chapter 8 before beginning to repair the fault.
2. All test equipment must be 'approved' and connected to the test circuits by recommended test probes as described by the Health and Safety Executive (HSE) Guidance Note GS 38 which were discussed in Chapter 8 and shown in Fig 8.7. The test equipment used must also be 'proved' on a known supply or by means of a proving unit such as that shown in Fig. 8.9.
3. Isolation devices must be 'secured' in the 'off' position as shown in Fig. 8.10. The key is retained by the person working on the isolated equipment.
4. Warning notices must be posted.
5. All relevant safety and functional tests must be completed before restoring the supply.

Live testing

The Electricity at Work Act tells us that it is 'preferable' that supplies be made dead before work commences (Regulation 4(3)). However, it does acknowledge that some work, such as fault finding and testing, may require the electrical equipment to remain energized. Therefore, if the fault finding and testing can only be successfully carried out 'live', then the person carrying out the fault diagnosis must:

- be trained so that he understands the equipment and the potential hazards of working live and can, therefore, be deemed to be 'competent' to carry out the activity;
- only use approved test equipment;
- set up barriers and warning notices so that the work activity does not create a situation dangerous to others.

Note that while live testing may be required in order to find the fault, live repair work must not be carried out. The individual circuit or item of equipment must first be isolated.

Definition

Isolation means the disconnection and separation of the electrical equipment from every source of electrical energy in such a way that this disconnection and separation is secure.

Secure isolation of electrical supply

The Electricity at Work Regulations are very specific in describing the procedure to be used for isolation of the electrical supply. Regulation 12(1) tells us that isolation means the disconnection and separation of the electrical equipment from every source of electrical energy in such a way that this disconnection and separation is secure. Regulation 4(3) tells us that we must also prove the conductors dead before work commences and that the test instrument used for this purpose must itself be proved immediately before and immediately after testing the conductors. To isolate an individual circuit or item of equipment successfully, competently and safely we must follow a procedure such as that given by the flow diagram in Fig. 8.11. Start at the top and work your way down the flowchart. When you get to the heavy-outlined amber boxes, pause and ask yourself whether everything is satisfactory up to this point. If the answer is yes, move on. If no, go back as indicated by the diagram.

Faulty equipment: to repair or replace?

Having successfully diagnosed the cause of the fault we have to decide if we are to repair or replace the faulty component or piece of equipment.

In many cases the answer will be straightforward and obvious, but in some circumstances the solution will need to be discussed with the customer. Some of the issues which may be discussed are as follows:

- What is the cost of replacement? Will the replacement cost be prohibitive? Is it possible to replace only some of the components? Will the labour costs of the repair be more expensive than a replacement? Do you have the skills necessary to carry out the repair? Would the repaired piece of equipment be as reliable as a replacement?
- Is a suitable replacement available within an acceptable time? These days, manufacturers carry small stocks to keep costs down.
- Can the circuit or system be shut down to facilitate a repair or replacement?
- Can alternative or temporary supplies and services be provided while replacements or repairs are carried out?

Selecting test equipment

The HSE has published Guidance Notes (GS 38) which advise electricians and other electrically competent people on the selection of suitable test probes, voltage indicating devices and measuring instruments. This is because they consider suitably constructed test equipment to be as vital for personal safety as the training and practical skills of the electrician. In the past, unsatisfactory test probes and voltage indicators have frequently been the cause of accidents, and therefore all test probes must now incorporate the following features:

1. The probes must have finger barriers or be shaped so that the hand or fingers cannot make contact with the live conductors under test.
2. The probe tip must not protrude more than 2 mm, and preferably only 1 mm, be spring-loaded and screened.
3. The lead must be adequately insulated and coloured so that one lead is readily distinguished from the other.

4. The lead must be flexible and sufficiently robust.
5. The lead must be long enough to serve its purpose but not too long.
6. The lead must not have accessible exposed conductors even if it becomes detached from the probe or from the instrument.
7. Where the leads are to be used in conjunction with a voltage detector they must be protected by a fuse.

A suitable probe and lead is shown in Fig. 8.7.

GS 38 also tells us that where the test is being made simply to establish the presence or absence of a voltage, the preferred method is to use a proprietary test lamp or voltage indicator which is suitable for the working voltage, rather than a multimeter. Accident history has shown that incorrectly set multimeters or makeshift devices for voltage detection have frequently caused accidents. Figure 8.8 shows a suitable voltage indicator. Test lamps and voltage indicators are not fail-safe, and therefore GS 38 recommends that they should be regularly proved, preferably before and after use, as described previously in the flowchart for a safe isolation procedure.

The IEE Regulations (BS 7671) also specify the test voltage or current required to carry out particular tests satisfactorily. All testing must, therefore, be carried out using an 'approved' test instrument if the test results are to be valid. **The test instrument must also carry a calibration certificate, otherwise the recorded results may be void.** Calibration certificates usually last for a year. Test instruments must, therefore, be tested and recalibrated each year by an approved supplier. This will maintain the accuracy of the instrument to an acceptable level, usually within 2% of the true value.

Modern digital test instruments are reasonably robust, but to maintain them in good working order they must be treated with care. An approved test instrument costs equally as much as a good-quality camera; it should, therefore, receive the same care and consideration.

CONTINUITY TESTER

To measure accurately the resistance of the conductors in an electrical installation we must use an instrument which is capable of producing an open circuit voltage of between 4 and 24V a.c. or d.c., and deliver a short-circuit current of not less than 200mA (Regulation 612.2.1). The functions of continuity testing and insulation resistance testing are usually combined in one test instrument.

INSULATION RESISTANCE TESTER

The test instrument must be capable of detecting insulation leakage between live conductors and between live conductors and earth. To do this and comply with Regulation 612.3 the test instrument must be capable of producing a test voltage of 250, 500 or 1000V and deliver an output current of not less than 1mA at its normal voltage.

EARTH FAULT LOOP IMPEDANCE TESTER

The test instrument must be capable of delivering fault currents as high as 25 A for up to 40 ms using the supply voltage. During the test, the instrument does an Ohm's law calculation and displays the test result as a resistance reading.

RCD TESTER

Where circuits are protected by an RCD we must carry out a test to ensure that the device will operate very quickly under fault conditions and within the time limits set by the IEE Regulations. The instrument must, therefore, simulate a fault and measure the time taken for the RCD to operate. The instrument is, therefore, calibrated to give a reading measured in milliseconds to an in-service accuracy of 10%.

If you purchase good-quality 'approved' test instruments and leads from specialist manufacturers they will meet all the Regulations and Standards and therefore give valid test results. However, to carry out all the tests required by the IEE Regulations will require a number of test instruments and this will represent a major capital investment in the region of £1000.

The specific tests required by the IEE Regulations: BS 7671 are described in detail in Chapter 8 of this book under the sub-heading 'Inspection and Testing Techniques'.

Electrical installation circuits usually carry in excess of 1 A and often carry hundreds of amperes. Electronic circuits operate in the milliampere or even microampere range. The test instruments used on electronic circuits must have a *high impedance* so that they do not damage the circuit when connected to take readings. All instruments cause some disturbance when connected into a circuit because they consume some power in order to provide the torque required to move the pointer. In power applications these small disturbances seldom give rise to obvious errors, but in electronic circuits, a small disturbance can completely invalidate any readings taken. We must, therefore, choose our electronic test equipment with great care.

So far in this chapter, I have been considering standard electrical installation circuits wired in conductors and cables using standard wiring systems. However, you may be asked to diagnose and repair a fault on a system that is unfamiliar to you or outside your experience and training. If this happens to you I would suggest that you immediately tell the person ordering the work or your supervisor that it is beyond your knowledge and experience. I have said earlier that fault diagnosis can only be carried out successfully by someone with a broad range of experience and a thorough knowledge of the installation or equipment that is malfunctioning. The person ordering the work will not think you a fool for saying straightaway that the work is outside your experience. It is better to be respected for your honesty than to attempt something that is beyond you at the present time and which could create bigger problems and waste valuable repair time.

Let us now consider some situations where special precautions or additional skills and knowledge may need to be applied.

Special situations

OPTICAL FIBRE CABLES

The introduction of fibre-optic cable systems and digital transmissions will undoubtedly affect future cabling arrangements and the work of the electrician. Networks based on the digital technology currently being used so successfully by the telecommunications industry are very likely to become the long-term standard for computer systems. Fibre-optic systems dramatically reduce the number of cables required for control and communications systems, and this will in turn reduce the physical room required for these systems. Fibre-optic cables are also immune to electrical noise when run parallel to mains cables and, therefore, the present rules of segregation and screening may change in the future. There is no spark risk if the cable is accidentally cut and, therefore, such circuits are intrinsically safe.

Optical fibre cables are communication cables made from optical-quality plastic, the same material from which spectacle lenses are manufactured. The energy is transferred down the cable as digital pulses of laser light as against current flowing down a copper conductor in electrical installation terms. The light pulses stay within the fibre-optic cable because of a scientific principle known as 'total internal refraction' which means that the laser light bounces down the cable and when it strikes the outer wall it is always deflected inwards and, therefore, does not escape out of the cable, as shown in Fig. 9.3.

The cables are very small because the optical quality of the conductor is very high and signals can be transmitted over great distances. They are cheap to produce and lightweight because these new cables are made from high-quality plastic and not high-quality copper. Single-sheathed cables are often called 'simplex' cables and twin-sheathed cables 'duplex', that is, two simplex cables together in one sheath. Multicore cables are available containing up to 24 single fibres.

Fibre-optic cables look like steel wire armour (SWA) cables (but of course are lighter) and should be installed in the same way and given the same level of protection as SWA cables. Avoid tight-radius bends if possible and kinks at all costs. Cables are terminated in special joint boxes which ensure cable ends are cleanly cut and butted together to ensure the continuity of the light pulses. Fibre-optic cables are Band I circuits when used for data

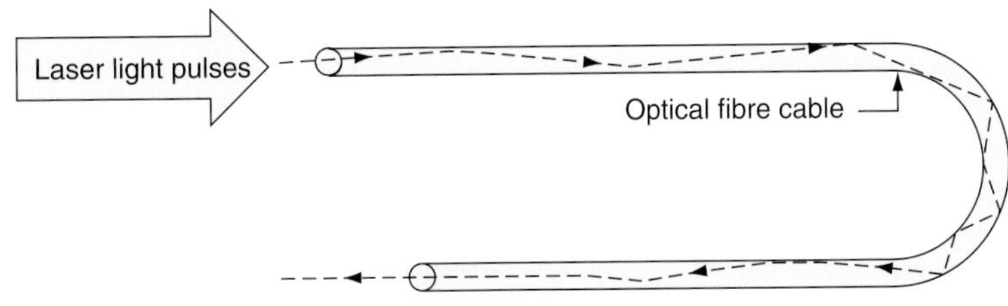

FIGURE 9.3
Digital pulses of laser light down an optical fibre cable.

transmission and must therefore be segregated from other mains cables to satisfy the IEE Regulations.

The testing of fibre-optic cables requires that special instruments be used to measure the light attenuation (i.e. light loss) down the cable. Finally, when working with fibre-optic cables, electricians should avoid direct eye contact with the low-energy laser light transmitted down the conductors.

Definition

Static electricity is a voltage charge which builds up to many thousands of volts between two surfaces when they rub together.

Definition

Static charge builds up between any two insulating surfaces or between an insulating surface and a conducting surface, but it is not apparent between two conducting surfaces.

Antistatic precautions

Static electricity is a voltage charge which builds up to many thousands of volts between two surfaces when they rub together. A dangerous situation occurs when the static charge has built up to a potential capable of striking an arc through the airgap separating the two surfaces.

Static charges build up in a thunderstorm. A lightning strike is the discharge of the thunder cloud, which might have built up to a voltage of 100 MV, to the general mass of earth which is at 0 V. Lightning discharge currents are of the order of 20 kA, hence the need for lightning conductors on vulnerable buildings in order to discharge the energy safely.

Static charge builds up between any two insulating surfaces or between an insulating surface and a conducting surface, but it is not apparent between two conducting surfaces.

A motor car moving through the air builds up a static charge which sometimes gives the occupants a minor shock as they step out and touch the door handle.

Static electricity also builds up in modern offices and similar carpeted areas. The combination of synthetic carpets, man-made footwear materials and dry air conditioned buildings contribute to the creation of static electrical charges building up on people moving about these buildings. Individuals only become aware of the charge if they touch earthed metalwork, such as a stair banister rail, before the static electricity has been dissipated. The effect is a sensation of momentary shock.

The precautions against this problem include using floor coverings that have been 'treated' to increase their conductivity or that contain a proportion of natural fibres that have the same effect. The wearing of leather soled footwear also reduces the likelihood of a static charge persisting as does increasing the humidity of the air in the building.

A nylon overall and nylon bed sheets build-up static charge which is the cause of the 'crackle' when you shake them. Many flammable liquids have the same properties as insulators, and therefore liquids, gases, powders and paints moving through pipes build-up a static charge.

Petrol pumps, operating theatre oxygen masks and car spray booths are particularly at risk because a spark in these situations may ignite the flammable liquid, powder or gas.

So how do we protect ourselves against the risks associated with static electrical charges? I said earlier that a build-up of static charge is not apparent between two conducting surfaces, and this gives a clue to the solution. Bonding surfaces together with protective equipotential bonding conductors prevents a build-up of static electricity between the surfaces. If we use large-diameter pipes, we reduce the flow rates of liquids and powders and, therefore, we reduce the build-up of static charge. Hospitals use cotton sheets and uniforms, and use protective equipotential bonding extensively in operating theatres. Rubber, which contains a proportion of graphite, is used to manufacture antistatic trolley wheels and surgeons' boots. Rubber constructed in this manner enables any build-up of static charge to 'leak' away. Increasing humidity also reduces static charge because the water droplets carry away the static charge, thus removing the hazard.

Avoiding shutdown of IT equipment

Every modern office now contains computers, and many systems are linked together or networked. Most computer systems are sensitive to variations or distortions in the mains supply and many computers incorporate *filters which produce high-protective conductor currents* of around 2 or 3 mA. This is clearly not a fault current, but is typical of the current which flows in the circuit protective conductor of IT equipment under normal operating conditions. IEE Regulations 543.7.1 and 4 deals with the earthing requirements for the installation of equipment having high-protective conductor currents. IEE Guidance Note 7 recommends that IT equipment should be connected to double sockets as shown in Fig. 9.4.

CLEAN SUPPLIES

Supplies to computer circuits must be 'clean' and 'secure'. Mainframe computers and computer networks are sensitive to mains distortion or interference, which is referred to as 'noise'. Noise is mostly caused by switching an inductive circuit which causes a transient spike, or by brush gear making contact with the commutator segments of an electric motor. These distortions in the mains supply can cause computers to 'crash' or provoke errors and are shown in Fig. 9.5.

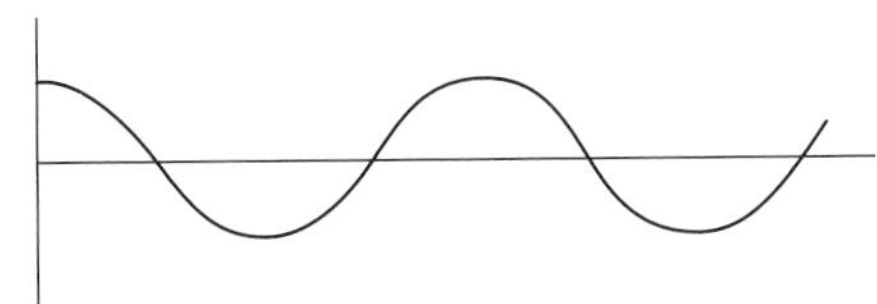

Clean supply

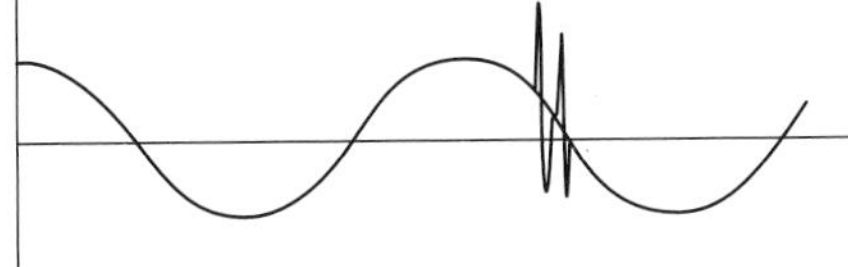

Spikes, caused by an over voltage transient surging through the mains

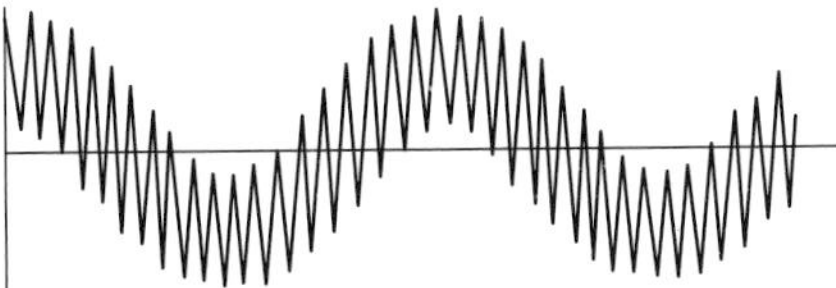

'Noise': unwanted electrical signals picked up by power lines or supply cords

FIGURE 9.5
Distortions in the a.c. mains supply.

Distribution fuse board

PE

Separate connections

Double socket outlets must have two terminals for protective conductors. One terminal to be used for each protective conductor, of a minimum size of 1.5 mm²

Ring final circuit supplying twin socket outlets (total protective conductor current exceeding 10 mA)

FIGURE 9.4
Recommended method of connecting IT equipment to socket outlets.

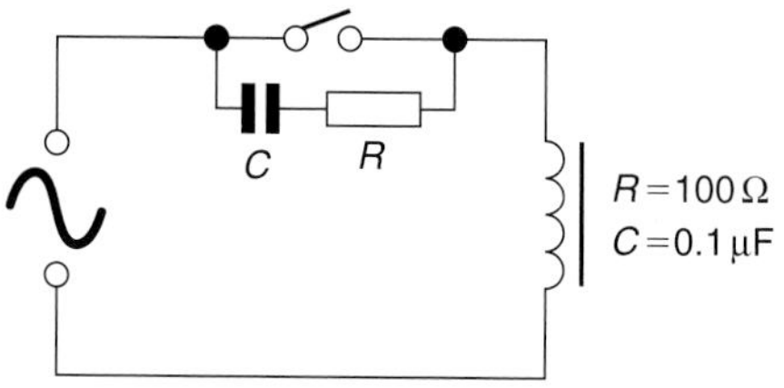

FIGURE 9.6
A simple noise suppressor.

To avoid this, a 'clean' supply is required for the computer network. This can be provided by taking the ring or radial circuits for the computer supplies from a point as close as possible to the intake position of the electrical supply to the building. A clean earth can also be taken from this point, which is usually one core of the cable and not the armour of an SWA cable, and distributed around the final wiring circuit. Alternatively, the computer supply can be cleaned by means of a filter such as that shown in Fig. 9.6.

SECURE SUPPLIES

The mains electrical supply in the United Kingdom is extremely reliable and secure. However, the loss of supply to a mainframe computer or computer network for even a second can cause the system to 'crash', and hours or even days of work can be lost.

One solution to this problem is to protect 'precious' software systems with an uninterruptible power supply (UPS). A UPS is essentially a battery supply electronically modified to provide a clean and secure a.c. supply. The UPS is plugged into the mains supply and the computer systems are plugged into the UPS.

Definition

A UPS is essentially a battery supply electronically modified to provide a clean and secure a.c. supply.

A **UPS** to protect a small network of, say, six PCs is physically about the size of one PC hard drive and is usually placed under or at the side of an operator's desk.

It is best to dedicate a ring or radial circuit to the UPS and either to connect the computer equipment permanently or to use non-standard outlets to discourage the unauthorized use and overloading of these special supplies by, for example, kettles.

Finally, remember that most premises these days contain some computer equipment and systems. Electricians intending to isolate supplies for testing or modification should *first check and then check again* before they finally isolate the supply in order to avoid loss or damage to computer systems.

Damage to electronic devices by 'overvoltage'

The use of electronic circuits in all types of electrical equipment has increased considerably over recent years. Electronic circuits and components can now be found in leisure goods, domestic appliances, motor starting and control circuits, discharge lighting, emergency lighting, alarm circuits and special-effects lighting systems. All electronic circuits are low-voltage circuits carrying very small currents.

Electrical installation circuits usually carry in excess of 1 A and often carry hundreds of amperes. Electronic circuits operate in the milliampere or even microampere range. The test instruments used on electronic circuits must have a *high impedance* so that they do not damage the circuit when connected to take readings.

The use of an insulation resistance test as described by the IEE Regulations (described in Chapter 8 of this book), must be avoided with any electronic equipment. The working voltage of this instrument can cause total devastation to modern electronic equipment. When carrying out an insulation resistance test as part of the prescribed series of tests for an electrical installation, all electronic equipment must first be disconnected or damage will result.

Any resistance measurements made on electronic circuits must be achieved with a battery-operated ohmmeter, high impedance to avoid damaging the electronic components.

Definition

Induction heating processes use high-frequency power to provide very focused heating in industrial processes.

Risks associated with high frequency or large capacitive circuits

Induction heating processes use high-frequency power to provide very focused heating in industrial processes.

The induction heater consists of a coil of large cross-section. The work-piece or object to be heated is usually made of ferrous metal and is placed inside the coil. When the supply is switched on, eddy currents are induced into the work-piece and it heats up very quickly so that little heat is lost to conduction and convection.

The frequency and size of the current in the coil determines where the heat is concentrated in the work-piece:

- the higher the current the greater is the surface penetration;
- the longer the current is applied the deeper the penetration;
- the higher the frequency the less is the depth of heat penetration.

For shallow penetration, high frequency, high current, short time application is typically used for tool tempering. Other applications are brazing and soldering industrial and domestic gas boiler parts.

When these machines are not working they look very harmless but when they are working they operate very quietly and there is no indication of the intense heat that they are capable of producing. Domestic and commercial microwave ovens operate at high frequency. The combination of risks of high frequency and intense heating means that before any maintenance, repair work or testing is carried out, the machine must first be securely isolated and no one should work on these machines unless they have received additional training to enable them to do so safely.

Industrial wiring systems are very inductive because they contain many inductive machines and circuits, such as electric motors, transformers, welding plants and discharge lighting. The inductive nature of the industrial load causes the current to lag behind the voltage and creates a bad power factor. Power factor is the percentage of current in an alternating current circuit that can be used as energy for the intended purpose. A power factor of say 0.7 indicates that 70% of the current supplied is usefully employed by the industrial equipment.

An inductive circuit, such as that produced by an electric motor, induces an electromagnetic force which opposes the applied voltage and causes the current waveform to lag the voltage waveform. Magnetic energy is stored up in the load during one half cycle and returned to the circuit in the next half cycle. If a capacitive circuit is employed, the current leads the voltage since the capacitor stores energy as the current rises and discharges it as the current falls. So here we have the idea of a solution to the problem of a bad power factor created by inductive industrial loads. Power Factor and Power Factor Improvement was discussed in Chapter 4 of this book.

Definition

The *power factor* of the consumer is governed entirely by the electrical plant and equipment that is installed and operated within the consumer's buildings.

The **power factor** at which consumers take their electricity from the local electricity supply authority is outside the control of the supply authority. The power factor of the consumer is governed entirely by the electrical plant and equipment that is installed and operated within the consumer's buildings. Domestic consumers do not have a bad power factor because they use very little inductive equipment, most of the domestic load is neutral and at unity power factor.

Electricity supply authorities discourage the use of equipment and installations with a low-power factor because they absorb part of the capacity of the generating plant and the distribution network to no useful effect. They, therefore, penalize industrial consumers with a bad power factor through a maximum demand tariff, metered at the consumer's intake position. If the power factor falls below a datum level of between 0.85 and 0.9 then extra charges are incurred. In this way industrial consumers are encouraged to improve their power factor.

Definition

Power factor improvement of most industrial loads is achieved by connecting capacitors to either:

- individual items of equipment or,
- banks of capacitors.

Power factor improvement of most industrial loads is achieved by connecting capacitors to either:

- individual items of equipment or,
- banks of capacitors may be connected to the main busbars of the installation at the intake position.

The method used will depend upon the utilization of the installed equipment by the industrial or commercial consumer. If the load is constant then banks of capacitors at the mains intake position would be indicated. If the load is variable then power factor correction equipment could be installed adjacent to the machine or piece of equipment concerned.

Power factor correction by capacitors is the most popular method because of the following:

- They require no maintenance.
- Capacitors are flexible and additional units may be installed as an installation or system is extended.
- Capacitors may be installed adjacent to individual pieces of equipment or at the mains intake position. Equipment may be placed on the floor or fixed high up and out of the way.

Capacitors store charge and must be disconnected before the installation or equipment is tested in accordance with Section 6 of the IEE Regulations BS 7671.

Small power factor correction capacitors as used in discharge lighting often incorporate a high-value resistor connected across the mains terminals. This discharges the capacitor safely when not in use. Banks of larger capacity capacitors may require discharging to make them safe when not in use. To discharge a capacitor safely and responsibly it must be discharged slowly over a period in excess of five 'time-constants' through a suitable discharge resistor. Capacitors and time-constants were discussed earlier in this book in Chapter 4 under the sub-heading 'Electrostatics'.

Presence of storage batteries

Definition

Since an emergency occurring in a building may cause the mains supply to fail, the *emergency lighting* should be supplied from a source which is independent from the main supply.

Since an emergency occurring in a building may cause the mains supply to fail, the **emergency lighting** should be supplied from a source which is independent from the main supply. A battery's ability to provide its output instantly makes it a very satisfactory source of standby power. In most commercial, industrial and public service buildings housing essential services, the alternative power supply would be from batteries, but generators may also be used. Generators can have a large capacity and duration, but a major disadvantage is the delay of time while the generator runs up to speed and takes over the load. In some premises a delay of more than 5 seconds is considered unacceptable, and in these cases a battery supply is required to supply the load until the generator can take over.

The emergency lighting supply must have an adequate capacity and rating for the specified duration of time (IEE Regulation 313.2). BS 5266 and BS EN 1838 states that after a battery is discharged by being called into operation for its specified duration of time, it should be capable of once again operating for the specified duration of time following a recharge period of not longer than 24 hours. The duration of time for which the emergency lighting should operate will be specified by a statutory authority but is normally 1–3 hours. The British Standard states that escape lighting should operate for a minimum of 1 hour. Standby lighting operation time will depend upon financial considerations and the importance of continuing the process or activity within the premises after the mains supply has failed.

The contractor installing the emergency lighting should provide a test facility which is simple to operate and secure against unauthorized interference. The emergency lighting installation must be segregated completely from any other wiring, so that a fault on the main electrical installation cannot damage the emergency lighting installation (IEE Regulation 528.1).

The batteries used for the emergency supply should be suitable for this purpose. Motor vehicle batteries are not suitable for emergency lighting applications, except in the starter system of motor-driven generators. The fuel supply to a motor-driven generator should be checked. The battery room of a central battery system must be well ventilated and, in the case of a motor-driven generator, adequately heated to ensure rapid starting in cold weather.

The British Standard recommends that the full load should be carried by the emergency supply for at least 1 hour in every 6 months. After testing, the emergency system must be carefully restored to its normal operative state. A record should be kept of each item of equipment and the date of each test by a qualified or responsible person. It may be necessary to produce the record as evidence of satisfactory compliance with statutory legislation to a duly authorized person.

Self-contained units are suitable for small installations of up to about 12 units. The batteries contained within these units should be replaced about every 5 years, or as recommended by the manufacturer.

Definition

Storage batteries are secondary cells. A secondary cell has the advantage of being rechargeable. If the cell is connected to a suitable electrical supply, electrical energy is stored on the plates of the cell as chemical energy. When the cell is connected to a load, the chemical energy is converted to electrical energy.

Storage batteries are secondary cells. A secondary cell has the advantage of being rechargeable. If the cell is connected to a suitable electrical supply, electrical energy is stored on the plates of the cell as chemical energy. When the cell is connected to a load, the chemical energy is converted to electrical energy.

A lead-acid cell is a secondary cell. Each cell delivers about 2V, and when six cells are connected in series a 12V battery is formed.

A lead-acid battery is constructed of lead plates which are deeply ribbed to give maximum surface area for a given weight of plate. The plates are assembled in groups, with insulating separators between them. The separators are made of a porous insulating material, such as wood or ebonite, and the whole assembly is immersed in a dilute sulphuric acid solution in a plastic container.

The capacity of a cell to store charge is a measure of the total quantity of electricity which it can cause to be displaced around a circuit after being fully charged. It is stated in ampere-hours, abbreviation Ah, and calculated at the 10-hour rate which is the steady load current which would completely discharge the battery in 10 hours. Therefore, a 50 Ah battery will provide a steady current of 5 A for 10 hours.

MAINTENANCE OF LEAD-ACID BATTERIES

- The plates of the battery must always be covered by dilute sulphuric acid. If the level falls, it must be topped up with distilled water.
- Battery connections must always be tight and should be covered with a thin coat of petroleum jelly.
- The specific gravity or relative density of the battery gives the best indication of its state of charge. A discharged cell will have a specific gravity of 1.150, which will rise to 1.280 when fully charged. The specific gravity of a cell can be tested with a hydrometer.
- To maintain a battery in good condition it should be regularly trickle-charged. A rapid charge or discharge encourages the plates to buckle, and may cause permanent damage. Most batteries used for standby supplies today are equipped with constant voltage chargers. The principle of these is that after the battery has been discharged by it being called into operation, the terminal voltage will be depressed and this

enables a relatively large current (1–5 A) to flow from the charger to recharge the battery. As the battery becomes more fully charged its voltage will rise until it reaches the constant voltage level where the current output from the charger will drop until it is just sufficient to balance the battery's internal losses. The main advantage of this system is that the battery controls the amount of charge it receives and is therefore automatically maintained in a fully charged condition without human intervention and without the use of any elaborate control circuitry.

- The room used to charge the emergency supply storage batteries must be well ventilated because the charged cell gives off hydrogen and oxygen, which are explosive in the correct proportions.

Check your Understanding

When you have completed these questions, check out the answers at the back of the book.
Note: more than one multiple choice answer may be correct.

1. To diagnose and find electrical faults the ideal person will:
   a. first isolate the whole system
   b. use a logical and methodical approach
   c. carry out the relevant tests recommended by Part 6 of the IEE Regulations
   d. recognize his own limitations and seek help and guidance where necessary.
2. The 'symptoms' of an electrical fault might be:
   a. there is a complete loss of power
   b. nothing unusual is happening
   c. there is a local or partial loss of power
   d. the local isolator switch is locked off.
3. A fault is not a natural occurrence, it is not planned and occurs unexpectedly. It may be caused by:
   a. regular maintenance
   b. negligence
   c. misuse
   d. abuse.
4. The main lighting in a room having only one entrance would probably be controlled by:
   a. pull switch
   b. intermediate switch
   c. two-way switch
   d. one-way switch.
5. The main lighting in a room having two entrances would probably be controlled by:
   a. pull switch
   b. intermediate switch
   c. two-way switch
   d. one-way switch.
6. Electrical test equipment must always:
   a. work on a.c. and d.c. supplies
   b. have an in date calibration certificate
   c. incorporate a range selector switch
   d. incorporate probes which comply with GS 38.

7. A final circuit feeding socket outlets with a rated current of less than 20 A and used by ordinary persons for general use must always have:
   a. overcurrent protection
   b. 30 mA RCD protection
   c. splash proof protection.
   d. incorporate an industrial type socket outlet
8. State three symptoms of an electrical fault.
9. State three causes of electrical faults.
10. Make a list of 10 places where faults might occur on an electrical system.
11. List four steps involved in fault finding.
12. State five requirements for safe working procedures when fault finding.
13. State four factors that might influence the decision to either repair or replace faulty equipment.
14. State four safety features you would look for before selecting test equipment.
15. State the advantages and dangers associated with optical fibre cables.
16. State briefly the meaning of 'static electricity'. What action is taken to reduce the build-up of a static charge on an electrical system.
17. State the problems which an unexpected mains failure of IT equipment would create.
18. Use a sketch to describe 'clean supplies', 'spikes' and 'noise' on IT supplies.
19. Briefly describe what we mean by secure supplies and UPS with regard to IT equipment.
20. Emergency supplies are often provided by storage batteries that are secondary cells. What is the advantage of a secondary cell in these circumstances.

Restoring systems to working order

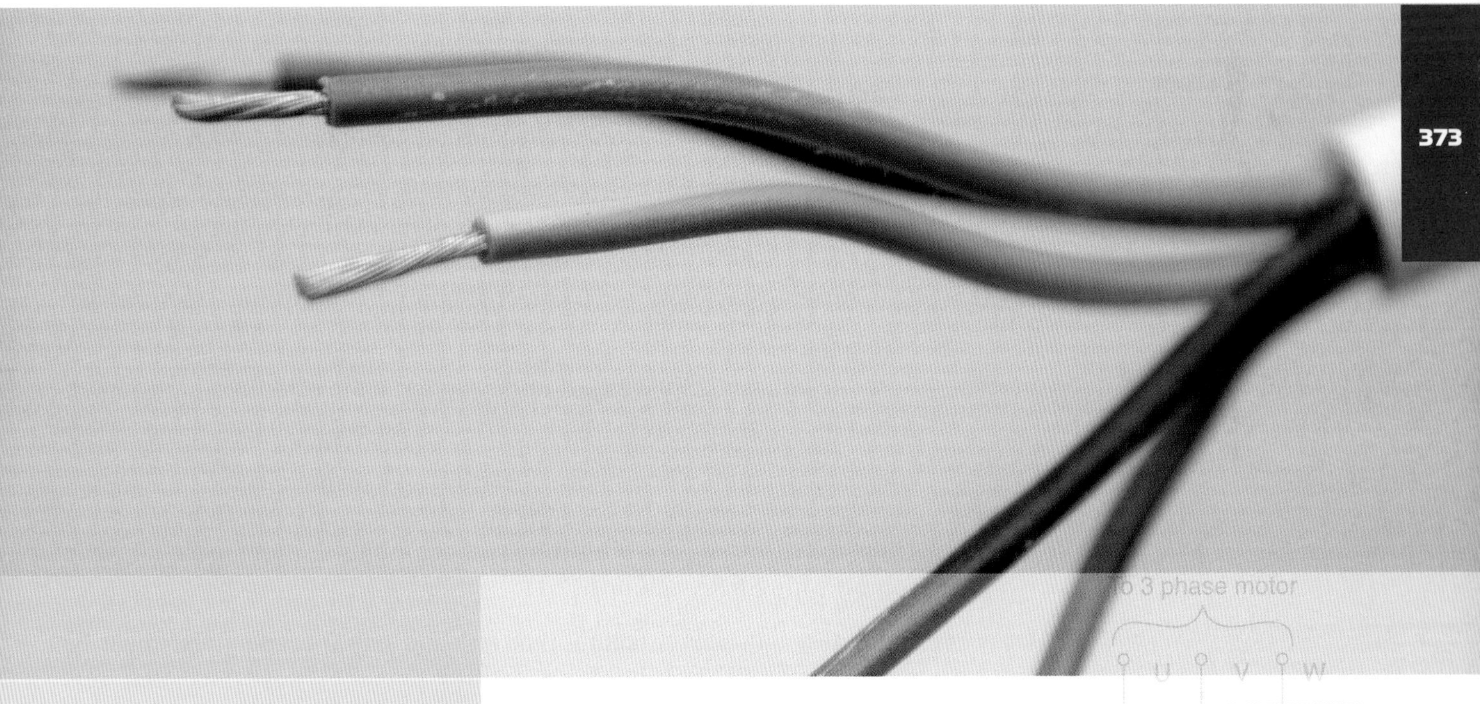

Installation (Buildings and structure): fault diagnosis and rectification – Outcome 2

Underpinning knowledge: When you have completed this chapter you should be able to:

- state the requirements of testing and commissioning
- state six reasons for carrying out functional tests
- state four assemblies that the IEE Regulations instruct us to carry out functional tests upon
- state the reasons for restoring the fabric of the building structure following fault diagnosis and repair work
- state procedures for dealing with waste material disposal
- describe the handover, completion process

Testing and commissioning

When the fault has been identified and repaired as described in Chapter 9, the circuit, system or equipment must be inspected, tested and functional checks carried out as required by IEE Regulations Chapter 61.

The purpose of inspecting and testing the repaired circuit, system or equipment is to confirm the electrical integrity of the system before it is re-energized.

The tests recommended by Part 6 of the IEE Regulations are:

1. Test the continuity of the protective conductors including the protective equipotential and supplementary bonding conductors.
2. Test the continuity of all ring final circuit conductors.
3. Test the insulation resistance between live conductors and earth.
4. Test the polarity to verify that single pole control and protective devices are connected in the line conductor only.
5. Test the earth electrode resistance where the installation incorporates an earth electrode as a part of the earthing system.

The supply may now be connected and the following tests carried out:

6. Test the polarity using an approved test lamp or voltage indicator.
7. Test the earth fault loop impedance where the protective measures used require a knowledge of earth fault loop impedance.

These tests *where relevant* must be carried out in the order given above to comply with IEE Regulation 612.1.

If any test indicates a failure, that test and any preceding test must be repeated after the fault has been rectified. This is because the earlier tests may have been influenced by the fault.

The above tests are described in Chapter 8 of this book, in Part 6 of the IEE Regulations and in Guidance Note 3 published by the IEE.

Safety First

Testing

The tests indicated in Part 6 of the Regulations:

- must be carried out in the order given so that essential safety features are proved first.
- IEE Regulations 612.1.

Functional testing (IEE Regulation 612.13)

Following the carrying out of the relevant tests described above we must carry out functional testing to ensure that:

- the circuit, system or equipment works correctly;
- it works as it did before the fault occurred;
- it continues to comply with the original specification;
- it is electrically safe;
- it is mechanically safe;
- it meets all the relevant regulations in particular the IEE Regulations (BS 7671).

IEE Regulation 612.13 tells us to check the effectiveness of the following assemblies to show that they are properly mounted, adjusted and installed:

- Residual current devices (RCDs)
- Switchgear
- Control gear
- Controls and interlocks.

Restoration of the building structure

If the structure, or we sometimes call it the fabric of the building, has been damaged as a result of your electrical repair work, it must be made good before you hand the installation, system or equipment back to the client.

Where a wiring system passes through elements of the building construction such as floors, walls, roofs, ceilings, partitions or cavity barriers, the openings remaining after the passage of the wiring system must be sealed according to the degree of fire resistance demonstrated by the original building material (IEE Regulation 527.2).

You should always make good the structure of the building using appropriate materials ***before*** you leave the job so that the general building structural performance and fire safety are not reduced. If additionally, there is a little cosmetic plastering and decorating to be done, then who actually will carry out this work is a matter of negotiation between the client and the electrical contractor.

Safety First

Fire

If your electrical activities cause damage to the fabric of the building then:

- the openings remaining must be sealed according to the degree of fire resistance demonstrated by the original building material.
- IEE Regulation 527.2.

Disposal of waste

Having successfully diagnosed the electrical fault and carried out the necessary repairs OR having completed any work in the electrotechnical industry, we come to the final practical task, leaving the site in a safe and clean condition and the removal of any waste material. This is an important part of your company's 'good customer relationships' with the client. We also know from Chapter 1 of this book that we have a 'duty of care' for the waste that we produce as an electrical company (see Chapter 1, under the sub-heading Controlled Waste Regulations 1998).

We have also said many times in this book that having a good attitude to health and safety, working conscientiously and neatly, keeping passageways clear and regularly tidying up the workplace is the sign of a good and competent craftsman. But what do you do with the rubbish that the working environment produces? Well:

- All the packaging material for electrical fittings and accessories usually goes into either your employer's skip or the skip on site designated for that purpose.
- All the off-cuts of conduit, trunking and tray also go into the skip.
- In fact, most of the general site debris will probably go into the skip and the waste disposal company will take the skip contents to a designated local council land fill area for safe disposal.

- The part coils of cable and any other reusable leftover lengths of conduit, trunking or tray will be taken back to your employer's stores area. Here it will be stored for future use and the returned quantities deducted from the costs allocated to that job.
- What goes into the skip for normal disposal into a land fill site is usually a matter of common sense. However, some substances require special consideration and disposal. We will now look at asbestos and large quantities of used fluorescent tubes which are classified as 'special waste' or 'hazardous waste'.

Asbestos is a mineral found in many rock formations. When separated it becomes a fluffy, fibrous material with many uses. It was used extensively in the construction industry during the 1960s and 1970s for roofing material, ceiling and floor tiles, fire resistant board for doors and partitions, for thermal insulation and commercial and industrial pipe lagging.

In the buildings where it was installed some 40 years ago, when left alone, it does not represent a health hazard, but those buildings are increasingly becoming in need of renovation and modernization. It is in the dismantling and breaking up of these asbestos materials that the health hazard increases. Asbestos is a serious health hazard if the dust is inhaled. The tiny asbestos particles find their way into delicate lung tissue and remain embedded for life, causing constant irritation and eventually, serious lung disease.

Working with asbestos materials is not a job for anyone in the electrotechnical industry. If asbestos is present in situations or buildings where you are expected to work, it should be removed by a specialist contractor before your work commences. Specialist contractors, who will wear fully protective suits and use breathing apparatus, are the only people who can safely and responsibly carry out the removal of asbestos. They will wrap the asbestos in thick plastic bags and store them temporarily in a covered and locked skip. This material is then disposed of in a special land fill site with other toxic industrial waste materials and the site monitored by the local authority for the foreseeable future.

There is a lot of work for electrical contractors in many parts of the country, updating and improving the lighting in government buildings and schools. This work often involves removing the old fluorescent fittings, hanging on chains or fixed to beams and installing a suspended ceiling and an appropriate number of recessed modular fluorescent fittings. So what do we do with the old fittings? Well, the fittings are made of sheet steel, a couple of plastic lampholders, a little cable, a starter and ballast. All of these materials can go into the ordinary skip. However, the fluorescent tubes contain a little mercury and fluorescent powder with toxic elements, which cannot be disposed of in the normal land fill sites. New Hazardous Waste Regulations were introduced in July 2005 and under these Regulations lamps and tubes are classified as hazardous. While each lamp contains only a small amount of mercury, vast numbers of lamps and tubes are disposed of in the United Kingdom every year resulting in a significant environmental threat.

The environmentally responsible way to dispose of fluorescent lamps and tubes is to recycle them.

The process usually goes like this:

- Your employer arranges for the local electrical wholesaler to deliver a plastic waste container of an appropriate size for the job.
- Expired lamps and tubes are placed whole into the container, which often has a grating inside to prevent the tubes breaking when being transported.
- When the container is full of used lamps and tubes, you telephone the electrical wholesaler and ask them to pick up the filled container and deliver it to one of the specialist recycling centres.
- Your electrical company will receive a 'Duty of Care Note' and full recycling documents which ought to be filed safely as proof that the hazardous waste was recycled safely.
- The charge is approximately 50p for each 1800mm tube and this cost is passed on to the customer through the final account.

Safety First

Waste

- Clean up before you leave the job.
- Put waste in the correct skip.
- Recycle used lamps and tubes.
- Get rid of all waste responsibly.

The Control of Substances Hazardous to Health (COSHH) Regulations and the Controlled Waste Regulations 1998 have encouraged specialist companies to set up businesses dealing with the responsible disposal of toxic waste material. Specialist companies have systems and procedures, which meet the relevant regulation, and they would usually give an electrical company a certificate to say that they had disposed of a particular waste material responsibly. The system is called 'Waste Transfer Notes'. The notes will identify the type of waste taken by whom and its final place of disposal. The person handing over the waste material to the waste disposal company will be given a copy of the notes and this must be filed in a safe place, probably in the job file or a dedicated file. It is the proof that your company has carried out its duty of care to dispose of the waste responsibly. The cost of this service is then passed on to the customer. These days, large employers and local authorities insist that waste is disposed of properly.

The Environmental Health Officer at your local Council Offices will always give advice and point you in the direction of specialist companies dealing with toxic waste disposal.

Hand over to the client

Handing over the repaired circuit, system or equipment is an important part of the fault diagnosis and repair process. You are effectively saying to the client 'here is your circuit, system or equipment, it is now safe to use and it works as it should work'.

The client will probably be interested in the following:

- What has been done to identify and repair the fault?
- The possible reasons why the fault occurred and recommendations which will prevent a reoccurrence of the problem.
- A demonstration of the operation of the circuit, system or equipment to show that the fault has been fully rectified.
- Finally, the handing over of certificates of test results and manufacturer's instructions, if new equipment has been installed.

Check your Understanding

When you have completed these questions, check out the answers at the back of the book.
Note: more than one multiple choice answer may be correct.

1. The IEE Regulation test for 'continuity of conductors' tests:
   a. line conductors
   b. neutral conductors
   c. CPCs
   d. protective equipotential bonding conductors.
2. The IEE Regulation test for 'continuity of ring final circuit conductors' tests:
   a. line conductors
   b. neutral conductors
   c. CPCs
   d. protective equipotential bonding conductors.
3. The IEE Regulation test for 'insulation resistance' tests:
   a. the effectiveness of conductors' insulation
   b. the live conductors to earth
   c. the earth conductors to the earth electrode
   d. that protective devices are connected in the line conductor.
4. The IEE Regulation test for 'Polarity' tests:
   a. the earth loop impedance
   b. the effective operation of RCDs
   c. the effective operation of controls and switchgear
   d. that protective devices are connected in the line conductor.
5. The IEE Regulation test for 'Functional testing' tests:
   a. the earth loop impedance
   b. the effective operation of RCDs
   c. the effective operation of controls and switchgear
   d. that protective devices are connected in the line conductor.
6. Use bullet points to list very briefly the requirements of testing and commissioning an installation following the repair of a fault.
7. Use bullet points to list very briefly the six reasons for carrying out functional testing following the repair of a fault.

8. The IEE Regulation 612.13 advises us to check the effectiveness of four assemblies to show that they are properly mounted, adjusted and installed. Name them.
9. State the reasons why it is important to make good any damage to the fabric of a building as a result of your electrotechnical activities. State how you would make good damage to a brick wall and a concrete floor where a 100 × 100 mm trunking passes through.
10. Very briefly state the responsibilities of an electrotechnical company with regard to the disposal of waste material.
11. State what we mean by 'ordinary waste' and 'hazardous waste' and give examples of each.
12. Very briefly describe the system of 'Waste Transfer Notes'.
13. Very briefly describe four points you would discuss with a client when handing over a repaired faulty system, circuit or piece of equipment.

Answers to Check your understanding

Chapter 1

1. a c
2. b, d
3. a, d
4. a, c, d
5. a
6. d
7. c
8. b
9. d
10. c
11. b
12. a

13 to 23 – answers in the text of Chapter 1.

Chapter 2

1. c
2. b
3. a, b
4. c
5. b
6. d
7. d
8. c
9. b
10. a
11. a
12. c
13. b, c

14 to 26 – answers in the text of Chapter 2.

Chapter 3

1. b
2. c
3. d
4. a

5 to 11 – answers in the text of Chapter 3.

Chapter 4

1. b
2. b
3. c
4. b
5. c
6. b
7. c
8. c
9. d
10. b
11. a
12. b
13. c
14. a, b

15 to 30 – answers in the text of Chapter 4.

Chapter 5

1. d	**2.** c	**3.** b
4. a	**5.** c, d	**6.** a, b
7. d	**8.** b	**9.** a
10. a, c		

11 to 25 – answers in the text of Chapter 5.

Chapter 6

1. b, c, d	**2.** a	**3.** a
4. b, c, d	**5.** b	**6.** c
7. a	**8.** b, c, d	**9.** c

10 to 22 – answers in the text of Chapter 6

Chapter 7

1. b	**2.** c, d	**3.** a
4. b	**5.** a	**6.** b
7. a	**8.** b	

9 to 20 – answers in the text of Chapter 7.

Chapter 8

1. b	**2.** c	**3.** c, d
4. b, c	**5.** c	**6.** d
7. b	**8.** d	**9.** b
10. d	**11.** a, d	**12.** c

13 to 25 – answers in the text of Chapter 8.

Chapter 9

1. b, c, d	**2.** a, c	**3.** b, c, d
4. d	**5.** c	**6.** b, d
7. a, b		

8 to 20 – answers in the text of Chapter 9

Chapter 10

1. c, d	**2.** c	**3.** a, b
4. d	**5.** b, c	

6 to 13 – answers in the text of Chapter 10.

Appendix A: Obtaining information and electronic components

For local suppliers, you should consult your local telephone directory. However, the following companies distribute electrical and electronic components throughout the United Kingdom. In most cases, telephone orders received before 5 p.m. can be dispatched the same day.

Electromail (R.S. mail order business), P.O. Box 33, Corby, Northants NN17 9EL, Telephone: 011536 204555. Website: RS www.com

Farnell Electronic Components, Canal Road, Leeds LS12 2TU, Telephone: 0113 636311. Email: Sale@farnellinone.co.uk

Maplin Electronics, Valley Road. Wombwell S73 OBS, Telephone: 01226 751155. Website: www.maplin.co.uk

Rapid Electronics Ltd, Heckworth Close, Severalls Industrial Estate, Colchester, Essex CO4 4TB, Telephone: 01206 751166. Fax: 01206 751188. Email: sales@rapidelec.co.uk

R.S. Components Ltd, P.O. Box 99, Corby, Northants NN17 9RS, Telephone: 01536 201234. Website: RS www.com

Verospeed Electronic Components, Boyatt Wood, Eastleigh, Hants SO5 4ZY, Telephone: 02380 644555.

Appendix B: Abbreviations, symbols and codes

Abbreviations used in electronics for multiples and sub-multiples		
T	tera	10^{12}
G	giga	10^{9}
M	mega or meg	10^{6}
k	kilo	10^{3}
d	deci	10^{-1}
c	centi	10^{-2}
m	milli	10^{-3}
µ	micro	10^{-6}
n	nano	10^{-9}
p	pico	10^{-12}

Suffixes used with semiconductor devices

Many semiconductor devices are available with suffix letters after the part number, that is, BC108B, C106D, TIP31C.

The suffix is used to indicate a specific parameter relevant to the device – some examples are shown below.

THYRISTORS, TRIACS, POWER RECTIFIERS

Suffix indicates voltage rating, for example, TIC 106D indicates device has a 400V rating. Letters used are:

Q = 15 V	B = 200 V	M = 600 V
Y = 30 V	C = 300 V	S = 700 V
F = 50 V	D = 400 V	N = 800 V
A = 100 V	E = 500 V	

SMALL SIGNAL TRANSISTORS

Suffix indicates h_{FE} range, for example, BC108C

A = h_{FE} of 125–260

B = h_{FE} of 240–500

C = h_{FE} of 450–900

POWER TRANSISTORS

Suffix indicates voltage, for example, TIP32C

No suffix = 40V

A = 60V

B = 80V

C = 100V

D = 120V

Resistor and capacitor letter and digit code (BS 1852)

Resistor values are indicated as follows:					
0.47Ω 1Ω 4.7Ω 47Ω	marked	R47 1R0 4R7 47R	100Ω 1kΩ 10kΩ 10MΩ	marked	100R 1K0 10K 10M

A letter following the value shows the tolerance.

F = ±1%; G = ±2%; J = ±5%; K = ±10%; M = ±20%;
R33M = 0.33 Ω ±20%; 6K8F = 6.8kΩ ± 1%.

Capacitor values are indicated as:

0.68pF 6.8pf 1000pF	marked	p68 6p8 1n0	6.8nf 1000nF 6.8μF	marked	6n8 1μ0 6μ8

Tolerance is indicated by letters as for resistors. Values up to 999pF are marked in pF, from 1000pf to 999 000pF (=999nF) as nF (1000pF = 1nF) and from 1000nF (=1μF) upwards as μF.

Some capacitors are marked with a code denoting the value in pF (first two figures) followed by a multiplier as a power of ten (3 denotes 10^3). Letters denote tolerance as for resistors but C = ±0.25pF.

For example 123J = 12pF × 10^3 ± 5% = 12,000pF (or 0.12μF).

Glossary of terms

Activities Activities are represented by an arrow, the tail of which indicates the commencement, and the head the completion of the activity.

Advantages of a d.c. machine One of the advantages of a d.c. machine is the ease with which the speed may be controlled.

Appointed person An appointed person is someone who is nominated to take charge when someone is injured or becomes ill, including calling an ambulance if required. The appointed person will also look after the first aid equipment, including re-stocking the first aid box.

As-fitted drawings When the installation is completed a set of drawings should be produced which indicate the final positions of all the electrical equipment.

Asphyxiation Asphyxiation is a condition caused by lack of air in the lungs leading to suffocation. Suffocation may cause discomfort by making breathing difficult or it may kill by stopping the breathing.

Assembly point The purpose of an assembly point is to get you away from danger to a place of safety where you will not be in the way of the emergency services.

Bar chart There are many different types of bar chart used by companies but the object of any bar chart is to establish the sequence and timing of the various activities involved in the contract as a whole.

Basic protection ***Basic Protection*** is provided by the insulation of live parts in accordance with Section 416 of the IEE Regulations.

Bill of quantities The size and quantity of all the materials, cables, control equipment and accessories. This is called a '**bill of quantities**'.

Block diagram A block diagram is a very simple diagram in which the various items or pieces of equipment are represented by a square or rectangular box.

Bonding conductor A protective conductor providing equipotential bonding.

Bonding The linking together of the exposed or extraneous metal parts of an electrical installation.

BS 5750/ISO 9000 certificate A BS 5750/ISO 9000 certificate provides a framework for a company to establish quality procedures and identify ways of improving its particular product or service. An essential part of any quality system is accurate record-keeping and detailed documentation which ensures that procedures are being followed and producing the desired results.

Cable tray Cable tray is a sheet-steel channel with multiple holes. The most common finish is hot-dipped galvanized but PVC-coated tray is also available. It is used extensively on large industrial and commercial installations for supporting MI and SWA cables which are laid on the cable tray and secured with cable ties through the tray holes.

Cables Most cables can be considered to be constructed in three parts: the *conductor* which must be of a suitable cross-section to carry the load current; the *insulation*, which has a colour or number code for identification; and the *outer sheath* which may contain some means of providing protection from mechanical damage.

Cage rotor The solid construction of the cage rotor used in many a.c. machines makes them almost indestructible.

Calibration certificates Calibration certificates usually last for a year. Test instruments must, therefore, be tested and recalibrated each year by an approved supplier.

Capacitance The property of a pair of plates to store an electric charge is called its *capacitance*.

Capacitive reactance (X_C) *Capacitive reactance* (X_C) is the opposition to an a.c. current in a capacitive circuit. It causes the current in the circuit to lead ahead of the voltage.

Capacitor By definition, a capacitor has a capacitance (C) of one farad (symbol F) when a p.d. of one volt maintains a charge of one coulomb on that capacitor.

Cartridge fuse The cartridge fuse breaks a faulty circuit in the same way as a semi-enclosed fuse, but its construction eliminates some of the disadvantages experienced with an open-fuse element.

CFLs CFLs (Compact Fluorescent Lamps) are miniature fluorescent lamps designed to replace ordinary GLS lamps.

Circuit diagram A circuit diagram shows most clearly how a circuit works.

Circuit Protective Conductor (CPC) A protective conductor connecting exposed conductive parts of equipment to the main earthing terminal.

Clean air act The Clean Air Act applies to all small and medium sized companies operating furnaces, boilers, or incinerators.

Competent person A competent person is anyone who has the necessary technical skills, training and expertise to safely carry out the particular activity.

Conductor A *conductor* is a material in which the electrons are loosely bound to the central nucleus and are, therefore, free to drift around the material at random from one atom to another.

Conduit A conduit is a tube, channel or pipe in which insulated conductors are contained.

Copper losses and iron losses As they have no moving parts causing frictional losses, most transformers have a very high efficiency, usually better than 90%. However, the losses which do occur in a transformer can be grouped under two general headings: copper losses and iron losses.

Copper losses Copper losses occur because of the small internal resistance of the windings.

Critical path Critical path is the path taken from the start event to the end event which takes the longest time.

Dangerous occurrence Dangerous occurrence is a 'near miss' that could easily have led to serious injury or loss of life. Near miss accidents occur much more frequently than injury accidents and are, therefore, a good indicator of hazard, which is why the HSE collects this data.

Daywork Daywork is one way of recording **variations to a contract**, that is, work done which is outside the scope of the original contract. It is **extra** work.

Delivery note By signing the delivery note the person is saying 'yes, these items were delivered to me as my company's representative on that date and in good condition and I am now responsible for these goods'.

Designer The designer of any electrical installation is the person who interprets the electrical requirements of the customer within the regulations.

Detail drawings and assembly drawings These are additional drawings produced by the architect to clarify some point of detail.

Direct current motors Direct current motors are classified by the way in which the field and armature windings are connected, which may be in series or in parallel.

Discharge lamps Discharge lamps do not produce light by means of an incandescent filament but by the excitation of a gas or metallic vapour contained within a glass envelope.

Disconnection and separation We must ensure the disconnection and separation of electrical equipment from every source of supply and that this disconnection and separation is secure.

Dummy activities Dummy activities are represented by an arrow with a dashed line.

'Duty holder' 'Duty holder', someone who has a duty of care for health, safety and welfare matters on site.

Duty of care Everyone has a *duty of care* but not everyone is a *duty holder*. The person who exercises 'control over the whole system, equipment and conductors' and is the Electrical Company's representative on site, is *the duty holder*.

Earth The conductive mass of the earth. Whose electrical potential is taken as zero.

Earthing The act of connecting the exposed conductive parts of an installation to the main protective earthing terminal of the installation.

Eddy currents Eddy currents are circulating currents created in the core material by the changing magnetic flux. These are reduced by building up the core of thin slices or laminations of iron and insulating the separate laminations from each other.

Efficacy The performance of a lamp is quoted as a ratio of the number of lumens of light flux which it emits to the electrical energy input which it consumes. Thus *efficacy* is measured in lumens per watt; the greater the efficacy the better is the lamp's performance in converting electrical energy into light energy.

Electricity at work act The Electricity at Work Act tells us that it is 'preferable' that supplies be made dead before work commences (Regulation 4(3)).

Emergency lighting Since an emergency occurring in a building may cause the mains supply to fail, the emergency lighting should be supplied from a source which is independent from the main supply.

Emergency switching Emergency switching involves the rapid disconnection of the electrical supply by a single action to remove or prevent danger.

Employees Employees have a duty to care for their own health and safety and that of others who may be affected by their actions.

Employer Under the Health and Safety at Work Act an employer has a duty to care for the health and safety of employees.

Environmental conditions Environmental conditions include unguarded or faulty machinery.

Event An event is a point in time, a milestone or stage in the contract when the preceding activities are finished.

Exit routes Exit routes are usually indicated by a green and white 'running man' symbol. Evacuation should be orderly, do not run but walk purposefully to your designated assembly point.

Exposed conductive parts This is the metalwork of an electrical appliance or the trunking and conduit of an electrical system.

Extraneous conductive parts This is the structural steelwork of a building and other service pipes such as gas, water, radiators and sinks.

Fault protection *Fault Protection* is provided by protective equipotential bonding and automatic disconnection of the supply (by a fuse or MCB) in accordance with IEE Regulations 411.3 to 6.

Fault A fault is not a natural occurrence; it is an unplanned event which occurs unexpectedly.

Fire extinguishers Fire extinguishers remove heat from a fire and are a first response for small fires.

Fire Fire is a chemical reaction which will continue if fuel, oxygen and heat are present.

First aid First aid is the initial assistance or treatment given to a casualty for any injury or sudden illness before the arrival of an ambulance, doctor or other medically qualified person.

First aider A first aider is someone who has undergone a training course to administer first aid at work and holds a current first aid certificate.

Flexible conduit Flexible conduit is made of interlinked metal spirals often covered with a PVC sleeving.

Float time, slack time or time in hand Float time, slack time or time in hand is the time remaining to complete the contract after completion of a particular activity.

Fluorescent lamp A fluorescent lamp is a linear arc tube, internally coated with a fluorescent powder, containing a low-pressure mercury vapour discharge.

FP 200 cable FP 200 cable is similar in appearance to an MI cable in that it is a circular tube, or the shape of a pencil, and is available with a red or white sheath. However, it is much simpler to use and terminate than an MI cable.

Functional switching Functional switching involves the switching on or off or varying the supply of electrically operated equipment in normal service.

Fuse element The fuse element is encased in a glass or ceramic tube and secured to end-caps which are firmly attached to the body of the fuse so that they do not blow off when the fuse operates.

GLS lamps GLS lamps produce light as a result of the heating effect of an electrical current. Most of the electricity goes to producing heat and a little to producing light. A fine tungsten wire is first coiled and coiled again to form the incandescent filament of the GLS lamp.

Hazardous malfunction if a piece of equipment was to fail in its function, that is fail to do what it is supposed to do and, as a result of this failure have the potential to cause harm, then this would be defined as a hazardous malfunction.

HBC cartridge fuses As the name might imply, these HBC (High Breaking Capacity) cartridge fuses are for protecting circuits where extremely high fault currents may develop such as on industrial installations or distribution systems.

Human errors Human errors include behaving badly or foolishly, being careless and not paying attention to what you should be doing at work.

Hysteresis loops Some materials magnetize easily, and some are difficult to magnetize. Some materials retain their magnetism, while others lose it. The result will look like the graphs shown in Fig. 4.21 and are called *hysteresis loops*.

Impedance The total opposition to current flow in an a.c. circuit is called impedance and given the symbol Z.

Improvement notice An **improvement notice** identifies a contravention of the law and specifies a date by which the situation is to be put right.

Induction heating processes Induction heating processes use high-frequency power to provide very focused heating in industrial processes.

Induction motor rotor There are two types of induction motor rotor – the wound rotor and the cage rotor.

Inductive reactance (X_L) *Inductive reactance* (X_L) is the opposition to an a.c. current in an inductive circuit. It causes the current in the circuit to lag behind the applied voltage.

Insulator An *insulator* is a material in which the outer electrons are tightly bound to the nucleus and so there are no free electrons to move around the material.

Inverse square law

The illumination of a surface follows the *inverse square law,* where $E = \frac{I}{d^2}$ (lx)

Investors in people

Investors in People is a National Quality Standard that focuses on the needs of the people working within an organization. It recognizes that a company or business is investing some of its profits in its workforce in order to improve the efficiency and performance of the organization. The objective is to create an environment where what people can do and are motivated to do, matches what the company needs them to do to improve.

Iron losses

Iron losses are made up of *hysteresis loss* and *eddy current loss.* The hysteresis loss depends upon the type of iron used to construct the core and consequently core materials are carefully chosen.

Isolation

Isolation means the disconnection and separation of the electrical equipment from every source of electrical energy in such a way that this disconnection and separation is secure.

Isolator

An isolator is a mechanical device that is operated manually and is provided so that the whole of the installation, one circuit or one piece of equipment, may be cut off from the live supply.

Job sheet or job card

A job sheet or job card such as that shown in Fig. 7.12 carries information about a job which needs to be done, usually a small job.

Layout drawings or site plan

These are scale drawings based upon the architect's site plan of the building and show the positions of the electrical equipment which is to be installed.

Low smoke and fume cables

Low smoke and fume cables give off very low smoke and fumes if they are burned in a burning building. Most standard cable types are available as LSF cables.

Lumen method

When designing interior lighting schemes the method most frequently used depends upon a determination of the total flux required to provide a given value of illuminance at the working place. This method is generally known as the *lumen method.*

Manual handling

Manual handling is lifting, transporting or supporting loads by hand or by bodily force.

Metallic trunking

Metallic trunking is formed from mild steel sheet, coated with grey or silver enamel paint for internal use or a hot-dipped galvanized coating where damp conditions might be encountered.

MI cable

An MI cable has a seamless copper sheath which makes it waterproof and fire- and corrosion-resistant. These characteristics often make it the only cable choice for hazardous or high-temperature installations.

Michael Faraday

Michael Faraday demonstrated on 29 August 1831 that electricity could be produced by magnetism. He stated that *'When a conductor cuts or is cut by a magnetic field an emf is induced in that conductor. The amount of induced emf is proportional to the rate or speed at which the magnetic field cuts the conductor'.*

Mini-trunking Mini-trunking is very small PVC trunking, ideal for surface wiring in domestic and commercial installations such as offices.

Motor starter The purpose of the motor starter is not to start the machine, as the name implies, but to reduce heavy starting currents and provide overload and no-volt protection in accordance with the requirements of Regulations 552.

Mutual inductance A mutual inductance of 1 H exists between two coils when a uniformly varying current of 1 A/s in one coil produces an emf of 1 V in the other coil.

Network diagram A network diagram can be used to co-ordinate all the interrelated activities of the most complex project.

Ohm's law Ohm's law, which says that the current passing through a conductor under constant temperature conditions is proportional to the potential difference across the conductor.

Overload current An overload current can be defined as a current which exceeds the rated value in an otherwise healthy circuit.

Paper insulated lead covered steel wire armour cables Paper insulated lead covered steel wire armour cables are only used in systems above 11 kV. Very high-voltage cables are only buried underground in special circumstances when overhead cables would be unsuitable, for example, because they might spoil a view of natural beauty.

Permit-to-work procedure The permit-to-work procedure is a type of 'safe system to work' procedure used in specialized and potentially dangerous plant process situations.

Power factor improvement Power factor improvement of most industrial loads is achieved by connecting capacitors to either:

- individual items of equipment or,
- banks of capacitors

Power factor The power factor of the consumer is governed entirely by the electrical plant and equipment that is installed and operated within the consumer's buildings.

PPE PPE is defined as all equipment designed to be worn, or held, to protect against a risk to health and safety.

Prohibition notice A **prohibition notice** is used to stop an activity which the inspector feels may lead to serious injury.

Protective equipotential bonding This is equipotential bonding for the purpose of safety.

Public nuisance A public nuisance is 'an act unwarranted by law or an omission to discharge a legal duty which materially affects the life, health, property, morals or reasonable comfort or convenience of Her Majesty's subjects'.

Pulsating field Once rotation is established, the pulsating field in the run winding is sufficient to maintain rotation and the start winding is disconnected by a centrifugal switch which operates when the motor has reached about 80% of the full load speed.

PVC insulated steel wire armour cables PVC insulated steel wire armour cables are used for wiring underground between buildings, for main supplies to dwellings, rising sub-mains and industrial installations. They are used where some mechanical protection of the cable conductors is required.

Quality Quality generally refers to the level of excellence, but in the business sense it means meeting the customer's expectations regarding performance, reliability and durability.

RCD An RCD is a type of circuit breaker that continuously compares the current in the line and neutral conductors of the circuit.

Rectification Rectification is the conversion of an a.c. supply into a unidirectional or d.c. supply.

Resistance In any circuit, *resistance* is defined as opposition to current flow.

Resistivity The *resistivity* (symbol ρ – the Greek letter 'rho') of a material is defined as the resistance of a sample of unit length and unit cross-section.

Resistor All materials have some resistance to the flow of an electric current but, in general, the term *resistor* describes a conductor specially chosen for its resistive properties.

Safety Officer *The Safety Officer* will be the specialist member of staff, having responsibility for health and safety within the company. He or she will report to the senior manager responsible for health and safety.

Schematic diagrams A schematic diagram is a diagram in outline of, for example, a motor starter circuit.

Semi-enclosed fuse The semi-enclosed fuse consists of a fuse wire, called the fuse element, secured between two screw terminals in a fuse carrier.

Shock protection Protection from electric shock is provided by Basic Protection and Fault Protection.

Short circuit A short circuit is an overcurrent resulting from a fault of negligible impedance connected between conductors.

Single PVC insulated conductors Single PVC insulated conductors are usually drawn into the installed conduit to complete the installation.

Skirting trunking A trunking manufactured from PVC or steel and in the shape of a skirting board is frequently used in commercial buildings such as hospitals, laboratories and offices.

Socket outlets Socket outlets provide an easy and convenient method of connecting portable electrical appliances to a source of supply.

Space factor The ratio of the space occupied by all the cables in a conduit or trunking to the whole space enclosed by the conduit or trunking is known as the *space factor*.

Special waste Special waste is covered by the Special Waste Regulations 1996 and is waste that is potentially hazardous or dangerous and which may, therefore, require special precautions during handling, storage, treatment or disposal. Examples of special waste are asbestos, lead-acid batteries, used engine oil, solvent-based paint, solvents, chemical waste and pesticides.

Static charge Static charge builds up between any two insulating surfaces or between an insulating surface and a conducting surface, but it is not apparent between two conducting surfaces.

Static electricity Static electricity is a voltage charge which builds up to many thousands of volts between two surfaces when they rub together.

Statutory nuisance 'A statutory nuisance must materially interfere with the enjoyment of one's dwelling. It is more than just irritating or annoying and does not take account of the undue sensitivity of the receiver'.

Steel wire armoured PVC insulated cables Steel wire armoured PVC insulated cables are now extensively used on industrial installations and often laid on cable tray.

Switching for mechanical maintenance The switching for mechanical maintenance requirements is similar to those for isolation except that the control switch must be capable of switching the full load current of the circuit or piece of equipment.

Team working Team working is about working with other people.

Test instruments and test leads The test instruments and test leads used by the electrician for testing an electrical installation must meet all the requirements of the relevant regulations.

The safety representative *The Safety Representative* will be the person who represents a small section of the workforce on the Safety Committee. The role of the Safety Representative will be to bring to the Safety Committee the health and safety concerns of colleagues and to take back to colleagues, information from the Committee.

Three-phase supply If a three-phase supply is connected to three separate windings equally distributed around the stationary part or stator of an electrical machine, an alternating current circulates in the coils and establishes a magnetic flux.

Time sheet A time sheet is a standard form completed by each employee to inform the employer of the actual time spent working on a particular contract or site.

Transformer A transformer is an electrical machine which is used to change the value of an alternating voltage. They vary in size from miniature units used in electronics to huge power transformers used in power stations.

Transformers rating Transformers are rated in kVA (kilovolt-amps) rather than power in watts because the output current and power factor will be affected by the load connected to the transformer.

Trunking A trunking is an enclosure provided for the protection of cables which is normally square or rectangular in cross-section, having one removable side. Trunking may be thought of as a more accessible conduit system.

Universal' motor A series motor will run on both a.c. or d.c. and is, therefore, sometimes referred to as a 'universal' motor.

UPS A UPS is essentially a battery supply electronically modified to provide a clean and secure a.c. supply.

Visual inspection The aim of the visual inspection is to confirm that all equipment and accessories are undamaged and comply with the relevant British and European Standards, and also that the installation has been securely and correctly erected.

Wiring diagram A wiring diagram or connection diagram shows the detailed connections between components or items of equipment.

Written messages A lot of communications between and within larger organizations take place by completing standard forms or sending internal memos.

Index